# 18th Asia Pacific Bioinformatics Conference (APBC 2020)



Seoul, South Korea
18 - 20 August 2020

**Editors:**

**Dokyun Na**
**Wenying Yan**
**Yi-Ping Phoebe Chen**

ISBN: 978-1-7138-1953-0

**Printed from e-media with permission by:**

Curran Associates, Inc.
57 Morehouse Lane
Red Hook, NY 12571

proceedings.com

**Some format issues inherent in the e-media version may also appear in this print version.**



Printed by Curran Associates, Inc. (2020)

For further information, please contact BioMed Central Ltd
at the address below. The collection is available online at
https://www.biomedcentral.com/search?query=APBC+2020&searchType=publisherSearch

BioMed Central
The Campus, 4 Crinan Street
London N1 9XW
United Kingdom

Phone: +44 (0) 20 3192 2009

Email: ORSupport@spingernature.com
Web: www.biomedcentral.com

**Additional copies of this publication are available from:**

Curran Associates, Inc.
57 Morehouse Lane
Red Hook, NY 12571 USA
Phone: 845-758-0400
Fax: 845-758-2633
Email: curran@proceedings.com
Web: www.proceedings.com

# TABLE OF CONTENTS

Abe *et al. BMC Bioinformatics* 2020, **21**(Suppl 13):393
https://doi.org/10.1186/s12859-020-03671-7

BMC Bioinformatics

RESEARCH

Open Access

# Model-based clustering for flow and mass cytometry data with clinical information

Ko Abe[1], Kodai Minoura[1,2], Yuka Maeda[3], Hiroyoshi Nishikawa[2,3] and Teppei Shimamura[1*]

*From* The 18th Asia Pacific Bioinformatics Conference
Seoul, Korea. 18–20 August 2020

*Correspondence:
shimamura@med.nagoya-u.ac.jp
[1]Division of Systems Biology, Nagoya University Graduate School of Medicine, 65 Tsurumai-cho, Showa-ku, 4668550 Nagoya, Japan
Full list of author information is available at the end of the article

## Abstract

**Background:** High-dimensional flow cytometry and mass cytometry allow systemic-level characterization of more than 10 protein profiles at single-cell resolution and provide a much broader landscape in many biological applications, such as disease diagnosis and prediction of clinical outcome. When associating clinical information with cytometry data, traditional approaches require two distinct steps for identification of cell populations and statistical test to determine whether the difference between two population proportions is significant. These two-step approaches can lead to information loss and analysis bias.

**Results:** We propose a novel statistical framework, called LAMBDA (Latent Allocation Model with Bayesian Data Analysis), for simultaneous identification of unknown cell populations and discovery of associations between these populations and clinical information. LAMBDA uses specified probabilistic models designed for modeling the different distribution information for flow or mass cytometry data, respectively. We use a zero-inflated distribution for the mass cytometry data based the characteristics of the data. A simulation study confirms the usefulness of this model by evaluating the accuracy of the estimated parameters. We also demonstrate that LAMBDA can identify associations between cell populations and their clinical outcomes by analyzing real data. LAMBDA is implemented in R and is available from GitHub (https://github.com/abikoushi/lambda).

**Keywords:** Flow cytomety, Mass cytometory, Bayesian mixture model, Stochastic EM algorithm

## Background

The recent development of high-dimensional flow cytometry and mass cytometry (CyTOF) allows for characterizing cell types and states by detecting the expression levels of pre-defined sets of surface and intracellular proteins at single cell resolution [1]. For an individual subject, the modern flow cytometry data consist of 20 or more protein measurements from millions of cells from the subject. The recent mass cytometry systems use antibodies tagged with heavy metal isotopes which reduce signal interference due to spectral overlap and autofluorescence and enable the detection of more than 40 proteins per cell [2]. This high-dimensional cytometry data contains useful information to diagnose diseases such as leukemia [3] and HIV [4], as well as to predict clinical outcomes such as the response to cancer immune-therapies [5].

A key challenge in the analysis of high-dimensional cytometry data is to identify unknown cell populations that relate as prognostic factors to clinical outcomes of interest. Traditional analysis is done by manual gating which suffers not only from the need to detect unknown cell populations, but also from the need to ensure reproducibility [6]. This lack of reproducibility has two subjective causes. The first concerns the order in which pairs of markers are explored. This order often allows some degree of freedom in the gating process, so it might lead to the selection of different cells by alternative gating strategies. The second concerns the boundaries of the gates being used. There is considerable diversity between operators in terms of gating strategy, in which some experts gate strictly and others gate generously. The subjective nature of manual gating allows for the production of too wide a variety of results to be accurately reproducible.

As an alternative to manual gating, researchers have developed several computational methods, including Citrus [7], cydar [8], and diffcyt [9], to infer cell populations or states associated with an outcome variable in high-dimensional cytometry data. However, these methods require two steps: a first step in which cell populations are identified using a clustering algorithm, and a second step in which the summary statistics of the identified cell populations are concatenated into a clinical outcome of interest which can lead to information loss and analysis bias. Furthermore, these methods do not consider the distinctive features of the expression values of mass cytometry data as opposed to flow cytometry data. Mass cytometry data is marked by a zero-inflated distribution (Fig. 1). That is, proteins can be either 'on' or 'off,' in which either a positive expression measure is recorded or the recorded expression is zero or negligible, and where a very high proportion of the data entries are zero. On the other hand, in flow cytometry, the boundary between 'on' and 'off' is more ambiguous, which leads to a bimodal Gaussian distribution (Fig. 1). Therefore, lack of consideration for this distribution difference in the existing methods masks the underlying difference in cell populations and gives rise to a misleading conclusion in both basic and clinical research.

To address the aforementioned problems, we propose a new probabilistic approach for identifying unknown cell populations associated with clinical outcomes of interest which we have named LAMBDA (Latent Allocation Model with Bayesian Data Analysis). The contributions of our proposed method are summarized as follows:

- Our method is a one-step procedure that directly uses cytometry data at the single cell level to simultaneously discover cell populations and to identify the associations of these populations with clinical outcomes of interest. Our model can also be used to find relationships between cell populations and a single clinical outcome as well as relationships between cell populations and multiple clinical outcomes.
- Our method is based on correctly specified probabilistic models that are designed for modeling the different distribution information of flow and mass cytometry data respectively. In the case of flow cytometry, LAMBDA assumes that the data is

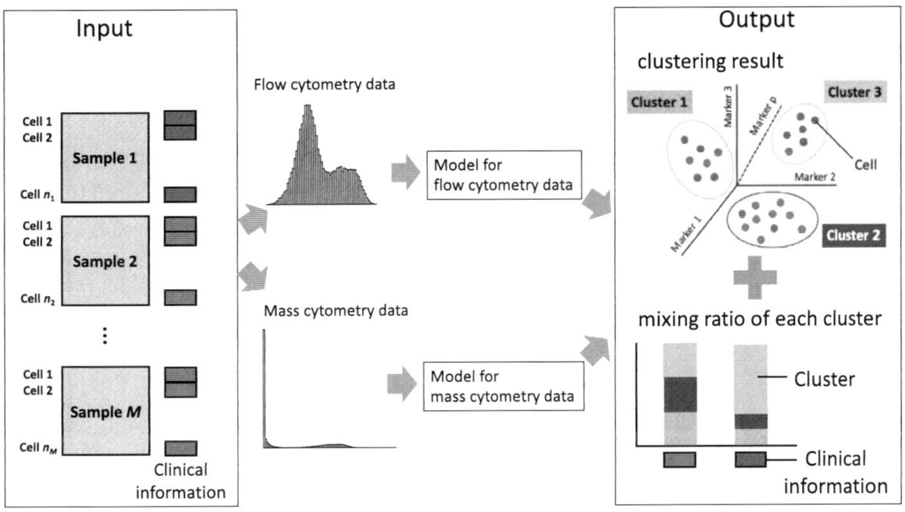


**Fig. 1** Conceptual diagram. LAMBDA estimates composition ratio of clusters depends on the clinical information

generated from a mixture of multivariate normal distributions, each of which represents an unknown cell population. On the other hand, in the case of mass cytometry, LAMBDA assumes that the data is generated from a mixture of zero-inflated distributions that represent censoring of expression below a substantial limit of detection. In both models, the compositions of cell populations are assumed to vary with clinical outcomes.

- We provide a simple and efficient learning procedure for the proposed model using a stochastic EM algorithm that reduces computational cost. LAMBDA is implemented in the R environment, which is available from https://github.com/abikoushi/LAMBDA.

From here we will explain the method and its implications in detail. Figure 1 shows a conceptual view of analysis by LAMBDA. The "Methods" section details the proposed model and algorithm. The "Results" section includes an analysis of the efficiency of LAMBDA using synthetic and real data. The "Conclusion" section summarizes the data presented here and describes the possibility for future expansion of this model.

## Methods

### Model for flow cytometry data

Suppose that we observe the flow cytometry dataset $\boldsymbol{y}_n \in \mathbb{R}^K$, $(n = 1, \ldots, N)$ and clinical information $\boldsymbol{x}_n \in \mathbb{R}^D$. The dataset includes $N$ cells, $K$ markers and $D$-dimensional clinical information. Our goal is to identify cells populations from the data. Furthermore we seek to understand how these cell populations change depending on the clinical information. LAMBDA is a model based clustering method. Let $L$ be the number of clusters. The data generative process of LAMBDA for **flow cytometry** data is defined as follows:

$$\boldsymbol{y}_n | w_n \sim \prod_{l=1}^{L} \text{Gaussian}\,(\boldsymbol{\mu}_l, \Sigma_l)^{w_{n,l}}$$

$$\boldsymbol{w}_n | \boldsymbol{x}_n \sim \text{Categorical}(\boldsymbol{\phi}_n)$$

$$
\begin{aligned}
\boldsymbol{\phi}_n &= \text{softmax}(\boldsymbol{x}_n \boldsymbol{\beta}) \\
\boldsymbol{\mu}_l &\sim \text{Gaussian}\left(\mathbf{0}, \frac{1}{\tau}\boldsymbol{\Sigma}_l\right) \\
\boldsymbol{\Sigma}_l^{-1} &\sim \text{Wishart}\,(\nu, \boldsymbol{\Lambda})
\end{aligned} \tag{1}
$$

where $w_{n,l}$ is $l$-th element of $\boldsymbol{w}_n$ and $D \times L$ matrix $\boldsymbol{\beta}$ is the effect of clinical information. For identifiability, the first column of $\boldsymbol{\beta}$ is always set as zero. Here, the softmax function is defined by $\text{softmax}(\boldsymbol{x}) = \frac{\exp(\boldsymbol{x})}{\sum_{k=1}^{K} \exp(x_k)}$ for vector $\boldsymbol{x} = (x_1, \ldots, x_K)^\top$ using an element-wise exponential function. Figure 2 shows a plate diagram of this data generating process. This model is a kind of conditional Gaussian mixture model [10]. However, the details of the estimation method are not described in detail in the publication, so we will describe them here.

### Parameter estimation for flow cytometry data

We find the maximum a posteriori probability (MAP) estimators, using an EM algorithm.

If the latent variable $w_{n,l}$ is given, the complete likelihood of this model is represented by the following formula:

$$
L^{(c)} = \prod_{n=1}^{N} \prod_{l=1}^{L} \phi_{n,l}^{w_{n,l}} \mathcal{N}(\boldsymbol{y}_n | \boldsymbol{\mu}_l, \boldsymbol{\Sigma}_l)^{w_{n,l}}. \tag{2}
$$

In the E-step, we calculate

$$
w_{n,l}^{(i)} = \frac{\phi_{n,l} \mathcal{N}\left(\boldsymbol{y}_n | \boldsymbol{\mu}_l^{(i-1)}, \boldsymbol{\Sigma}_l^{(i-1)}\right)}{\sum_{l=1}^{L} \phi_{n,l} \mathcal{N}\left(\boldsymbol{y}_n | \boldsymbol{\mu}_l^{(i-1)}, \boldsymbol{\Sigma}_l^{(i-1)}\right)}, \tag{3}
$$

where $\mathcal{N}(\boldsymbol{y}|\boldsymbol{\mu}, \boldsymbol{\Sigma})$ is the density function of the multivariate Gaussian distribution with mean $\boldsymbol{\mu}$ and covariance $\boldsymbol{\Sigma}$.

In the M-step, we update the parameters using:

$$
\mu_{k,l}^{(i)} = \frac{\sum_{n=1}^{N} w_{n,l}^{(i)} y_{n,k}}{\sum_{n=1}^{N} w_{n,l}^{(i)} + \tau} \tag{4}
$$

$$
\boldsymbol{\Sigma}_l = \frac{\sum_n w_{n,l}^{(i)} \left(\boldsymbol{y}_n - \boldsymbol{\mu}_l^{(i)}\right)\left(\boldsymbol{y}_n - \boldsymbol{\mu}_l^{(i)}\right)^\top + \tau \boldsymbol{\mu}_l^{(i)\top} \boldsymbol{\mu}_l^{(i)} + \boldsymbol{\Lambda}}{\sum_n w_{n,l}^{(i)} + \nu - K}, \tag{5}
$$

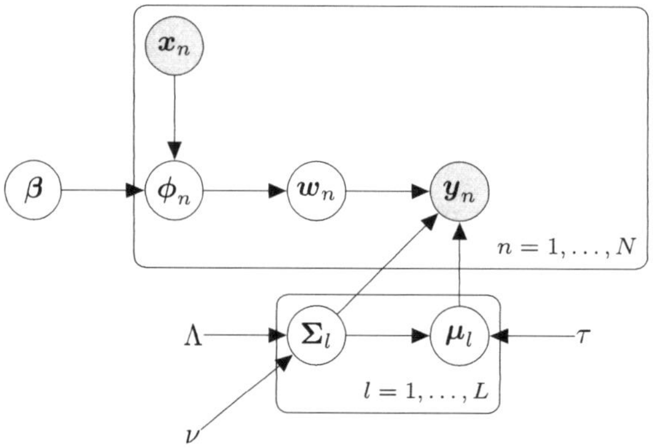


**Fig. 2** Plate diagram of the data generating process in LAMBDA for flow cytometry data. The white nodes indicate latent variables and the gray nodes indicate observed variables

Because closed form solutions for $\boldsymbol{\beta}$ are unavailable, we use Newton's method to obtain estimates. We obtain estimates of $\boldsymbol{\beta}$ by maximizing the following equation, with respect to $\boldsymbol{\beta}$:

$$Q(\boldsymbol{\beta}) = \sum_{n=1}^{N}\sum_{l=1}^{L}\left\{ w_{nl}^{(i)} \left( \sum_{d=1}^{D} x_{nd}\beta_{d,l} - \log\sum_{l=1}^{L}\exp\left(\sum_{d=1}^{D} x_{n,d}\beta_{d,l}\right)\right)\right\} \quad (6)$$

First order derivative of the function $Q(\boldsymbol{\beta})$ is represented by:

$$\nabla Q(\boldsymbol{\beta}) = \sum_{n=1}^{N}(\boldsymbol{w}_n - \text{softmax}(X_n\boldsymbol{\beta})) \otimes \boldsymbol{x}_n. \quad (7)$$

Second order derivative of the function $Q(\boldsymbol{\beta})$ is represented by:

$$\nabla^2 Q(\boldsymbol{\beta}) = \sum_{n=1}^{N}(P_n - \text{softmax}(X_n\boldsymbol{\beta})\text{softmax}(X_n\boldsymbol{\beta})^\top) \otimes \boldsymbol{x}_n\boldsymbol{x}_n^\top \quad (8)$$

where $\otimes$ denotes the Kronecker product and $P_n$ is defined as follows:

$$P_n = \begin{pmatrix} \phi_{n,1} & 0 & 0 & \cdots & 0 \\ 0 & \phi_{n,2} & 0 & \cdots & 0 \\ \vdots & & & & \vdots \\ 0 & 0 & 0 & \cdots & \phi_{n,L} \end{pmatrix}. \quad (9)$$

Thus, we update the estimate of $\boldsymbol{\beta}$ using:

$$\text{vec}(\boldsymbol{\beta}^{(i+1)}) = \text{vec}(\boldsymbol{\beta}^{(i)}) - \text{vec}(\nabla Q(\boldsymbol{\beta}))\left(\nabla^2 Q(\boldsymbol{\beta})\right)^{-1}, \quad (10)$$

where vec is the vec operator.

#### Model for mass cytometry data

In the same manner as the previous subsection, we observe a mass cytometry dataset, in this case $\boldsymbol{y}_n \in \mathbb{R}^K$, $(n = 1, \ldots, N)$ and clinical information $\boldsymbol{x}_n \in \mathbb{R}^D$. An important feature of mass cytometry data, is that a very high proportion of the data entries are zero. The ordinary mixture of the Gaussian model can not explain these zeroes. Here, LAMBDA steps in to properly assess the data. The data generative process of LAMBDA for **mass cytometry** data is defined as follows:

$$\begin{aligned} \boldsymbol{y}_n &= \begin{cases} z_{n,k} & z_{n,k} > 0 \\ 0 & z_{n,k} \leq 0 \end{cases} \\ \boldsymbol{z}_n | w_n &\sim \prod_{l=1}^{L} \text{Gaussian}\,(\boldsymbol{\mu}_l, \Sigma_l)^{w_{n,l}} \\ \boldsymbol{w}_n | \boldsymbol{x}_n &\sim \text{Categorical}(\boldsymbol{\phi}_n) \\ \boldsymbol{\phi}_n &= \text{softmax}(\boldsymbol{x}_n\boldsymbol{\beta}) \\ \boldsymbol{\mu}_l &\sim \text{Gaussian}\left(\mathbf{0}, \frac{1}{\tau}\boldsymbol{\Sigma}_l\right) \\ \boldsymbol{\Sigma}_l^{-1} &\sim \text{Wishart}\,(\nu, \boldsymbol{\Lambda}) \end{aligned} \quad (11)$$

Figure 3 shows a plate diagram of this data generating process.

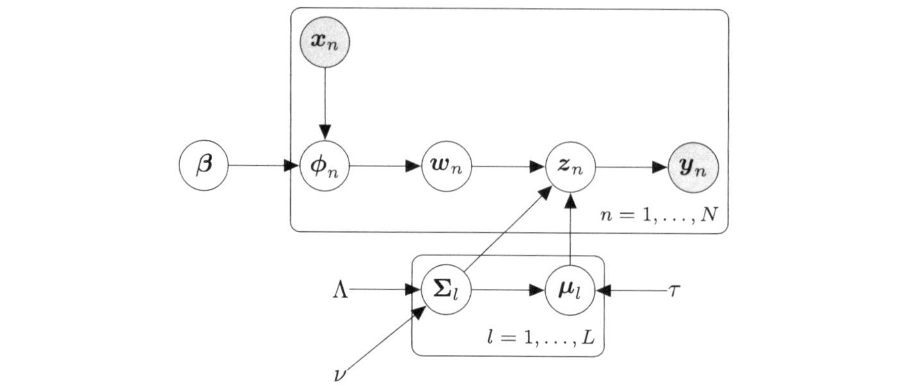


**Fig. 3** Plate diagram of the data generating process in LAMBDA for mass cytometry data

### Parameter estimation for mass cytometry data

Looking again at Eq. (11), if the latent variables $w_n$ and $z_n$ are given, the complete likelihood of this model is represented by the following formula:

$$L^{(c)} = \prod_{n=1}^{N}\prod_{l=1}^{L} \phi_{n,l}^{w_{n,l}} \mathcal{N}(\boldsymbol{z}_n|\boldsymbol{\mu}_l, \boldsymbol{\Sigma}_l)^{w_{n,l}}. \tag{12}$$

For mass cytometry data, we use a stochastic EM algorithm. In the E step, Monte Carlo samples $\tilde{\boldsymbol{w}}_n$ and $\tilde{\boldsymbol{z}}_n$ replace missing data $\boldsymbol{w}_n$ and $\boldsymbol{z}_n$.

If an arbitrary value for $\boldsymbol{z}_n$ is given, we can sample $\tilde{\boldsymbol{w}}_n$ from following categorical distribution:

$$\tilde{\boldsymbol{w}}_n \sim \text{Categorical}(\boldsymbol{\eta}_n) \tag{13}$$

where the $l$-th element of $\boldsymbol{\eta}_n$ is represented by the following formula:

$$\eta_{nl} = \frac{\phi_{n,l}\mathcal{N}(\tilde{\boldsymbol{z}}_n|\boldsymbol{\mu}_l, \boldsymbol{\Sigma}_l)}{\sum_{l=1}^{L}\phi_{n,l}\mathcal{N}(\tilde{\boldsymbol{z}}_n|\boldsymbol{\mu}_l, \boldsymbol{\Sigma}_l)}. \tag{14}$$

In contrast, if an arbitrary value for $\boldsymbol{w}_n$ is given, we can sample $\tilde{\boldsymbol{z}}_n$ from truncated Normal distribution. The steps of the Gibbs sampler for generating $\tilde{\boldsymbol{z}}_n$ are:

- If $y_{n,k} > 0$, $\tilde{z}_{n,k} = y_{n,k}$, othewise,
- let $\mu_k^{(n)} = \prod_{l=1}^{L} \mu_{k,l}^{\tilde{w}_{n,l}}$ and $\Sigma_{i,j}^{(n)} = \prod_{l=1}^{L} \Sigma_{i,j,l}^{w_{n,l}}$, where $\Sigma_{i,j,l}$ is $(i,j)$-element of $\boldsymbol{\Sigma}_l$
- let $\mu_{k,-k}^{(n)} = \mu_k^{(n)} - \Sigma_{k,k}^{(n)}(\boldsymbol{\Sigma}^{(n)})_{k,-k}^{-1}(z_{n,-k} - \boldsymbol{\mu}_{-k}^{(n)})$
- let $b = \Phi(0|\mu_{k,-k}^{(n)}, \Sigma_{k,k}^{(n)})$
- let $u \sim \text{Uniform}(0,1)$
- $\tilde{z}_{n,k} = \mu_{k,-k}^{(n)} + \sqrt{\Sigma_{k,k}^{(n)}}\Phi^{-1}(ub|0,1)$

where $\Phi(y|\mu, \Sigma)$ denotes the distribution function of a univariate Gaussian distribution with mean $\mu$ and variance $\Sigma$ and $x_{-i}$ is the set of all variables in $x$ except for the $i$-th variable.

In the M-step, by replacing $y_{n,k}$ and $w_{n,l}^{(i)}$ with samples $\tilde{z}_{n,k}$ and $\tilde{w}_{n,l}$, Eqs. (4), (5), and (10) can be used.

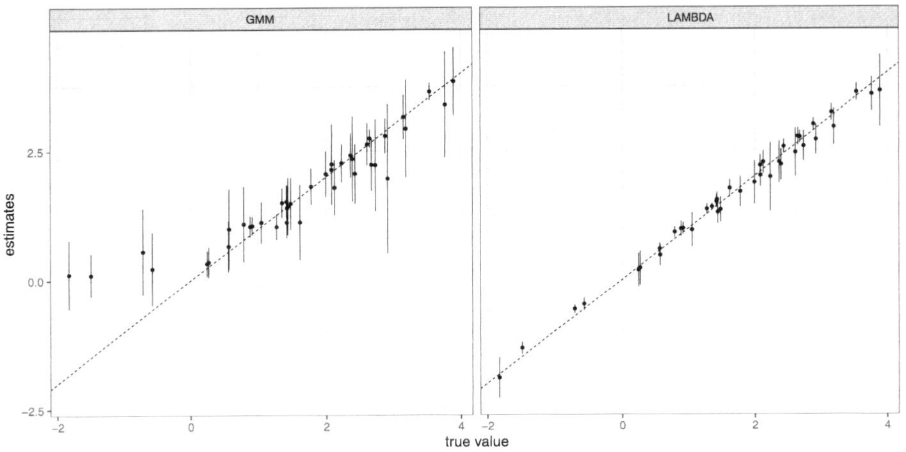


**Fig. 4** Simulation result of $\boldsymbol{\mu}$. The comparison true $\boldsymbol{\mu}$ and the mean of estimates $\hat{\boldsymbol{\mu}}$. The error bars indicates SE. Left panel show the result of GMM, right panel show the result of LAMBDA

#### Model selection

In fitting the model, it is important to choose an appropriate number for $L$. It is well known that the number of cluster $L$ with the lowest Bayesian information criterion (BIC) is an appropriate number. The BIC is defined as follows:

$$\mathrm{BIC} = -2\log(\mathcal{L}) + f\log(N), \tag{15}$$

where $\mathcal{L}$ is the likelihood and $f$ is the number of estimated parameters. However, for mass cytometry data, it requires a high computational cost to calculate the exact likelihood in the stochastic EM algorithm. Thus, in this article, we use BIC for flow cytometry data, and *elbow* method for mass cytometry data to choose $L$. Elbow method chooses a number of clusters that adding another cluster doesn't give a better fit to the data. The goodness of fit of the model to data is evaluated by the sum of squared error (SSE). SSE is defined by following:

$$\mathrm{SSE} = \sum_{n=1}^{N}\sum_{k=1}^{K}\left(y_{n,k} - \max\left(0, \mu_k^{(n)}\right)\right)^2. \tag{16}$$

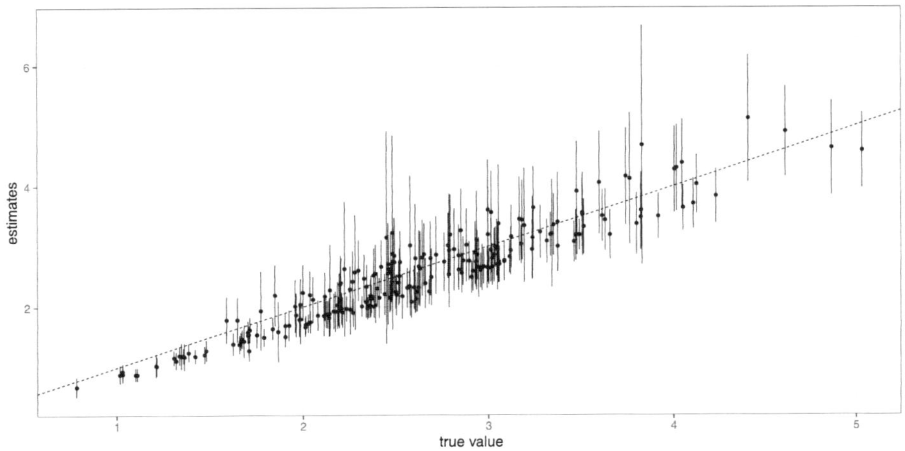


**Fig. 5** Simulation result of $\boldsymbol{\Sigma}$. The comparison true $\boldsymbol{\Sigma}$ and the mean of estimates $\hat{\boldsymbol{\Sigma}}$. The error bars indicates SE

**Table 1** Simulation result of mixture proportion

| cluster | 1 | 2 | 3 | 4 |
|---|---|---|---|---|
| | category 1 | | | |
| true value | 0.1 | 0.2 | 0.3 | 0.4 |
| mean (LAMBDA) | 0.10 | 0.20 | 0.30 | 0.40 |
| mean (GMM) | 0.10 | 0.20 | 0.31 | 0.39 |
| SE (LAMBDA) | 0.01 | 0.02 | 0.00 | 0.01 |
| SE (GMM) | 0.01 | 0.04 | 0.03 | 0.03 |
| | category 2 | | | |
| true value | 0.25 | 0.25 | 0.25 | 0.25 |
| mean (LAMBDA) | 0.24 | 0.26 | 0.25 | 0.25 |
| mean (GMM) | 0.25 | 0.26 | 0.24 | 0.24 |
| SE (LAMBDA) | 0.04 | 0.04 | 0.03 | 0.01 |
| SE (GMM) | 0.03 | 0.04 | 0.03 | 0.06 |

## Results

### Simulation study

To evaluate the standard error (SE) and the bias of the estimations, we conducted simulation experiments. The bias of $\hat{\theta}$ is defined by the difference between the true value and the estimated value ($E[\hat{\theta}] - \theta$). The synthetic data was naturally produced via the data generating process given by Eq 11. We set $K = 10$. The $\mu$ and $\Sigma$ were randomly generated.

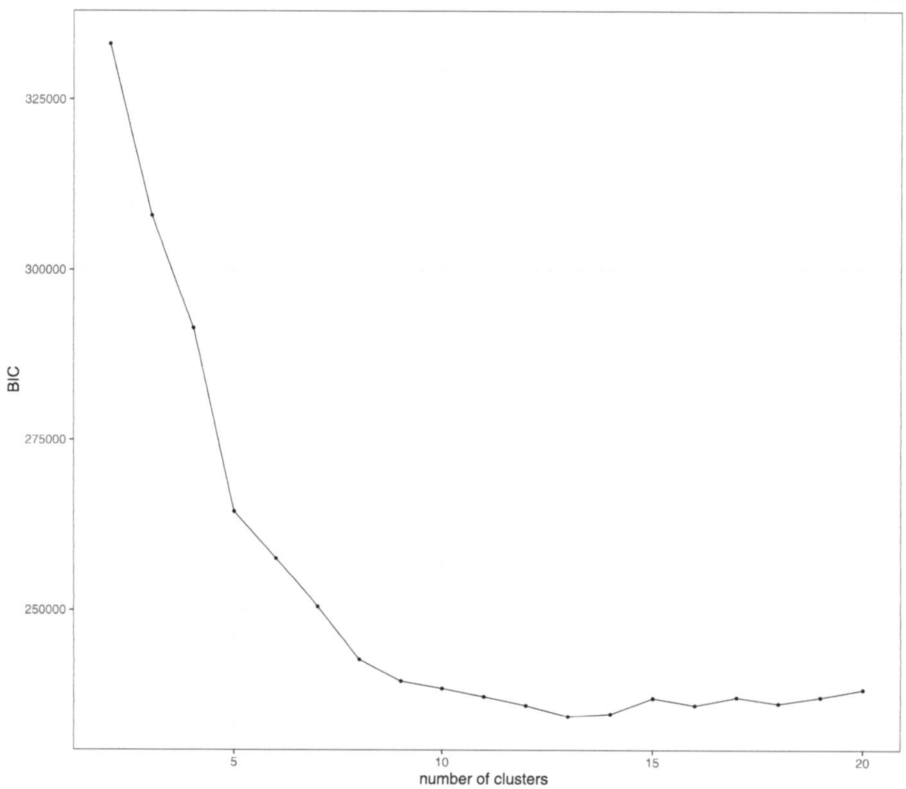


**Fig. 6** BIC. *x*-axis corresponds to the number of clusters. *y*-axis corresponds to the BIC

We used a multivariate normal distribution to generate the synthetic data, and values less than 0 were replaced by 0.

We estimated the parameters from 100 replicates of the experiment. We set the sample size $N = 2000$, the number of clusters $L = 4$, and the categories $D = 2$. One category has the mixture proportion $\boldsymbol{\phi}_1 = (0.1, 0.2, 0.3, 0.4)^\top$, the other has the mixture proportion $\boldsymbol{\phi}_2 = (0.25, 0.25, 0.25, 0.25)^\top$. To differentiate these two categories, we set $X = (\mathbf{1}, \boldsymbol{x})$, where $\mathbf{1}$ is a vector of ones. The variable $\boldsymbol{x}$ is a dummy variable to indicate the category. When estimating parameters, we set $\tau = 0.01$, $\nu = K + 2$, and $\Lambda$ is an identity matrix, which is equivalent to a weakly-informative prior distribution. To avoid the problem of label switching [11], the estimated parameters are rearranged as $\phi_{1,1} \leq \phi_{1,2} \leq \phi_{1,3} \leq \phi_{1,4}$.

Synthetic data was analyzed by LAMBDA along with ordinaly Gaussian mixture model (GMM) which cannot incorporate explanatory variables. using R package "mclust". We estimated the clusters by GMM and calculate the mixture proportion of the estimated clusters by category and the median for each cluster was used as the estimates of $\mu$.

The mean and SE for the estimated $\hat{\mu}$ and $\hat{\Sigma}$ are shown in Figs. 4 and 5, respectively. We observed that the points were arranged diagonally, indicating that the estimator of LAMBDA is unbiased. In contrast, GMM estimates often have a large bias. The mean and SE for the estimated $\hat{\phi}$ is shown in Table 1. In the case of synthetic data, the algorithm of

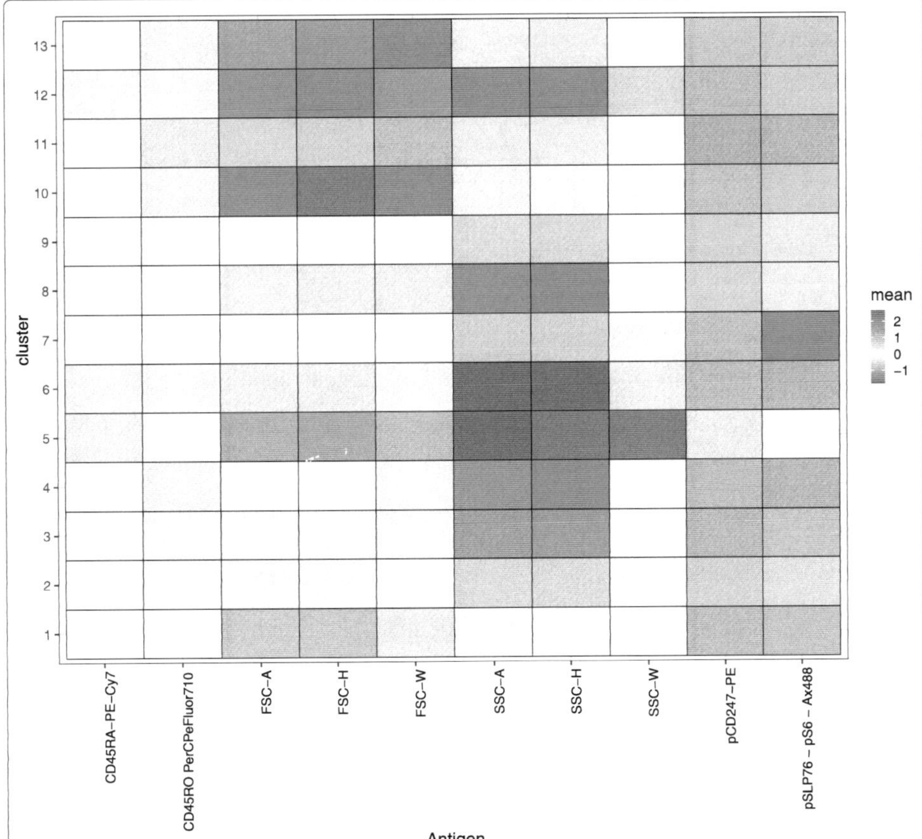


**Fig. 7** Estimated $\hat{\mu}$ for Landrigan's study. The *x*-axis corresponds to the markers, and the *y*-axis corresponds to the clusters. Red, white, and blue indicates high, middle, and low marker intensity, respectively

LAMBDA uses parameters estimated with small biases and is able to produce reasonable estimates.

### Results on real data

We applied LAMBDA to real world flow and mass cytometry data. When estimating parameters, we set $\tau = 0.01$, $\nu = K + 2$, and $\Lambda$ is an identity matrix.

For the case of flow cytometry we turn to Landrigan's study (https://community.cytobank.org/cytobank/experiments/35226), in which naive CD4+ T cells were purified and stimulated by anti-CD3 and anti-CD28 antibodies.

Five cases were tested: unstimulated, stimulated by only the anti-CD3 antibody, stimulated by both the anti-CD3 and anti-CD28 antibodies, and two cases with different dosages for the anti-CD3 antibody (0.3 $\mu$g/mL and 0.8 $\mu$g/mL). The purpose of this study is identifying the associations of cell populations with elapsed time from the stimulations start point.

Thus, we use time, dosage, anti-CD3, and anti-CD28 as the covariate $\boldsymbol{X}$. All variables are treated as dummy variables.

It is known that stimulation of CD3 triggers activation of naive CD4+ T cells, which accompany the phosphorylation of SLP76/S6 and CD247 (pSLP76/pS6, pCD246) [12].

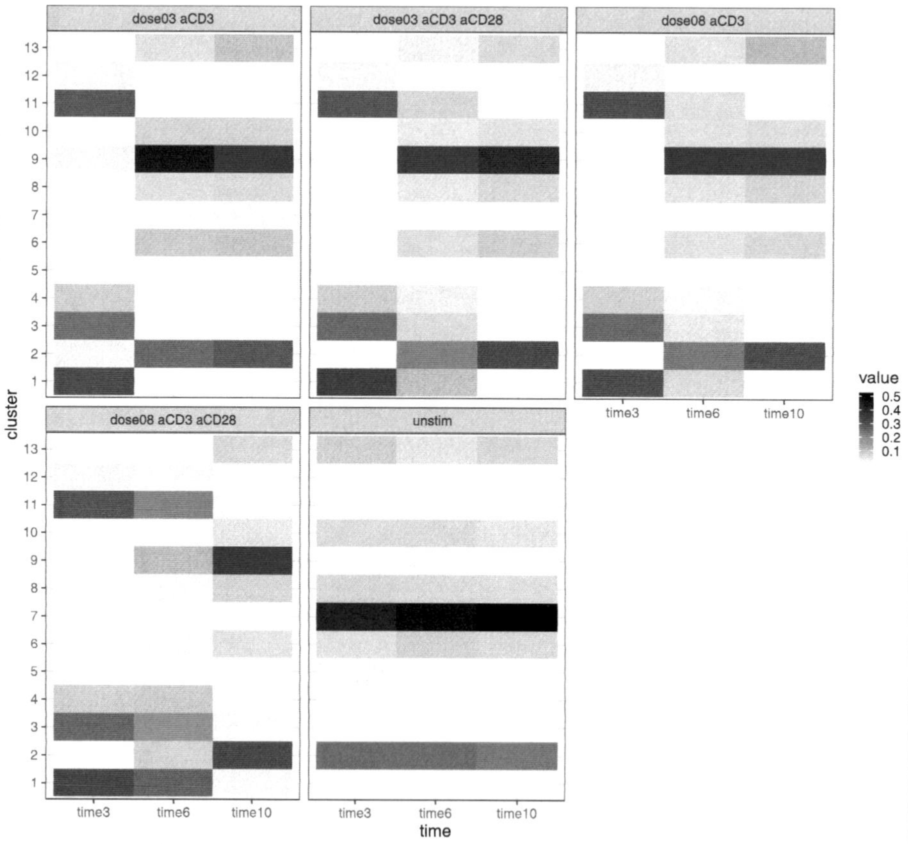


**Fig. 8** Estimated $\hat{\phi}$ for Landrigan's study. The $x$-axis corresponds to the timepoints, and the $y$-axis corresponds to the clusters

CD28 is the co-stimulatory factor that enhances and prolongs T cell activation [13]. Soon after activation, the levels of phosphorylated SLP/S6 and CD247 decrease by negative feedback. Then the cells become CD45RO+ memory T cells. BIC (Fig. 6) determined the setting of 13 clusters. Figure 7 shows the $\hat{\boldsymbol{\mu}}$. Also as shown in Fig. 7, clusters 1, 3, 4, 11, and 12 are the pSLP76/pS6+ pCD247+activated naive T cells, and clusters 2, 9, and 10 are pSLP76/pS6- pCD247- CD45RO+ memory T cells. The mixture proportion is shown in Fig. 8. While the mixture proportion remains stable over time in unstimulated cases, other cases show a high proportion of activated naive T cells at 3 min and their proportion decreases at 6 and 10 min by deactivation through negative feedback. Figure 8 shows that as activated T cells decrease, the memory T cell population increases, indicating the transformation of naive T cells to memory T cells. This shows that in the case of flow cytometry data the method is able to provide a reasonable interpretation of the cell population clusters.

We also applied LAMBDA to mass cytometry data from Jin's study [14]. This data is available in FlowRepository (https://flowrepository.org/) under Repository ID: FR-FCM-ZY6C. The purpose of this study is discovering cell population related to clinical responses. In the case of mass cytometry data, for determination of the number of clusters, we used the elbow method, which is performed by plotting the SSE within each cluster against the number of clusters. In this case, the elbow method determined 14 clusters (Fig. 9).

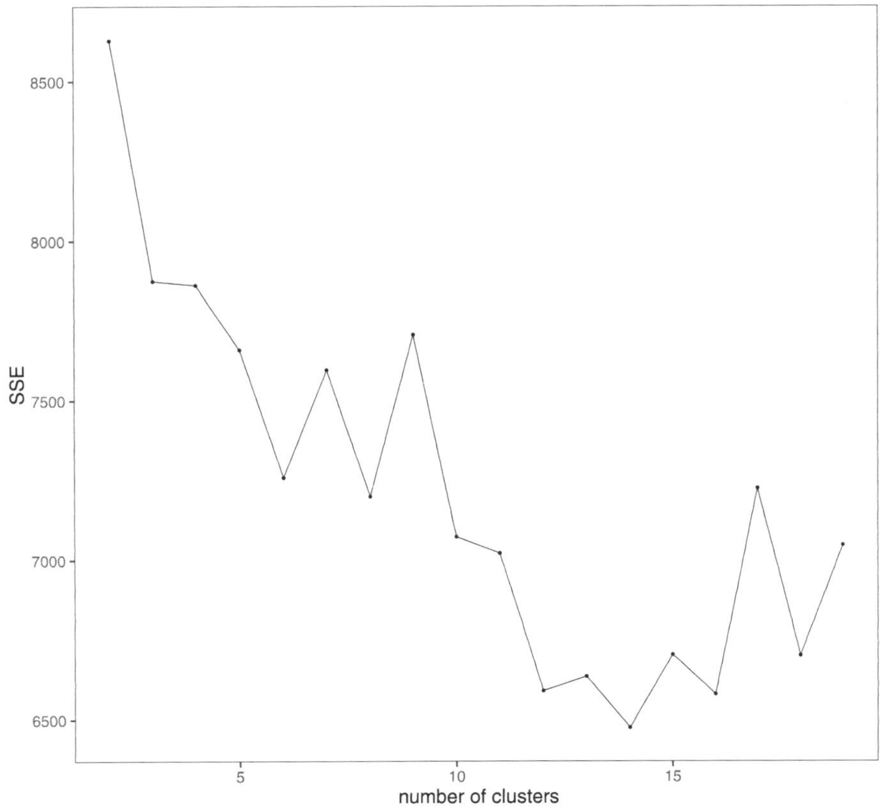


**Fig. 9** Elbow plot for Jin's study. Decrease in SSE saturated at 8 clusters

Arranged by severity, clinical responses include healthy donors (ND), partial response (PR), stable disease (SD), and progressive disease (PD). ND were used as the baseline for ICB samples. Figure 10 shows the estimated mixture proportion $\hat{\phi}$. We observed that the value of the mixture proportion for cluster 2 increases as cancer progresses from PR to PD. Figure 11 shows the estimated $\hat{\mu}$. We denote hi and lo that the marker is high and low level expressed respectively. Cluster 2 is characterized by CD8+, T-bet lo, EOMES hi, PD1hi, and Ki67 lo. T-bet lo, EOMES hi, PD1hi, and Ki67 lo are exhaustion markers of the T cell. "Exhaustion" refers to cases where a T cell becomes dysfunctional due to the long-term induction of various co-repressive molecules such as PD-1, CTLA-4, and TIM-3. Pauken & Wherry [15] reported that the CD8+ T cells of T-bet hi and EOMES lo become T-bet lo, EOMES hi, and PD1 hi through exhaustion. The marker KI67 indicate cell mass culturing. The exhausted T cells have a low expression level in KI67. Blackburn [16] reported that the cell populations of T-bet lo, EOMES hi, and PD1 hi are not activated by blocking the PD1 / PDL-1 pathway with immune checkpoint inhibitors. LAMBDA shows that the cluster 2 cell population is high in patients who underwent a PDL1 inhibitor treatment with a poor prognosis. This finding is consistent with Pauken and Blackburn's study, showing the effectiveness of LAMBDA in interpreting high dimensional mass cytometry data in real situations.

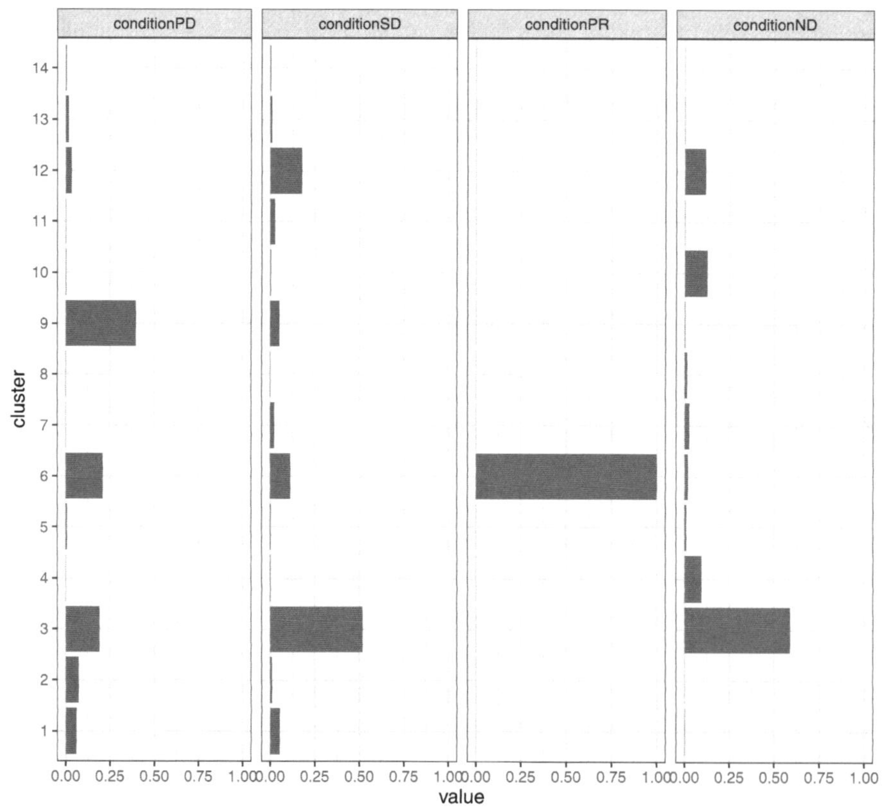


**Fig. 10** Estimated $\hat{\phi}$ for Jin's study. In the left figure, *x*-axis corresponds to the mixture proportion $\hat{\phi}$, and the *y*-axis corresponds to the clusters

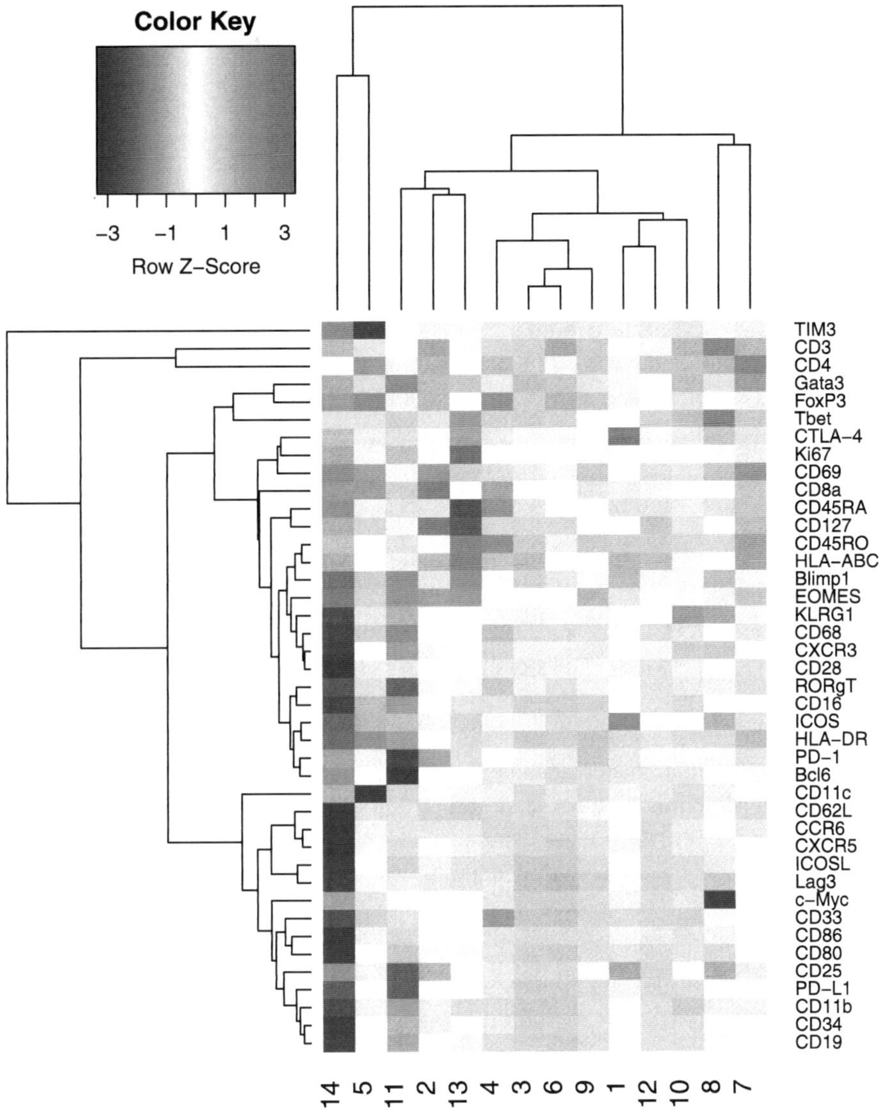


**Fig. 11** Heatmap for Jin's study. *x*-axis corresponds to the markers, and the *y*-axis corresponds to the clusters

Through this analysis we can see that LAMBDA is a method that can efficiently estimate various clusters within cell populations and identify the associations between these cell clusters and their clinical outcomes in cases of both flow and mass cytometry data.

## Discussion

LAMBDA should prove useful as it is described in this paper, but there is room for future study and improvement. Recently, with the development of next generation sequencing technologies, single cell sequencing was introduced to the field of biomedical research. Sequencing the DNA provides a higher resolution of cellular differences and a better understanding of the function of an individual cell in the context of its microenvironment. In this context, our future aim is the extension of LAMBDA for application to single cell

DNA data. This will allow us to understand the condition of the cell on a fundamental level, contributing to our overall understanding of biology and its processes.

## Conclusion

With the development of high dimensional flow and mass cytometry data, researches have been challenged with the need to properly identify and interpret data about cell populations. To meet this challenge, we proposed a statistical framework that uses flow and mass cytometry data to discover cell clusters and the associations between individual clusters and clinical information.

As described in the "Methods" section, this model uses a stochastic EM to estimate parameters. In terms of computation, this parameter-estimation procedure offers an advantage over procedures that use an ordinary EM algorithm. This is because, in an algorithm that uses ordinary EM, the computational cost is large due to calculating the high-dimensional conditional expectation. By contrast, our procedure involves a Gibbs sampling that substitutes for this requirement, significantly reducing the computational cost.

In addition to being computationally efficient, our framework also has useful properties from the perspective of data analysis. Usual methods of clustering are not able to support the inclusion of explanatory variables. However, LAMBDA can include any explanatory variables. This property allows LAMBDA to analyze experimental results with various settings. Because of this novel feature, we expect that LAMBDA will be efficiently applied to studies that seek an association between cell populations and clinical information, advancing our ability to predict disease and predict outcomes of treatment.

#### Abbreviations

BIC: Bayesian information criterion; EM: Expectation-maximization; GMM: Gaussian mixture model; HIV: Human immunodeficiency virus; MAP: Maximum a posteriori probability; SE: Standard error

#### Acknowledgements

Not applicable.

#### About this supplement

This article has been published as part of *BMC Bioinformatics Volume 21 Supplement 13, 2020: Selected articles from the 18th Asia Pacific Bioinformatics Conference (APBC 2020): bioinformatics*. The full contents of the supplement are available online at https://bmcbioinformatics.biomedcentral.com/articles/supplements/volume-21-supplement-13.

#### Authors' contributions

KA and TS designed the proposed algorithm. KM, YM and HN designed the experiments. All authors have read and approved the final manuscript.

#### Funding

This research was supported by JSPS Grant-in-Aid for Scientific Research on Innovative Areas [No. 15H05912, 18H04798, and 19H05210]. It was also supported by the Japan Agency for Medical Research and Development (AMED) under the Strategic Research Program for Brain Sciences [No. JP18dm0107087], and under Practical Research Project for Rare / Intractable Diseases [No. JP17ek0109281]. The super-computing resources were provided by Human Genome Center, University of Tokyo. Publication costs are funded by all of the funding. The funding bodies played no role in the design of the study and collection, analysis, and interpretation of data and in writing the manuscript.

#### Availability of data and materials

LAMBDA is implemented with R and is available from GitHub (https://github.com/abikoushi/LAMBDA).

#### Ethics approval and consent to participate

Not applicable.

#### Consent for publication

Not applicable.

#### Competing interests

The authors declare that they have no competing interests.

**Author details**
[1]Division of Systems Biology, Nagoya University Graduate School of Medicine, 65 Tsurumai-cho, Showa-ku, 4668550 Nagoya, Japan. [2]Division of Immunology, Nagoya University Graduate School of Medicine, 65 Tsurumai-cho, Showa-ku, 4668550 Nagoya, Japan. [3]Division of Cancer Immunology, Research Institute/EPOC, National Cancer Center, Chuo-ku tsukiji 5-1-1/Kashiwa-shi kashiwanoha 6-5-1, 1040045/2778577 Tokyo/Chiba, Japan.

Published: 17 September 2020

**References**

1. Spitzer MH, Nolan GP. Mass Cytometry: Single Cells, Many Features. Cell. 2016;165(4):780–91.
2. Bendall SC, Simonds EF, Qiu P, Amir E, Krutzik PO, Finck R, Bruggner RV, Melamed R, Trejo A, Ornatsky OI, Balderas RS, Plevritis SK, Sachs K, Pe'er D, Tanner SD, Nolan GP, et al. Single-cell mass cytometry of differential immune and drug responses across a human hematopoietic continuum. Sci (N Y). 2011;332(6030):687–96.
3. Rawstron AC, Kreuzer KA, Soosapilla A, Spacek M, Stehlikova O, Gambell P, Milani R, et al. Reproducible diagnosis of chronic lymphocytic leukemia by flow cytometry: An European Research Initiative on CLL (ERIC) & European Society for Clinical Cell Analysis (ESCCA) Harmonisation project. Cytom B Clin Cytom. 2018;94(1):121–8.
4. Abraham RS, Aubert G. Flow Cytometry, a Versatile Tool for Diagnosis and Monitoring of Primary Immunodeficiencies. Clin Vaccine Immunol. 2016;23(4):254–71. https://doi.org/10.1128/CVI.00001-16.
5. Smid M, Rodríguez-González FG, Sieuwerts AM, Salgado R, Prager-Van der Smissen WJ, van Der Vlugt-Daane M, Van de Vijver MJ, et al. Breast cancer genome and transcriptome integration implicates specific mutational signatures with immune cell infiltration. Nat Commun. 2016;7(1):2910.
6. Saeys Y, Van Gassen S, Lambrecht BN. Computational flow cytometry: helping to make sense of high-dimensional immunology data. Nat Rev Immunol. 2016;16(7):449.
7. Bruggner RV, Bodenmiller B, Dill DL, Tibshirani RJ, Nolan GP. Automated identification of stratifying signatures in cellular subpopulations. Proc Natl Acad Sci. 2014;111(26):E2770–7.
8. Lun A, Lun MA, BiocParallel D, biocViews FlowCytometry M. Package 'cydar'. 2017.
9. Weber LM, Nowicka M, Soneson C, Robinson MD. diffcyt: Differential discovery in high-dimensional cytometry via high-resolution clustering. BioRxiv. 2018349738.
10. Fahey MT, Thane CW, Bramwell GD, Coward WA. Conditional Gaussian mixture modelling for dietary pattern analysis. J R Stat Soc Ser A Stat Soc. 2007;170(1):149–66.
11. Stephens M. Dealing with label switching in mixture models. J R Stat Soc Ser B Stat Methodol. 2000;62(4):795–809.
12. Gaud G, Lesourne R, Love PE. Regulatory mechanisms in T cell receptor signalling. Nat Rev Immunol. 2018;18(8): 485–97.
13. Alegre ML, Frauwirth KA, Thompson CB. T-cell regulation by CD28 and CTLA-4. Nat Rev Immunol. 2001;1(3):220.
14. Jin G, Xue G, Wang RS, Wu LY, Lance M, Lu Y, Zhang W. Single-Cell Modeling of CD8+ T Cell Exhaustion Predicts Response to Cancer Immunotherapy. bioRxiv. 2018459867.
15. Pauken KE, Wherry EJ. Overcoming T cell exhaustion in infection and cancer. Trends Immunol. 2015;36(4):265–76.
16. Blackburn SD, Shin H, Freeman GJ, Wherry EJ. Selective expansion of a subset of exhausted CD8 T cells by $\alpha$ PD-L1 blockade. Proc Natl Acad Sci. 2008;105(39):15016–21.

## Publisher's Note

Zhong et al. BMC Bioinformatics 2020, 21(Suppl 13):385
https://doi.org/10.1186/s12859-020-03672-6

BMC Bioinformatics

RESEARCH Open Access

# Twadn: an efficient alignment algorithm based on time warping for pairwise dynamic networks

Yuanke Zhong[1†], Jing Li[2†], Junhao He[1], Yiqun Gao[1], Jie Liu[1], Jingru Wang[1], Xuequn Shang[1] and Jialu Hu[1,3*]

*From* The 18th Asia Pacific Bioinformatics Conference
Seoul, Korea. 18-20 August 2020

*Correspondence:
jhu@nwpu.edu.cn
†Yuanke Zhong and Jing Li contributed equally to this work.
[1]School of Computer Science, Northwestern Polytechnical University, West Youyi Road 127, 710072 Xi'an, China
[3]Centre of Multidisciplinary Convergence Computing, School of Computer Science, Northwestern Polytechnical University, 1 Dong Xiang Road, 710129 Xi'an, China
Full list of author information is available at the end of the article

**Abstract**

**Background:** Network alignment is an efficient computational framework in the prediction of protein function and phylogenetic relationships in systems biology. However, most of existing alignment methods focus on aligning PPIs based on static network model, which are actually dynamic in real-world systems. The dynamic characteristic of PPI networks is essential for understanding the evolution and regulation mechanism at the molecular level and there is still much room to improve the alignment quality in dynamic networks.

**Results:** In this paper, we proposed a novel alignment algorithm, Twadn, to align dynamic PPI networks based on a strategy of time warping. We compare Twadn with the existing dynamic network alignment algorithm DynaMAGNA++ and DynaWAVE and use area under the receiver operating characteristic curve and area under the precision-recall curve as evaluation indicators. The experimental results show that Twadn is superior to DynaMAGNA++ and DynaWAVE. In addition, we use protein interaction network of Drosophila to compare Twadn and the static network alignment algorithm NetCoffee2 and experimental results show that Twadn is able to capture timing information compared to NetCoffee2.

**Conclusions:** Twadn is a versatile and efficient alignment tool that can be applied to dynamic network. Hopefully, its application can benefit the research community in the fields of molecular function and evolution.

**Keywords:** PPI network, Dynamic network, Network alignment, Dynamic time warping

## Background

In recent years, due to the rapid development of biotechnology, we can obtain a large amount of biological data, such as: gene expression data, methylation data, protein interaction network data and so on [1]. Protein is a substance closely related to life and various forms of life activities. It plays a vital role in almost all life activities. Therefore, research on proteins plays a crucial role in our biological research. Protein is not a single biological function. It usually interacts with other proteins to perform certain biological functions [2–5]. All protein interactions form a protein-protein interaction (PPI) network. Most of networks are dynamic in real-world systems. For instance, PPI could change over time, and online professional network will also evolve over time [6]. A large number of PPIs are transient interactions, which briefly exists in only certain cellular context related with cell types, cell cycle stages etc. However, most of network alignment (NA) methods are designed for static networks [7], since static networks were used to model complex real-world systems. The aim of NA on PPI networks is to find an optimal node mapping that can indicate similar biological meanings between matched proteins. However, these networks actually change over time. The dynamic characteristic of PPI networks is essential for the understanding of evolution and regulation mechanism at the molecular level. Some pioneer works [8] attempt to improve NA quality using dynamic network model on evolving systems. This new computational framework can use dynamic characteristic as a supplementary information in the measure of node similarity, whereas it also suffers from the lack of high-confidence dynamic networks of real-world systems. Network study consists of a lot of parts, such as KF-finder [9], which can identify key factors from host-microbial networks in cervical cancer, besides, detection of network motif [10] is also a major search of network. In this paper, we focus on NA, which can be used to predict protein function by transferring functional knowledge from a well-studied species to a poorly-studied species.

There are two categories of alignment methods according to the target regions of networks: global alignment and local alignments. Global alignment is to find one global node mapping for compared networks [11], while local alignment aims to identify multiple conserved subregions which reflect putative functional modules of biological systems [12]. Alignments of two networks are called pairwise network alignments, those of three or more are termed as multiple network alignments. In this paper, we aim to address the global alignment problem of two dynamic networks. IsoRank was originally proposed to solve pairwise global alignment. It was intuitively guided by the assumption that one protein is a good match for another protein in the other compared network if their neighborhood topologies and sequences are similar. Many more alignment tools were developed to improve the algorithm performance of existing methods over the past decade. Among these, there are NETAL [13], H-GRAAL [14], MAGNA [15], MAGNA++ [16], which can provide one-to-one global node mapping for two compared networks. To find protein match sets for multiple species, IsoRankN [17], NetCoffee [11], SMETANA [18] and multiMAGNA++ [19] were used to find one global node mapping for multiple PPI networks. All these algorithms focused on aligning protein pairs based on static networks, although these networks evolve over time. DynaMAGNA++ [8] and DynaWAVE [20] were recently proposed to make up this deficiency. DynaMAGNA++ is the first dynamic NA algorithm, which was adapted from the MAGNA++ method. DynaMAGNA++ takes two measures (node conservation and edge conservation) to capture functionally conserved proteins. However, there is still much room to improve the alignment quality in dynamic networks. It is still a challenge to solve the alignment problem for dynamic networks.

To overcome these issues, we proposed a novel NA algorithm based on a technique termed as dynamic time warping (DTW) to align dynamic PPI networks across species. A

5-tuple-feature vector was calculated on each node of each time snapshot. A target scoring function was used to evaluate the quality of NA, which integrates both topology and sequence information. Then, the alignment problem is transformed into an optimization problem. Simulated annealing was applied to iteratively search for a near-optimal global node mapping between two compared networks.

## Methods

The Twadn algorithm returns the optimal alignment results over two given dynamic networks. A dynamic network can be seen as a series of static networks based on a time sequence. So the structure feature of each static network can be extracted by a traditional static NA. In our program, one of our previous work NetCoffee2 [21] was applied to extract the topological feature of each node in the network. Then we can get a sequence of features of each vertex in the dynamic network. Simulated annealing algorithm was used to search for a near-optimal solution. Twadn's algorithm framework is shown in Fig 1. and it has four major steps: 1) perform pair-wise sequence alignment for all pairs of proteins, and select the similar pairs which are statistically significant; 2) extract the topological feature of each vertex in each of the static network using NetCoffee2; 3) calculate the dynamic time warping similarity of all pairs of proteins; 4) use simulated annealing algorithm to find an optimal NA.

### Sequence-based similarity

We use the open source tool BLASTP [22] to sequence alignment of all proteins in the network, and obtain the sequence similarity scores e-value and bit-score for each protein pair. Considering that the amino acid sequence that affects protein function may be just a functional region of the sequence, we use the e-value parameter for preliminary filtering and select those protein pairs with e-value less than 1e-7 as the e-value can affect the coverage of predicted homologous proteins by the NetCoffee2 algorithm. Note that $\Omega$ denotes the candidates of homology proteins. Given a protein pair $u$ and $v$, the sequence similarity $s_h(u, v)$ can be calculated in the following formula:

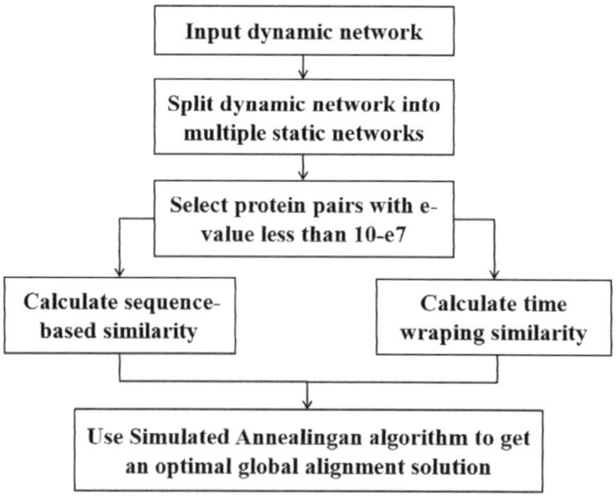


**Fig. 1** Algorithm framework of Twadn

$$s_h(u,v) = \frac{\varepsilon(u,v) - \varepsilon_{min}(u,v)}{\Delta\varepsilon} \quad (1)$$

Here, $\varepsilon(u,v)$ can be $-log(e - value)$ or bit-score of the protein pair $u$ and $v$, $\Delta\varepsilon$ is the largest difference between any two pairs of homolog in $\Omega$, and $\Delta\varepsilon = \varepsilon_{max}(u,v) - \varepsilon_{min}(u,v)$, which servers as a normalization factor. The similarity values calculated by this method are in the interval [0, 1], where 0 represents the least similar protein pair and 1 represents the most similar protein pair.

#### 5-tuple-feature vector of every vertex

Dynamic networks can be regarded as a series of static networks at many snapshots. Here, we attempt to construct a 5-tuple-feature $(\gamma, \sigma, \tau, \eta, \theta)$ for each node in the static network to represent local connectivity of its corresponding node. We denote the adjacent matrix of a netwrok $G$ as $M_{n\times n}$. Since $M$ is real and symmetric, it must has a major normalized eigenvector $K = (k_1, k_2, ..., k_n)$, which is the normalized eigenvector of the largest eigenvalue. Then, we use $k_i$, $1 \leq i \leq n$ as the reputation of the node $v_i$ while the greater the reputation is, the more important the node is. Therefore, we use $k_i$ as the first element of the 5-tuple-feature vector (i.e. $\gamma$) for node $v_i$. The set of neighbors of $v$ is denoted as $N_v$. Then, we use $|N_v|$ as the second element of the 5-tuple-feature vector (i.e. $\sigma$), the sum of the reputation of these nodes $\sum_{x\in N_v} k_x$ as the third element(i.e. $\tau$). Let us denote nodes that are 2-step away from $v$ as $N_v^2$ and all nodes in $N_v^2$ are not directly connected to $v$. Then, we use $\left|N_v^2\right|$ as the fourth element (i.e. $\eta$). The last element $\eta$ is calculated by the following formula:

$$\frac{1}{2}\sum_{x\in N_v^2} k_x p_{xv} \quad (2)$$

where $p_{xv}$ presents the number of the shortest paths from $x$ to $v$.

#### Dynamic time warping similarity

The input for calculating the dynamic time warping similarity between networks is two dynamic network $DN1$ and $DN2$, and output is a matrix $S$ in which each element $s_{ij}$ represents the time warping similarity between protein i from $DN1$ and protein j from $DN2$. Therefore, for each protein in $DN1$, we need to calculate it's time warp similarity with each protein in $DN2$. Here we carefully explain the calculation method of the time warping similarity between the proteins $P$ and $Q$ of $DN1$ and $DN2$, then through apply this method to all other protein pairs, we can obtain the result matrix $S$.

Since proteins from different dynamic network may have different snapshot number, the length of each sequence might be different. So, Euclidean distance is unable to be used in measuring the similarity of two given nodes as their differences in sequence length. Suppose $P$ and $Q$ appearing $m$ and $n$ times in their snapshots respectively, so their time sequences can be written as, $p = (p_1, p_2, ..., p_i, ..., p_m), q = (q_1, q_2, ..., q_j, ..., q_n)$. Here, $p_i$ and $q_j$ are the 5-tuple-feature vectors of time sequences $p$ and $q$ at its $i^{th}$ and $j^{th}$ appearance in the snapshot.

DTW is one of the algorithms for measuring similarity between two temporal sequences, which may vary in length and it's proved that DTW is a robust distance measure for time series, allowing similar shapes to match even if they are out of phase in the time index or different length of time series. So we apply DTW algorithm to calculate the time warping similarity between the sequences p and q.

To align two sequence using DTW, we construct a matrix $D_{m\times n}$ in which each matrix element value $d_{ij}$ represents distance between $p_i$ and $q_j$ $\left(i.e.\ d_{ij} = \sum_{k=1}^{5} \left(p_{ik} - q_{jk}\right)^2\right)$, which capture the time feature of protein in the network. Here, we define a warping path to describe the time correspondence between $p$ and $q$ as follow:

$$W = w_1, w_2, ..., w_k, \max(m, n) \le k \le m + n - 1 \tag{3}$$

The form of $w_k$ is $(i, j)$, which represents that this path passes through the lattice corresponding to $p_i$ and $q_j$. DTW is a typical optimization problem and its purpose is to find the warping path that minimizes the cumulative distance between two sequence as follow:

$$DTW(p, q) = \min_W \sum_{i=1}^{k} \delta(w_i) \tag{4}$$

Here, $\delta(w_k) = d_{ij}$ is the distance between two time series elements of $w_k$.

The selection of this path needs to meet the following constraints: 1)$w_1 = (1, 1)$ and $w_k = (m, n)$; 2) If $w_{k-1} = (i, j)$ and $w_k = \left(i', j'\right)$, then $i <= i' <= i + 1, j <= j' <= j + 1$. Then, the search for an optimal path can be transformed into a dynamic programming problem. We define the recurrence formula of the cumulative distance of each pair of proteins as follow:

$$\lambda(i, j) = d(p_i, q_j) + min\{\lambda(i - 1, j - 1), \lambda(i - 1, j), \lambda(i, j - 1)\} \tag{5}$$

where $\lambda(i, j)$ is the dynamic time warping distance of $p$ and $q$. With the warping path $W$, the node similarity of any two nodes $p$ and $q$ in the dynamic network can be calculated with the Gaussian function $s_t(p, q) = exp\left(-\frac{1}{2}[\lambda(m, n)]^2\right)$.

Figure 2 shows an example of calculate dynamic time warping similarity of each two nodes in dynamic network $DN1$ and $DN2$. The matrix (A) are the 5-tuple-feature sequence of node $a$ at five snapshots. Red numbers in matrix (B) constitute the warping path, which can be used to calculate the dynamic time warping distance of node $a$ and node $B$. The matrix (C) contains all dynamic time warping distance of each two nodes in $DN1$ and $DN2$. The matrix (D) contains all dynamic time warping similarity of each two nodes in $DN1$ and $DN2$.

### Simulated annealing

To find an optimal NA, Twadn used the simulated annealing technique to search for an approximately optimal solution, which maximizes the sequence similarity and dynamic time warping similarity with an objective function. The objective function writes as $f(A) = \sum_{m\in A} s_m$, where $A$ and $m$ refer to all possible match sets and a match set, respectively. A match set is a putative functional orthologs that could be a group of functionally related proteins. Suppose there is a match set $m = (m_1, m_2, ..., m_v)$, the alignment score of $m$ is:

$$s_m = \sum_{i,j,i\neq j} \alpha s_h(i, j) + \sum_{i,j,i\neq j} (1 - \alpha) s_t(i, j), i, j \in \{m_1, m_2, ..., m_v\} \tag{6}$$

where $s_h(i, j)$ and $s_t(i, j)$ is sequence-based similarity and dynamic time warping similarity of protein $i$ and $j$ described above. With this definition, an optimal global alignment

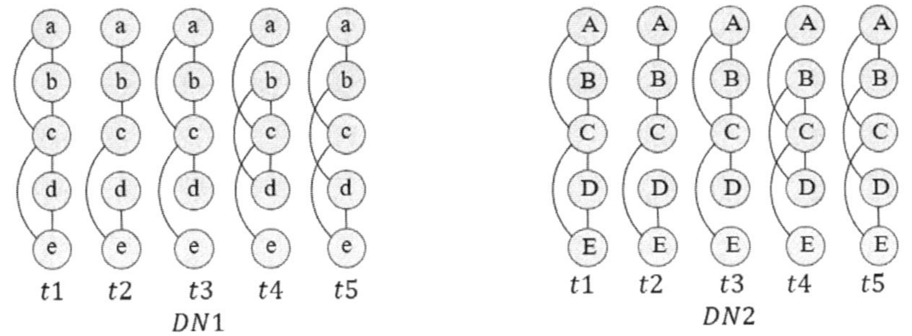


| | $\gamma$ | $\delta$ | $\tau$ | $\eta$ | $\theta$ |
|---|---|---|---|---|---|
| $a_{t1}$ | 0.64 | 2.00 | 1.64 | 2.00 | 0.64 |
| $a_{t2}$ | 0.50 | 1.00 | 0.86 | 2.00 | 0.50 |
| $a_{t3}$ | 0.74 | 2.00 | 1.74 | 2.00 | 0.42 |
| $a_{t4}$ | 0.42 | 1.00 | 1.00 | 0.00 | 0.00 |
| $a_{t5}$ | 1.00 | 2.00 | 2.00 | 2.00 | 1.92 |

(A)

| | $a_{t1}$ | $a_{t2}$ | $a_{t3}$ | $a_{t4}$ | $a_{t5}$ |
|---|---|---|---|---|---|
| $B_{t1}$ | $0.00^{0.00}$ | $0.53^{0.53}$ | $0.01^{0.54}$ | $1.17^{1.71}$ | $0.08^{1.79}$ |
| $B_{t2}$ | $0.01^{0.22}$ | $0.30^{0.30}$ | $0.21^{0.51}$ | $0.52^{1.03}$ | $0.32^{1.35}$ |
| $B_{t3}$ | $0.01^{0.23}$ | $0.57^{0.79}$ | $0.00^{0.30}$ | $1.16^{1.46}$ | $0.09^{1.12}$ |
| $B_{t4}$ | $0.01^{0.24}$ | $0.57^{0.80}$ | $0.00^{0.30}$ | $1.16^{1.46}$ | $0.09^{1.21}$ |
| $B_{t5}$ | $0.08^{0.32}$ | $0.76^{1.00}$ | $0.09^{0.39}$ | $1.47^{1.77}$ | $0.00^{1.21}$ |

(B)

| | $a$ | $b$ | $c$ | $d$ | $e$ |
|---|---|---|---|---|---|
| $A$ | 0.00 | 1.21 | 6.41 | 1.56 | 0.90 |
| $B$ | 1.21 | 0.00 | 4.45 | 0.90 | 1.45 |
| $C$ | 6.41 | 4.45 | 0.00 | 6.52 | 6.39 |
| $D$ | 1.56 | 0.90 | 6.52 | 0.00 | 1.47 |
| $E$ | 0.90 | 1.45 | 6.39 | 1.47 | 0.00 |

(C)

$e^{-\frac{1}{2}x^2}$ ⟹

| | $a$ | $b$ | $c$ | $d$ | $e$ |
|---|---|---|---|---|---|
| $A$ | 1.00 | 0.48 | 0.00 | 0.30 | 0.67 |
| $B$ | 0.48 | 1.00 | 0.00 | 0.67 | 0.35 |
| $C$ | 0.00 | 0.00 | 1.00 | 0.00 | 0.00 |
| $D$ | 0.30 | 0.67 | 0.00 | 1.00 | 0.34 |
| $E$ | 0.67 | 0.35 | 0.00 | 0.34 | 1.00 |

(D)

**Fig. 2** Dynamic time warping similarity between dynamic network *DN*1 and *DN*2

solution could be solved by maximizing a target function:

$$A^* = arg\max_A f(A) = arg\max_A \sum_{m \in A} s_m \tag{7}$$

## Result

To test our method, Twadn was evaluated on both simulated dynamic networks and real-world dynamic networks. We use a simulated dynamic network to compare Twadn, DynaMAGNA++ and DynaWAVE and evaluate the quality of alignment results in terms of area under the precision-recall curve(AUPR) and area under the ROC curve(AUROC). To show the ability of characterizing the time information in dynamic network, Twadn and NetCoffee2 were implemented on two real-world dynamic networks. We aggregated a dynamic network into a static network. In the static version, the network has the same set of nodes as the dynamic network, and a static edge will exist between two nodes if there is at least one edge between the same two nodes in the dynamic network. This kind of dynamic network aggregation was commonly applied in other time series network analyses, such as [23].

### Evaluation using synthetic dynamic network

#### *Model of synthetic dynamic network generation*

To simulate a network that mimics the evolution of protein-protein interaction networks, we generated the simulated network using a scale free gene duplication network model

[20]. First, a small seed network was given with two connected nodes. Then, we add a node and some edges to network at each step, which simulates the gene duplication and divergence mechanism during the evolution of PPI networks:

**Duplication:** A node $i$ is selected at random. A new node $i'$, with a link to all the neighbors of $i$, is created. With probability $p$ a link between $i$ and $i'$ is established.

**Divergence:** For each of the nodes $j$ linked to $i$ and $i'$, we choose randomly one of the two links $(i,j)$ or $(i',j)$ and remove it with probability $q$.

### *Evaluation measures*

A good NA approach should be able to produce high quality alignments between networks that are similar, and to produce low quality alignments between networks that are dissimilar [8]. It assumes that networks originating from a model with a same set of parameters should be more similar than these from different parameters [20]. We generate 20 dynamic networks using two sets of parameters $p = 0.3, q = 0.7$ and $p = 0.7, q = 0.6$. Each model generates 10 dynamic networks. We align all possible pairs of the synthetic networks by using Twadn, DynaMAGNA++ and DynaWAVE. Alignment quality of the $C_{20}^2 = 190$ pairs of synthetic dynamic networks can reflect the alignment power of the three algorithms. Alignment tools can classify all pairs of networks based on the alignment score into two categories, similar pairs and dissimilar pairs. With a given threshold $\rho$, these network pairs with a score $s > \rho$ will be categorized into similar pairs, others will go to dissimilar pairs.

Afterwards, we can compare the performance of all the alignment algorithms using precision-recall and receiver operating characteristic (ROC). The precision is the fraction of network pairs which are true positive among all pairs which have a score $s > \rho$. The recall is the fraction of network pairs which are true positive among all pairs which are true (similar) network pairs. A precision-recall curve can be plotted by adjusting the threshold $\rho$ from 0 to the maximum observed alignment score. The area under the precision-recall curve (AUPR) and F-score are two commonly used measures for the performance of classification methods. The measure F-score is the harmonic mean of precision and recall which can be calculated by following function:

$$F = 2 \times \frac{precision \times recall}{precision + recall} \tag{8}$$

We used the F-score at the point that precision and recall are equal, termed as F-score$_{cross}$ and the score when F-score gets the maximum, termed as F-score$_{max}$. We also evaluated the binary classification according to the receiver operating characteristic (ROC) curve, which was created by plotting the true positive rate (TPR) against the false positive rate (FPR). Here, TPR is the same as recall. FPR is the fraction of false positive network pairs among all false pairs (i.e. dissimilar network pairs). The ROC curve can be plotted by adjusting the criterion $\rho$. The area under the ROC curve (AUROC) can be calculated after we got the plot.

### *Performance on synthetic dynamic networks*

For synthetic dynamic networks, we aim to develop a NA tool to distinguish these similar network pairs from the other. So we use AUPR, F-score$_{cross}$, F-score$_{max}$, and AUROC to evaluate the quality of the alignment algorithm. As shown in Table 1, the performance

**Table 1** Network discrimination performance of DynaMAGNA++, DynaWAVE and Twadn. For biological synthetic networks, with respect to the area under the precision-recall curve (AUPR), F-score at which precision and recall cross and are thus equal (F-score$_{cross}$), maximum F-score (F-score$_{max}$), and the area under the ROC curve (AUROC). In each column, the best score is bolded

| algorithm | AUPR | F-score$_{cross}$ | F-score$_{max}$ | AUROC |
|---|---|---|---|---|
| DynaMAGNA++ | 0.467 | 0.489 | 0.642 | 0.507 |
| DynaWAVE | 0.600 | 0.556 | 0.642 | 0.594 |
| Twadn | **0.653** | **0.589** | **0.735** | **0.718** |

of Twadn outperforms all other alignment algorithms in terms of AUPR, F-score$_{cross}$, F-score$_{max}$, and AUROC, which are 0.653, 0.589, 0.735, and 0.718, respectively. DynaWAVE shows a better performance than DynaMAGNA++ in terms of AUPR, F-score$_{cross}$, and AUROC. From Fig. 3, it shows that the PR curve of DynaMAGNA++ starts from origin of the coordinate system. It means that DynaMAGNA++ failed to discriminate the one with the best alignment score. From Figs. 3 and 4, Twadn is the best aligner according to both the PR curve and the ROC curve over all.

### Evaluation using real-world dynamic networks

#### *Experimental design on real-world dynamic networks*

To show the alignment capabilities in real-world networks, we evaluated our algorithm on PPI networks. In contrast to static NA tools, Twadn is able to capture dynamic features of a node in the time axis, which would benefit the alignment quality. It's difficult to tell whether two biological networks come from a same evolution model or not because of the ambiguity of evolution model in biology. So we use a randomized (noisy) version of the network (see below). The larger randomized noise level, the more dissimilar the two dynamic network are. We have two randomized versions of the network. One randomizes only the temporal aspect of the network and another randomized both temporal aspect and structure aspect.

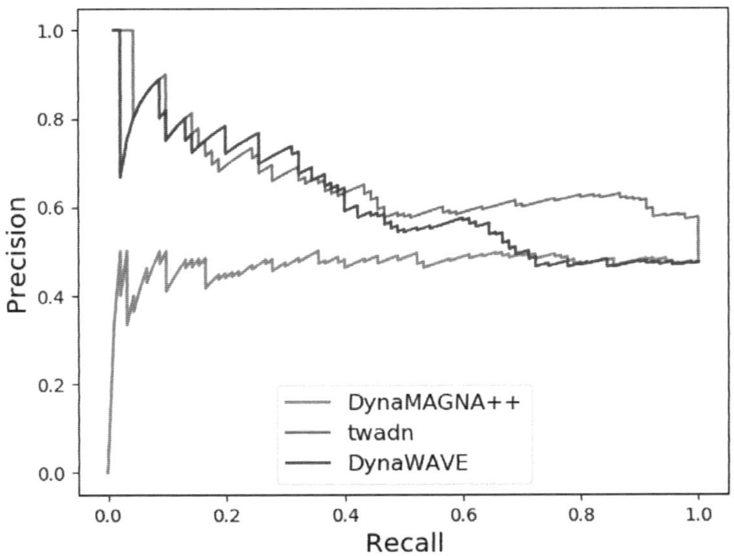


**Fig. 3** Network discrimination performance of DynaMAGNA++ and Twadn for biological synthetic networks with respect to precision-recall curve

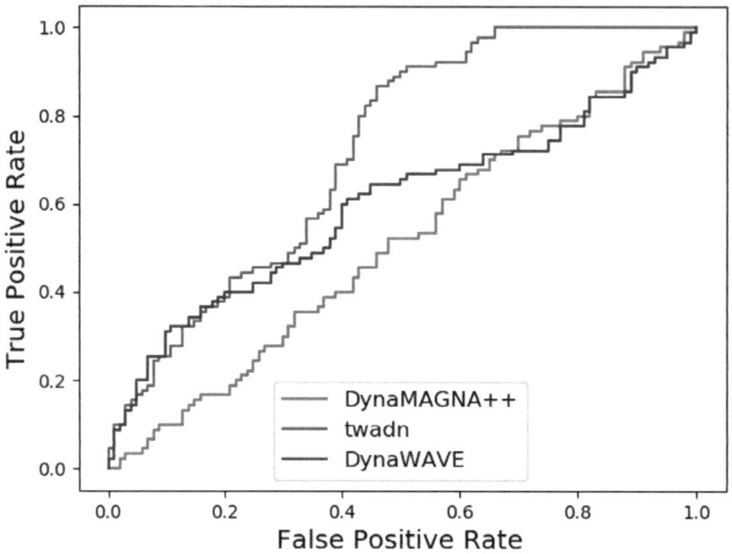


**Fig. 4** Network discrimination performance of DynaMAGNA++ and Twadn for biological synthetic networks with respect to ROC curve

Since there is a lack of available experimental dynamic molecular networks, we create a dynamic Drosophila melanogaster PPI network from an artificial temporal sequence of static PPI networks. Here, the static PPI network that are used as snapshots of the dynamic PPI network are all real-world networks, it is just their temporal sequence that is artificial. The sequence consists of seven static PPI network snapshots: at the first snapshot, network have 70% high confidence interactions of original network. at second snapshot, we add 5% high confidence interactions. Now the network have 75% interactions of original which have greater confidence value than the rest interactions. We continue to add 5% interaction until the network has hole 100% interaction of original network. Then we get a real-world dynamic network with seven static PPI network snapshots. Then we generate two randomized versions of network, as follow:

**Randomizes only temporal aspect of network**: Since the difference between dynamic NA and static NA is that the former accounts for the temporal aspect of the data more explicitly than the latter, we first randomizes only temporal aspect of network, which means randomized network will preserve as much structure as possible of dynamic network. They are only different from each other in time information. This way, the only difference observed between Twadn's and NetCoffee2's performance will be the consequence of considering the temporal aspect of the data. We randomize network with a parameter the noisy level $p$, the larger the $p$ value, the more noise is added. Given the noise level $p$ and $G = (g_1, g_2, ..., g_7), g_i = (V_i, E_i)$, for each $e_{ab}$ $(a \in V_i, b \in V_i)$ in $E_i$, with probability $p$, we arbitrarily select $e_{cd}$ $(c \in V_j, d \in V_j)$ from another snapshot $g_j (j \neq i)$ and swap this two interactions. The specific method of swap is to delete the edge $e_{ab}$ in the snapshot $g_i$ and add an edge $e_{cd}$. At the same time, delete $e_{cd}$ in the snapshot $g_j$ and increase $e_{ab}$. By randomizing only temporal aspect of network, an dynamic network can be aggregated into the same static network as its noisy version's.

**Randomizes both temporal aspect and structure aspect of network**: To observe the performance of Twadn and NetCoffee2 in different sets of noisy versions of the original network using a somewhat more flexible randomization scheme that does not conserve the structure of the flattened version of the original dynamic network, we randomize both temporal aspect and structure aspect of network. Given the noisy level $p$ and $G = (g_1, g_2, ..., g_7), g_i = (V_i, E_i)$, for each $e_{ab}$ $(a \in V_i, b \in V_i)$ in $E_i$, with probability $p$, we arbitrarily select $e_{cd}$ $(c \in V_j, d \in V_j)$ from another snapshot $g_j (j \neq i)$, if there is no edge connection between node $a$ and $d$, and there is no edge connection between node $b$ and node $c$, then connect $a$ and $d$, $b$ and $c$. If the process of adding noise creates a loop (i.e., an edge from a node to itself) or a multiple link (i.e., duplicate edge between the same nodes), then we undo it and re-randomly select another edge to do the above process. By randomizing both temporal aspect and structure aspect of network, the resulting compressed version of the static network is different from the original dynamic network.

#### Evaluation measures

For NA of Drosophila dynamic networks with increased noise, we know the true alignment results (the same protein can be considered a homologous protein). So we use the Alignment Score, which is the algorithm's objective function value (Eq. 3) and node correctness (NC) to measure the network alignment results and the NC can be calculate as follow:

$$NC = \frac{N_{correct}}{N_{all}} \tag{9}$$

where $N_{correct}$ represents the number of correctly aligned protein pairs and $N_{all}$ represents the count of all protein pairs. The greater the node-correctness is, the better the algorithm is. When noise is added to the dynamic network only on the timing information, the dynamic network will become more and more dissimilar to the original network as the level of noise increases, while the dynamic network will be compressed into the same static network. Then as the noise level increases, the comparison result of Twadn will become worse and worse, and the result of NetCoffee2 should not change much. When randomizes both temporal aspect and structure aspect of network, if the dynamic network is compressed to a static version, different static networks will be obtained. As noise level increase, the noise and original versions of dynamic and static networks become more and more dissimilar. Then if Twadn and NetCoffee2 are used to compare the dynamic network with the static network respectively, the comparison results of the two algorithms should be worse as the noise level increases.

#### Result

As Fig. 5 shows, the alignment score of Twadn decrease with the increase of noisy, while the alignment score of NetCoffee2 does not change when randomizing only temporal aspect. It is reasonable since the flattened version of noisy of dynamic network is the same as original network's. The alignment quality of NetCoffee2 and Twadn both decrease when randomizing both temporal aspect and structure aspect of network in Fig. 6. This illustrates that Twadn really capture the time information of dynamic network compared with NetCoffee2. In other hand, the node-correctness of Twadn is higher than NetCoffee2 in every noisy level, which means Twadn is superior to NetCoffee2. We think it is reasonable as Twadn capture the time information of dynamic network while NetCoffee2 does not.

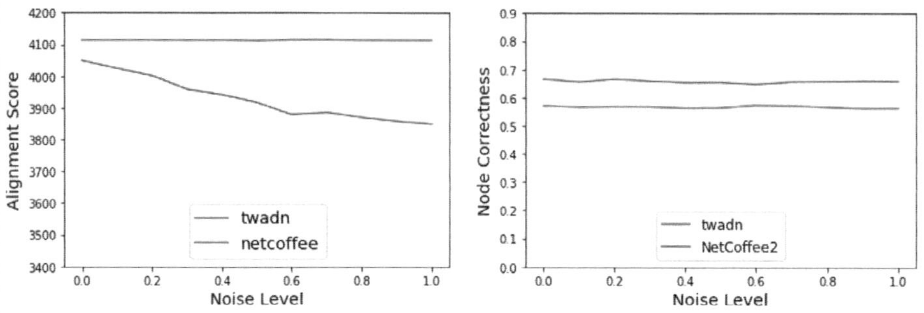


**Fig. 5** The performance of node correctness and alignment score when randomizing only temporal aspect of network

## Discussion

Although the experimental results have been able to achieve better results than existing algorithms. There are still many problems in the work done in this paper that can be further explored: 1) The algorithm temporarily does not support multiple dynamic networks for simultaneous alignment. However, according to the framework of the simulated annealing algorithm, it is hoped that multiple networks can be aligned at the same time in the future. 2) The algorithm is currently only used in PPI networks, and it is expected to be applied to other networks in the future, such as gene regulatory networks, metabolic networks, etc.

## Conclusion

NA is a very important computational framework for understanding molecular function and phylogenetic relationships. Although many NA methods have been developed in the last decade, most of these focused on aligning proteins in static PPI networks. All species and PPIs evolve in different speed. Therefore, there is an urgent demand to develop efficient computational tools to deal with these dynamic networks. To supplement this shortcoming, a novel method Twadn based on the dynamic model of networks was proposed, which can include the time information of molecular interactions. We construct a 5-tupe-feature vector and an optimal warping path to extract topology structures and evolving patterns of all nodes in networks. Twadn was applied in both synthetic datasets and real biological datasets. The synthetic dataset was generated based on a scale-free gene duplication model. Results show that Twadn is superior to DynaMAGNA++ and

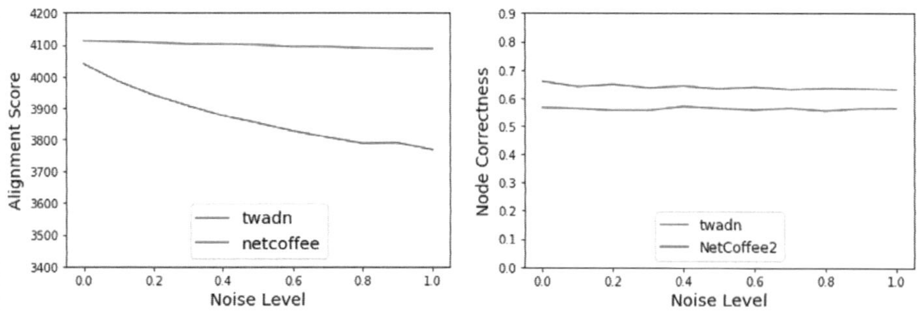


**Fig. 6** The performance of node correctness and alignment score when randomizing both temporal aspect and structure aspect of network

DynaWAVE in synthetic network. At the same time, in order to show that the Twadn can capture timing information compared to the static NA, we add timing information and noise to the Drosophila protein interaction network, and then run the Twadn and Net-Coffee2, experimental results in line with our expectations. Dynamic NA algorithm do capture timing information. It suggests that Twadn is a versatile and efficient alignment tool that can be applied to dynamic network. Hopefully, its application can benefit the research community in the fields of molecular function and evolution.

**Abbreviations**
AUPR: Area under the precision-recall curve; AUROC: Area under the receiver operating characteristic curve; DN: Dynamic network; DTW: Dynamic time warping; FPR: False positive rate; NA: Network alignment; NC: Node correctness; PPI: Protein-protein interaction; ROC: Receiver operating characteristic; TPR: True positive rate; Twadn: Time warping based alignment algorithm for pairwise dynamic networks

**About this supplement**
This article has been published as part of *BMC Bioinformatics Volume 21 Supplement 13, 2020: Selected articles from the 18th Asia Pacific Bioinformatics Conference (APBC 2020): bioinformatics*. The full contents of the supplement are available online at https://bmcbioinformatics.biomedcentral.com/articles/supplements/volume-21-supplement-13.

**Authors' contributions**
YKZ and JL designed the computational framework, JH and YQG implemented the program. JL, XS and JW performed all the analyses of the data; YKZ and JLiu wrote the manuscript; JLH is the major coordinator, who contributed a lot of time and efforts in the discussion of this project. All author(s) read and approved the final manuscript.

**Funding**
Publication costs were funded by the National Natural Science Foundation of China (Grant No. 61702420); This project has also been funded by the National Natural Science Foundation of China (Grant No. 61332014, 61702420 and 61772426); the China Postdoctoral Science Foundation (Grant No. 2017M613203); the Natural Science Foundation of Shaanxi Province (Grant No. 2017JQ6037); the Fundamental Research Funds for the Central Universities (Grant No. 3102018zy032); the Top International University Visiting Program for Outstanding Young Scholars of Northwestern Polytechnical University.

**Availability of data and materials**
The source code of Twadn is freely available at: https://github.com/screamer/twadn.

**Ethics approval and consent to participate**
Not applicable.

**Consent for publication**
Not applicable.

**Competing interests**
The authors declare that they have no competing interests.

**Author details**
[1]School of Computer Science, Northwestern Polytechnical University, West Youyi Road 127, 710072 Xi'an, China. [2]Xi'an Mingde Institute of Technology, Fenghe Campus, Fenghe Campus, 710124 Xi'an, China. [3]Centre of Multidisciplinary Convergence Computing, School of Computer Science, Northwestern Polytechnical University, 1 Dong Xiang Road, 710129 Xi'an, China.

Published: 17 September 2020

**References**
1. Zhong Y, Li J, Liu J, Zheng Y, Shang X, Hu J. Deep learning enables accurate alignment of single cell rna-seq data. In: Yoo I, Bi J, Hu J, editors. 2019 IEEE International Conference on Bioinformatics and Biomedicine (BIBM). New York: IEEE. p. 778–81. https://doi.org/10.1109/BIBM47256.2019.8982969.
2. Hu JL, Wang JR, Lin JA, Liu TW, Zhong YK, Liu J, Zheng Y, Gao YQ, He JH, Shang XQ. Md-svm: a novel svm-based algorithm for the motif discovery of transcription factor binding sites. BMC Bioinformatics. 2019;20:8. https://doi.org/10.1186/s12859-019-2735-3.
3. Hu JL, Gao YQ, Li J, Shang XQ. Deep learning enables accurate prediction of interplay between lncrna and disease. Front Genet. 2019;10:7. https://doi.org/10.3389/fgene.2019.00937.
4. Hu JL, Gao YQ, Li J, Zheng Y, Wang JR, Shang XQ. A novel algorithm based on bi-random walks to identify disease-related lncrnas. BMC Bioinformatics. 2019;20:569. https://doi.org/10.1186/s12859-019-3128-3.
5. Hu J, He J, Li J, Gao Y, Zheng Y, Shang X. A novel algorithm for alignment of multiple ppi networks based on simulated annealing. BMC Genomics. 2019;20(Suppl. 13):932. https://doi.org/10.1186/s12864-019-6302-0.
6. Oentaryo LEAXJSea RJ. Talent flow analytics in online professional network. Data Sci Eng. 2018;3(3):199–220. https://doi.org/10.1007/s41019-018-0070-8.

7. Hu JL, Reinert K. Localali: an evolutionary-based local alignment approach to identify functionally conserved modules in multiple networks. Bioinformatics. 2015;31(3):363–72. https://doi.org/10.1093/bioinformatics/btu652.
8. Vijayan V, Critchlow D, Milenkovic T. Alignment of dynamic networks. Bioinformatics. 2017;33(14):180–9. https://doi.org/10.1093/bioinformatics/btx246.
9. Hu JL, Gao YQ, Zheng Y, Shang XQ. Kf-finder: identification of key factors from host-microbial networks in cervical cancer. BMC Syst Biol. 2018;12(4):54. https://doi.org/10.1186/s12918-018-0566-x.
10. Hu JL, Shang XQ. Detection of network motif based on a novel graph canonization algorithm from transcriptional regulation networks. Molecules. 2017;22(12):9. https://doi.org/10.3390/molecules22122194.
11. Hu JL, Kehr B, Reinert K. Netcoffee: a fast and accurate global alignment approach to identify functionally conserved proteins in multiple networks. Bioinformatics. 2014;30(4):540–8. https://doi.org/10.1093/bioinformatics/btt715.
12. Kalaev M, Smoot M, Ideker T, Sharan R. Networkblast: comparative analysis of protein networks. Bioinformatics. 2008;24(4):594–6. https://doi.org/10.1093/bioinformatics/btm630.
13. Behnam N, Ahmadreza K, Somaye H, Seyed Shahriar A. Netal: a new graph-based method for global alignment of protein-protein interaction networks. Bioinformatics. 2013;29(13):1654–62. https://doi.org/10.1093/bioinformatics/btt202.
14. Paufique J, Madec PY, Kolb J, Kiekebusch MJ, Arsenault R, Siebenmorgen R, Downing M, Hibon P, Valenzuela JJ, Haguenauer P. Graal on the mountaintop. In: Adaptive Optics Systems V. vol. 9909. Bellingham: SPIE; 2016. p. 806–20. https://doi.org/10.1117/12.2232826.
15. Saraph V, Milenkovic T. Magna: Maximizing accuracy in global network alignment. Bioinformatics. 2014;30(20): 2931–40. https://doi.org/10.1093/bioinformatics/btu409.
16. Vijayan V, Saraph V, Milenkovic T. Magna++: Maximizing accuracy in global network alignment via both node and edge conservation. Bioinformatics. 2015;31(14):2409–11. https://doi.org/10.1093/bioinformatics/btv161.
17. Chung-Shou L, Kanghao L, Michael B, Rohit S, Bonnie B. Isorankn: spectral methods for global alignment of multiple protein networks. Bioinformatics. 2009;25(12):253–8. https://doi.org/10.1093/bioinformatics/btp203.
18. Sahraeian SM, Yoon BJ. Smetana: accurate and scalable algorithm for probabilistic alignment of large-scale biological networks. Plos ONE. 2013;8(7):67995. https://doi.org/10.1371/journal.pone.0067995.
19. Vijayan V, Milenković T. Multiple network alignment via multimagna++. IEEE/ACM Trans Comput Biol Bioinforma. 2018;15(5):1669–82.
20. Vijayan V, Milenkovic T. Aligning dynamic networks with dynawave. Bioinformatics. 2017;34(10):1795–8. https://doi.org/10.1093/bioinformatics/btx841.
21. Hu JL, He JH, Gao YQ, Zheng Y, Shang XQ. Netcoffee2: A novel global alignment algorithm for multiple ppi networks based on graph feature vectors. In: Huang DS, Jo KH, Zhang XL, editors. Intelligent Computing Theories and Application, Pt Ii, Lecture Notes in Computer Science, vol. 10955. Cham: Springer; 2018. p. 241–6.
22. Altschul SF, Gish W, Miller W, Myers EW, Lipman DJ. Basic local alignment search tool. J Mol Biol. 1990;215(3): 403–10. https://doi.org/10.1016/S0022-2836(05)80360-2.
23. Holme P. Modern temporal network theory: a colloquium. Eur Phys J B. 2015;88(9):234. https://doi.org/10.1140/epjb/e2015-60657-4.

## Publisher's Note

Wang et al. BMC Bioinformatics 2020, 21(Suppl 13):380
https://doi.org/10.1186/s12859-020-03673-5

BMC Bioinformatics

RESEARCH Open Access

# Biomedical document triage using a hierarchical attention-based capsule network

Jian Wang, Mengying Li, Qishuai Diao, Hongfei Lin, Zhihao Yang and YiJia Zhang*

*From* The 18th Asia Pacific Bioinformatics Conference
Seoul, Korea. 18-20 August 2020

*Correspondence: zhyj@dlut.edu.cn
Dalian University of Technology, The School of Computer Science and Technology, 116024 Dalian, China

**Abstract**

**Background:** Biomedical document triage is the foundation of biomedical information extraction, which is important to precision medicine. Recently, some neural networks-based methods have been proposed to classify biomedical documents automatically. In the biomedical domain, documents are often very long and often contain very complicated sentences. However, the current methods still find it difficult to capture important features across sentences.

**Results:** In this paper, we propose a hierarchical attention-based capsule model for biomedical document triage. The proposed model effectively employs hierarchical attention mechanism and capsule networks to capture valuable features across sentences and construct a final latent feature representation for a document. We evaluated our model on three public corpora.

**Conclusions:** Experimental results showed that both hierarchical attention mechanism and capsule networks are helpful in biomedical document triage task. Our method proved itself highly competitive or superior compared with other state-of-the-art methods.

**Keywords:** Biomedical document triage, Capsule network, Hierarchical attention mechanism, Biomedical literature

## Background

Biomedical natural language processing (BioNLP) has an important role in the framework for implementing precision medicine [1–3] . Biomedical document triage is an important task in BioNLP, and is the first step in the literature curation workflow [4, 5]. Biomedical document triage helps curators and researchers focus on the biomedical literature that contains information relevant to their tasks [6, 7].

In the past decade, biomedical document triage has been an important shared task in the BioCreative challenge community. For example, BioCreative II (IAS) [8] and III

(ACT) [9] focused on the classification of whether a given article contains protein interaction information. BioCreative VI (PM) [10] focused on identifying relevant PubMed citations describing genetic mutations affecting protein-protein interactions. Similarly, various methods have been proposed for the task of biomedical document triage [11]. The majority of these tasks can be divided into either machine learning-based methods or neural network-based methods.

As for machine learning-based methods for biomedical document triage, most depend on effective feature engineering including lexical and syntactic information. For example, Si L et al. [12] utilized logistic regression and support-vector machine algorithms to generate ranked lists of documents. Almeida H et al. [13] experimented with dataset sampling factors and a set of features, as well as three different machine learning algorithms including naive Bayes, support-vector machine and logistic model trees. Generally, machine learning-based methods are skill-dependent and labor-intensive, requiring lots of effort to design particular features.

Recently, neural network-based methods [14] have been successfully applied to biomedical documents. Kim et al. [15] reported on a series of experiments using convolutional neural networks (CNN) trained on top of pretrained word vectors for sentence-level classification tasks. Lai et al. [16] introduced a recurrent convolutional neural network for text classification, which combines CNN with a recurrent neural network (RNN). Some of the above-mentioned methods have been successfully applied to biomedical document triage [17–19]. Attention mechanisms, which can capture the relatively important parts of the input text, have been successfully applied in BioNLP [20, 21].In 2016, Yang et al. proposed a two-layer attention network for text classification [22]. That network would obtain the characteristics of both words and sentences within a document. In 2017, Hinton proposed the CapsNet network architecture [23], which was based on the traditional CNN but it modified some layers.

However, while these methods can automatically extract features to save time and energy in the documents triage task, they have limitations in dealing with long biomedical documents. In the BioCreative VI Precision Medicine Track of the triage task, even the top team received an F-score of less than 70 percent [10]. Those models mentioned above cannot effectively learn the latent feature representation from long biomedical texts. Considering that the word-level attention layer can capture the internal association in the sentence. The obtained vector can reflect the global feature with the information about all the words of the entire sentence. The sentence-level attention layer can capture the association feature between sentences in an article. Recent studies [23–25] indicates that the capsule network retains the advantages of CNN and improve its shortcomings. It can capture more information on spatial patterns aggregated at lower levels that contribute to representing higher level concepts. It forms a more effective recursive process to articulate what to be modeled when there is less training dataset. In general, document text is much longer than sentence text. In particular, some biomedical document texts contain several very complicated sentences including medical terms. In our study, we make full use of the complementarity of hierarchical attention and capsule network to construct our model, in which the attention mechanism can reduce the problem of dependence information loss in the long biomedical document text and the capsule network can capture more feature information at lower levels even there are complicated sentences in biomedical document. These recent advances in neural networks may improve the performance of biomedical document triage. Both attention mechanisms and capsule networks may be helpful in biomedical document triage.

In this paper, we propose a hierarchical attention-based capsule network model, which can effectively capture the important features across sentences and learn the comprehensive latent feature representation for the whole document text. Firstly, our model employed the dynamic route algorithm to accurately identify more features in the biomedical text and improve precision instead of using a max-pooling method where important information may be lost to filter features. Additionally, a hierarchical attention mechanism was used to capture valuable features at both the sentence-level and word-level to better deal with the long text of a document. Our method was evaluated on three public corpora including BioCreative VI Precision Medicine (PM) corpus, BioCreative II (IAS) corpus and BioCreative III (ACT) corpus. Experimental results suggest that the proposed model achieved state-of-the-art performance on all three corpora.

The rest of paper is organized as follows. In "Methods" we give a brief introduction of the biomedical document triage task and describe our proposed model in detail. Then, we present and discuss the experimental results on the three corpora in "Results and discussion". Finally, our conclusion and future plans are presented in "Conclusion" sections.

## Methods

### Biomedical document triage

Biomedical document triage is generally approached as the task of classifying whether a specific article contains information relevant to what is needed. In this paper, we choose three public corpora: BioCreative VI Precision Medicine (PM) corpus, BioCreative II (IAS) corpus and BioCreative III (ACT). All three corpora have been examined by biological curators and domain experts.

The PM corpus is provided by the BioCreative VI Precision Medicine Track task [10]. The PM corpus contains training and test sets that are stored in the 'JSON' format, Fig. 1 gives a negative example and a positive example of the PM corpus. The corpus file includes two types of passages. One is the title text and the other is the abstract text, the latter of which is marked by the label of 'infons'. Negative examples are annotated with a 'no' under the 'relevant' label, this annotation helps identify text that is relevant

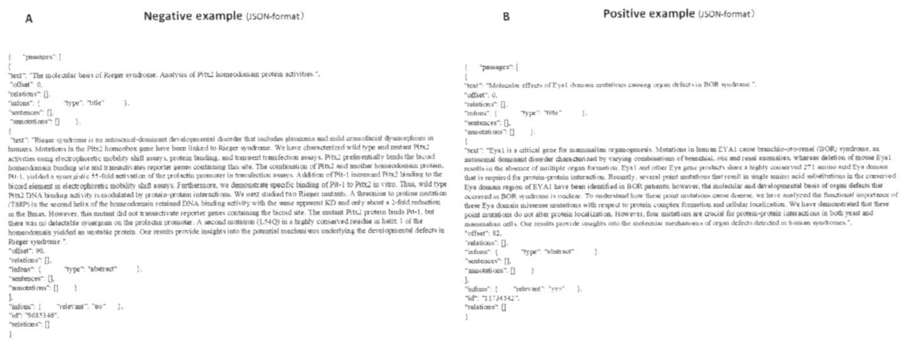


**Fig. 1** Negative examples and positive examples of JON-format from PM datset. **a** negative examples of JON-format from PM datset. **b** positive examples of JON-format from PM dataset

to our area of interest, genetic mutations affecting protein-protein interactions (PPIm). All selected articles were manually annotated by the official organization to answer these questions: Does this article describe experimentally verified protein-protein interactions? Are the database curated PPI pairs for this article mentioned in the abstract? Does this article describe a disease known mutation or a mutational analysis experiment? Is the PPI affected by the mutation? Then, based on the above annotations, articles are carefully categorized as 1)Positives, for articles specifically describing PPI influenced by genetic mutations, 2)Negatives, for articles describing both PPIs and genetic variation analysis with no inference of relation between them or containing PPI but no mutations or containing mutations but no PPI or mentioning neither [10]. In the training set of the PM corpus, there were 1729 relevant articles that describe the protein-protein interactions affected by genetic mutations and 2353 articles that were not relevant. Lastly, the PubMed Unique Identifier (PMID) (e.g. 9685346) of the passages is given under the 'id' label.

The IAS corpus and ACT corpus are a little different from the PM corpus. They are provided in the XML-format. The IAS corpus file includes training and test sets. The IAS training set corpus contained 3536 positive examples that were relevant to protein interactions and 1959 negative examples that were not relevant to protein interactions. The IAS test set contained 338 positive examples and 339 negative examples. The ACT corpus file includes training, development and test sets. There were 1140 positive examples and 1140 negative examples in the ACT training set, 682 positive examples and 3318 negative examples in the ACT development set, and 910 positive examples and 5090 negative examples in the ACT test set. Figure 2 gives an illustration of positive and negative examples of XML-like corpus format. The label of positive and negative is under the 'CURATIONRELEVANCE' label. The corpus in the XML-format also gives the PMID, title text, abstract text and so on under the relevant labels.

## The model architecture

An architecture schematic overview of our model is shown in Fig. 3. In general, our model consists of three main parts: the hierarchical attention layer, the convolution neural network (CNN) and the capsule network layer. The inputs of our model are text sequences. The word embedding generates the distributed representation vector including semantic information for each word. The hierarchical attention layer applies a sentence-level and a word-level multi-attention mechanism to capture the relatively important features based

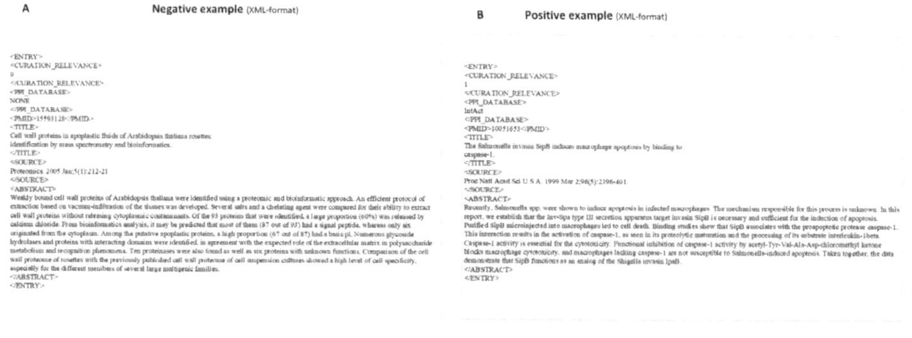
**A** Negative example (XML-format)

<ENTRY>
<CURATION_RELEVANCE>
0
</CURATION_RELEVANCE>
<PPI_DATABASE>
NONE
</PPI_DATABASE>
<PMID>15593128</PMID>
<TITLE>
Cell wall proteins in apoplastic fluids of Arabidopsis thaliana rosettes: identification by mass spectrometry and bioinformatics.
</TITLE>
<SOURCE>
Proteomics. 2005 Jan;5(1):212-21
</SOURCE>
<ABSTRACT>
Weakly bound cell wall proteins of Arabidopsis thaliana were identified using a proteomic and bioinformatic approach. An efficient protocol of extraction based on vacuum-infiltration of the tissues was developed. Several salts and a chelating agent were compared for their ability to extract cell wall proteins without releasing cytoplasmic contaminants. Of the 93 proteins that were identified, a large proportion (60%) was released by calcium chloride. From bioinformatics analysis, it may be predicted that most of them (87 out of 93) had a signal peptide, whereas only six originated from the cytoplasm. Among the putative apoplastic proteins, a high proportion (67 out of 87) had a basic pI. Numerous glycoside hydrolases and proteins with interacting domains were identified, in agreement with the expected role of the extracellular matrix in polysaccharide metabolism and recognition phenomena. Ten proteinases were also found as well as six proteins with unknown functions. Comparison of the cell wall proteome of rosettes with the previously published cell wall proteome of cell suspension cultures showed a high level of cell specificity, especially for the different members of several large multigenic families.
</ABSTRACT>
</ENTRY>

**B** Positive example (XML-format)

<ENTRY>
<CURATION_RELEVANCE>
1
</CURATION_RELEVANCE>
<PPI_DATABASE>
IntAct
</PPI_DATABASE>
<PMID>10051653</PMID>
<TITLE>
The Salmonella invasin SipB induces macrophage apoptosis by binding to caspase-1.
</TITLE>
<SOURCE>
Proc Natl Acad Sci U S A. 1999 Mar 2;96(5):2396-401
</SOURCE>
<ABSTRACT>
Recently, Salmonella spp. were shown to induce apoptosis in infected macrophages. The mechanism responsible for this process is unknown. In this report, we establish that the inv-Spa type III secretion apparatus target invasin SipB is necessary and sufficient for the induction of apoptosis. Purified SipB microinjected into macrophages led to cell death. Binding studies show that SipB associates with the proapoptotic protease caspase-1. This interaction results in the activation of caspase-1, as seen in its proteolytic maturation and the processing of its substrate interleukin-1beta. Caspase-1 activity is essential for the cytotoxicity. Functional inhibition of caspase-1 activity by acetyl-Tyr-Val-Ala-Asp-chloromethyl ketone blocks macrophage cytotoxicity, and macrophages lacking caspase-1 are not susceptible to Salmonella-induced apoptosis. Taken together, the data demonstrate that SipB functions as an analog of the Shigella invasin IpaB.
</ABSTRACT>
<ENTRY>

**Fig. 2** Negative examples and positive examples of XML-like format from IAS dataset. **a** negative examples of XML-like format from IAS dataset. **b** Positive examples of XML-like format from IAS dataset

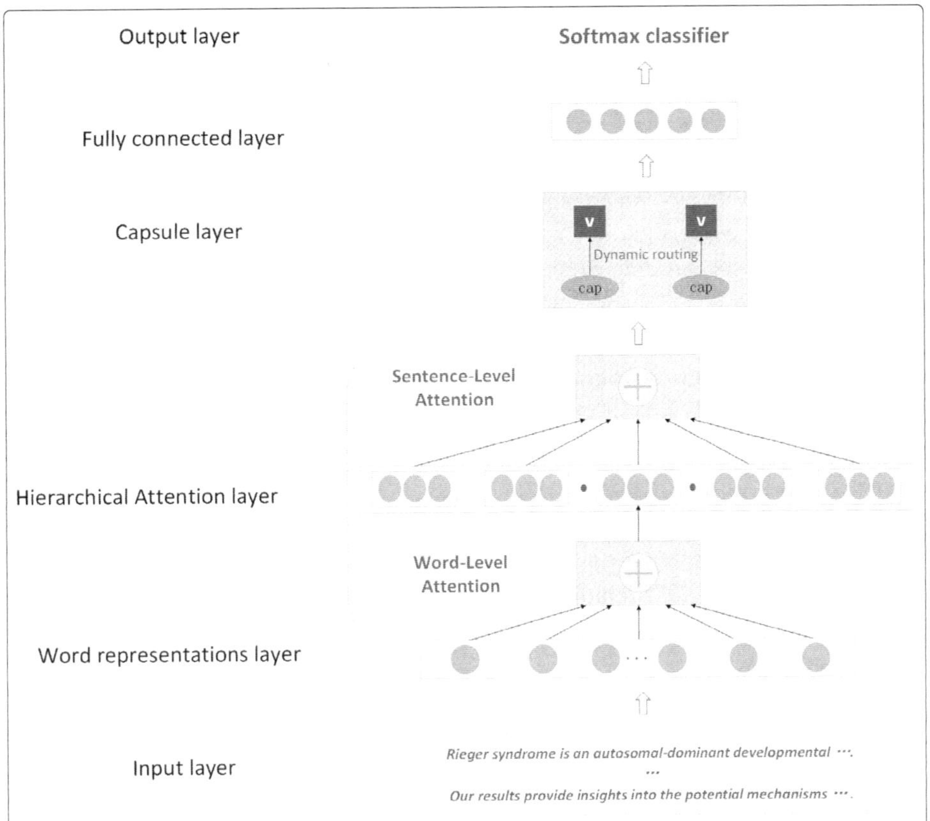


**Fig. 3** The schematic overview of our model

on the whole word representations from the two levels in the long text. After the hierarchical attention mechanism layer, a convolution layer is used to learn some local features from the long text. Importantly, in order to prevent losing significant features to the max-pooling operation of the CNN, we use the dynamic routing algorithm in the capsule layer and convert scalar feature output to vector feature output to learn more features. At last, we employ a fully connected layer and the *Softmax* function to implement document triage.

### Word representations

The distributed representation, also known as word embedding [26, 27], is based on the hypothesis that semantically similar words have similar semantics. In the field of BioNLP word embedding is widely used, it effectively captures the semantic information underlying each word. For this paper, we used the pre-trained word embedding downloaded from https://github.com/cambridgeltl/BioNLP-2016, which was trained on the PubMed Central Open Access subset (PMC) corpus.

### Hierarchical attention mechanism

Attention mechanisms have become an important part of some compelling sequence models and transduction models for various tasks, allowing modeling of dependencies without regard for their distance in the input or output sequences [22, 28, 29]. In our model, we combine word-level and sentence-level attention mechanisms to capture important features across sentences.

### *Multi-head attention mechanism*

The principle on which the multi-head attention mechanism functions is that applying the attention many times may learn more features than a single application. For attention mechanisms, the self-attention mechanism is a special case where the input and output are the same sequences in the Encoder-Decoder framework. The physical meaning in machine translation is a word alignment mechanism between the target word and the source word in the general attention mechanism, while the self-attention mechanism learns the internal connection or grammatical structure of the sentence. As an example, we consider the sentence "The animal didn't cross the street because it was too tired." What does "it" refer to, "street" or "animal?" This is a simple problem for humans, but it is not simple for an algorithm. When the model processes the word "it," the self-attention mechanism associates "it" with "animal." When the model processes each word, that is, when processing each position of the input sequence, the self-attention mechanism allows it to look at other locations in the input sequence to find ways to better encode each word. Figure 4 illustrates the calculation process of the self-attention mechanism.

An attention function is created by mapping a query and a set of key-value pairs to an output, where the key, values, query and output are all vectors. A weighted sum of the values is the output, where the weight assigned to each value is computed by a compatibility function of the query with the corresponding key. Our input was made of queries and keys of the dimension $d_k$, and values of the dimension $w$. The dot products of the query with all keys were computed and each was divided by $\sqrt{d_k}$. Finally, a *Softmax* function was be used to obtain the weights of the values, which is shown in the following formula.

$$Attention(Q, K, V) = softmax\left(\left(QK^T\right)/\sqrt{d_k}\right)V$$

where $Q$ is a matrix in which a set of queries is packed together. Similarly, $K$ and $V$ are the matrices in which the keys and values, respectively, are packed together.

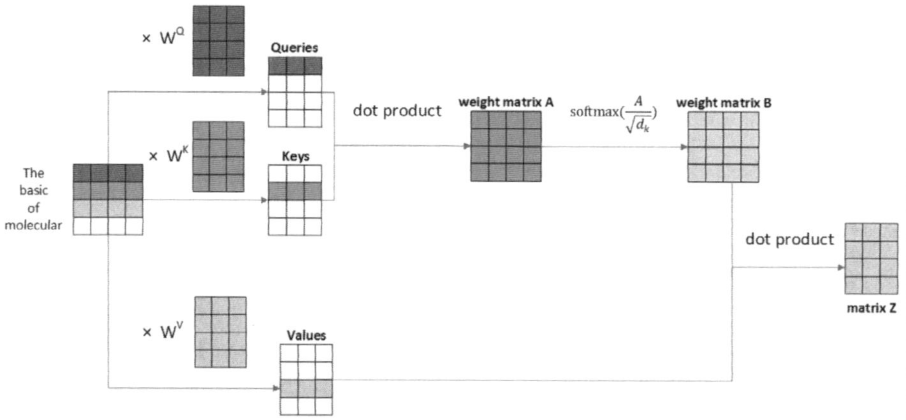


**Fig. 4** The self-attention mechanism calculation process. We can get three vectors that are a query vector, a key vector and a value vector for each word by multiplying the embedding word vector by the three matrices trained during the training. The size of these new vectors is 3, while the dimensions of the embedding word and encoder are 4. We evaluate the dependence between the words with the dot product operation

It is more effective to linearly project the queries, keys, and values many times with different dimensions for each than to perform a single attention function with the dimensional keys, values, and queries, which is the main idea of the multi-head attention mechanism. The model is allowed to jointly attend to information from different representation subspaces at different positions in the multi-attention mechanism, which is shown as follows:

$$head_i = Attention\left(QW_i^Q, KW_i^K, VW_i^v\right)$$

$$MultiHead(Q, K, V) = Concat\left(head_1, \ldots, head_h\right) W^o$$

where the $W_i^Q$, $W_i^K$ and $W_i^v$ are the parameter matrices, whose dimensions are $d_model \times d_k$, $d_model \times d_k$, and $d_model \times d_v$, respectively. The dimension of $W^o$ is $hd_v \times d_model$, where $h$ is the number of times a single attention calculation is performed. The matrices above all are initialized randomly. After training, each set of input vectors is projected into a different representation subspace.

The multi-head attention mechanism improves the performance of the attention layer by extending the model's ability to focus on different locations and provide multiple "representative subspaces." Our goal is that the attention model will learns different dependency information in different heads from the long biomedical texts so that we can further improve on the performance of the biomedical text classification task.

#### *Hierarchical attention mechanism*

Hierarchical attention is aimed to capture two basic types of features from a biomedical document structure, one being word-level features and the other being sentence-level features.

Since the time complexity of the attention mechanism is $O(n2)$, the amount of computation increases significantly when the input of the model is a longer text. On the other hand, there is little connection between words found in different sentences or long sentences. There are the word-to-sentence and sentence-to-document features in each text. Correspondingly, the representation of the sentence can be constructed first by the word, followed by a representation of the text that can be constructed by the sentence. Because different words and sentences have different information, not only can the information between the words be obtained, but the information between the sentences can be obtained with the two-levels of features. Hierarchical attention mechanisms can give words and sentences different weights to accommodate that fact the same words and sentences can have different roles in different texts. The problem that dependence information is lost when the input text is too long, which often occurs in the text classification task, can be solved by the hierarchical attention mechanism and we can get more features from words and sentences in the document in this way.

There are many sentences in each text and many words in each sentence, complicating the biomedical document triage task, this makes the training speed slow when using the self-attention mechanism directly. In our work, we mainly use hierarchical attention mechanisms based on self-attention. First, the self-attention mechanism at the word level is used to find the dependencies between words in the sentence, and then it is used at the sentence level to find the dependencies between the sentences in the document, which not only speeds up the training, but also establishes the characteristics of the word and sentence levels in the text. Our experimental result shows its effectiveness in improving the performance of the document triage task. We use the first part of the hierarchical attention model to process the words of each clause with the purpose of transforming the sentences into vectors. The second part deals with the sentence vector of the document to generate a new sentence vector using the self-attention calculations and perform subsequent convolution operations. Figure 5 gives the hierarchical mechanism architecture. We can see that the input text is split into words $(w_1, w_2, \ldots, w_n)$ that will be encoded by the recurrent neural network. Each word will get an attention weight $(a_1, a_2, \ldots, a_n)$ about the input text sequence. Then the output of the word-level attention will be used as the feature of the sentence-level attention for the sentence encoder. The sentence-level attention input is the document which is split into sentences $(s_1, s_2, \ldots, s_n)$, and we will calculate the attention weights $(b_1, b_2, \ldots, b_n)$ for the sentences. At last, we will get a feature vector with more information and the *Softmax* function will be used to normalize the result.

### Convolution neural network (CNN)

After the hierarchical attention layer, we use a slight variant of the traditional convolutional layer as the baseline to further capture local features for optimization of our model. Generally, each convolutional layer of the traditional CNN is serially connected, that is,

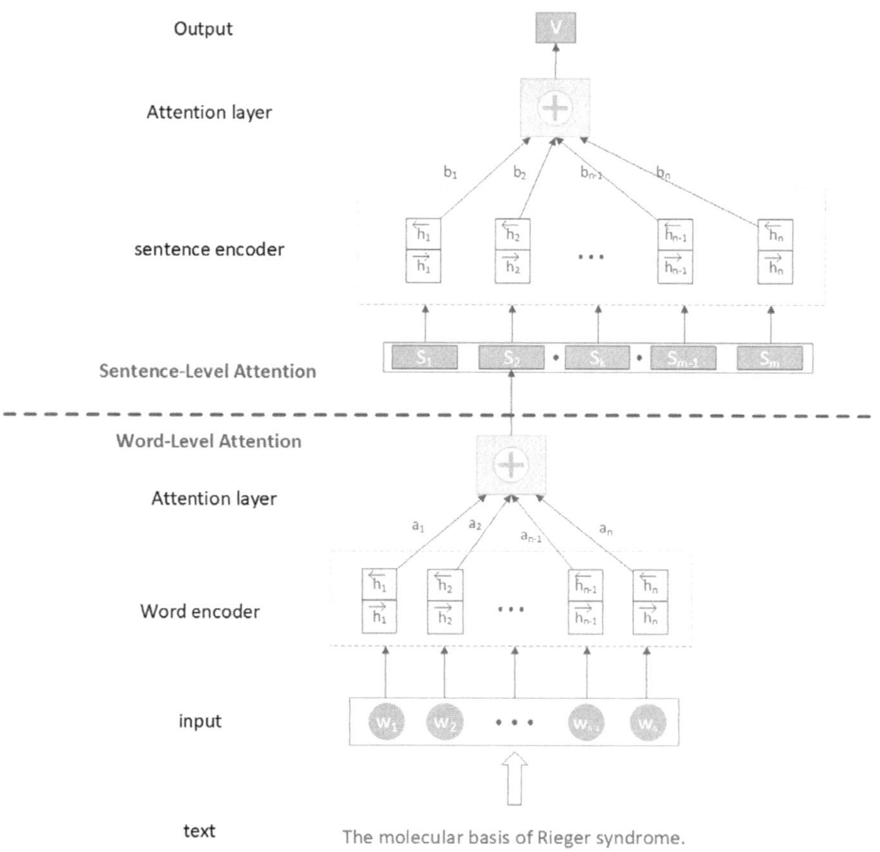


**Fig. 5** The hierarchical attention mechanism architecture. We connect the word-level and sentence-level features by attention mechanism to get a feature vector with more information and the Softmax function is used to normalize the result

the output of the upper layer is used as the input of the next layer, but in our experiment, each convolutional layer of the CNN is connected in parallel, and the output of each convolutional layer is spliced together as the output of the CNN. In this layer, let $x_i$ be the k-dimensional word vector corresponding to the $i - th$ word in the sentence. A sentence of length n is represented as follows. If the length is not n, we will pad it where necessary.

$$x_{(1:n)} = x_1, x_2, \ldots, x_n$$

$x_{(1:n)}$ denotes the concatenation operator of $x_1, x_2, \ldots, x_n$. The convolution operation can be seen as filtering features, which obtains local optimal features through the kernel function. Then these features are combined together to form new features. In this way, each layer is filtered out and the more significant features are passed to the higher layers to calculate as follows:

$$S_{((t))} = ReLU\left(Wx_{(t:t+w-1)} + b\right)$$

where $W$ is a transformation matrix, also known as the convolution kernel function. The input sequence is $\left[x_t, x_{(t+1)}, \ldots x_{(t+w-2)}, x_{(t+w-1)}\right]$, where the lowercase $w$ is the input window size. $ReLU$ is an activation function that is a non-linear unit function and $b$ is the bias vector. The filter is applied to each possible window of words in the sentence $x_{(t:t+w-1)}$ to produce the feature map S, which is the convolutional layer result.

### Capsule network

A CNN model can effectively capture local features, but cannot capture global features. In brief, a CNN generates different features through multiple convolution kernels, and the features are accumulated layer by layer, but in this process, the network loses important information: i.e. the spatial relationship between the features. To address this disadvantage of the CNN model, Hinton et al. proposed capsule networks [23]. An important concept of CNN is the pooling strategy, which downsamples the input vectors. In the text classification area, each convolution kernel can be used to detect the relevant meanings of consecutive words to generate text features. If a similar text feature reappears, the output value of this convolution kernel becomes larger, which is well preserved by the pooling strategy. The pooling can handle the translation change. When a feature moves, as long as it does not exceed the size of the pooled window, it will not be lost and will be detected, which can make the network position-invariant. However, the disadvantage of this method is that the pooling operation, such as max-pooling, retains only the most important features while losing a lot of information. The ideal pooling not only reduces the data dimension, but also retains various features and information so that each feature does not change through the pooling layer. Based on this idea, the capsule network replaces the scalar-output feature detectors of CNN with vector-output capsules and max-pooling with routing-by-agreement, it still likes to replicate learned knowledge across space. Unlike max-pooling, information about the precise position of the entity within the region is not thrown away in the capsule network.

The capsule network is well trained by a powerful dynamic routing mechanism that ensures the capsule's output reaches the appropriate parent node. The basic idea of the dynamic routing mechanism is to design a nonlinear mapping strategy, whose task it is to let the output reach the appropriate parent capsule. More specifically, the capsule output is sent to all of the parent capsules in the next layer, and then, for each possible parent capsule, the sub-capsules calculate their outputs by multiplying by the weight matrix. If this result is a large scalar product of the parent capsules output, then the connection between the sub-capsule and the parent capsule should be close, which is achieved by increasing the coupling coefficient of the sub-capsule to the parent capsule and reducing the coupling coefficient with other parent capsules. In theory, this dynamic routing protocol is more efficient than the routing method implemented by max-pooling. From a mathematical point of view, a non-linear "squashing" function is used to ensure that short vectors get shrunk to a near zero length, which is shown as follows.

$$v_j = \left(||s_j||^2\right) / \left(1 + ||s_j||^2\right) s_j / \left(||s_j||\right)$$

where $v_j$ is the vector output of capsule $j$ and $s_j$ is its total input. For all but the first layer of the capsules, the total input to a capsule $s_j$ is a weighted sum over all "prediction vectors" $\widehat{u}_{(j|i)}$ from the capsules in the layer below and is produced by multiplying the output $u_i$ of a capsule in the layer by the weight matrix $W_{(ij)}$, which is shown as follows.

$$s_j = \Sigma_i c_{(ij)} \widehat{u}_{(j|i)}$$

$$\widehat{u}_{(j|i)} = W_{(ij)} u_i$$

where the $c_ij$ are coupling coefficients that are determined by the iterative dynamic routing process. The calculation method of $c_ij$ is shown as follows.

$$c_{(ij)} = exp\left(z_{(ij)}\right) / \Sigma_i exp\left(z_{(ik)}\right)$$

The coupling coefficients between capsule $i$ and all the capsules in the layer above are determined by a routing *Softmax* whose initial logits $z_ij$ are the log prior probabilities that capsule $i$ should be coupled to capsule $j$. The coupling coefficients sum to 1.

Our capsule network architecture is shown in Fig. 6. It can be seen that we first use the convolutional layer to extract some primary features and next use a primary capsule layer to capture more features and convert the scalar output to vector output to be the

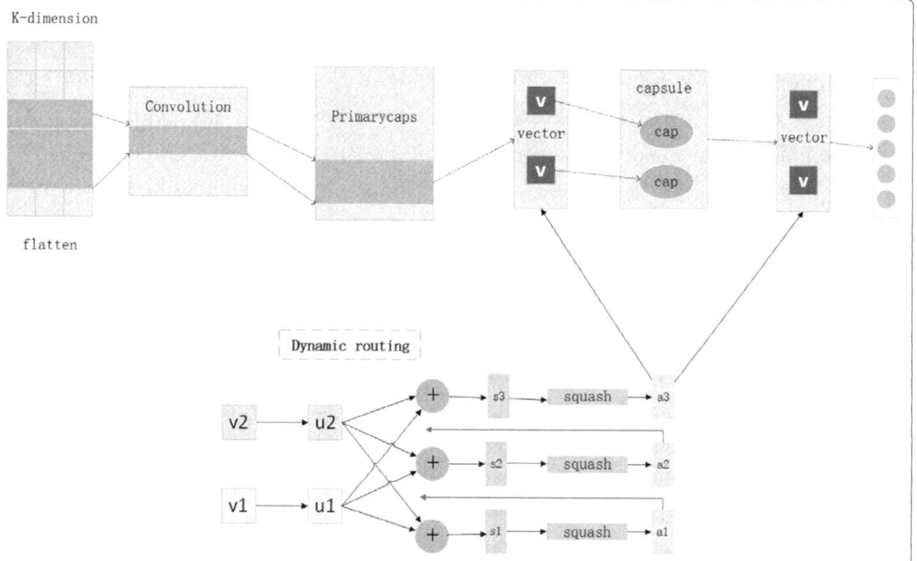


**Fig. 6** The capsule network model architecture

input of the next capsule layer which can get accuracy features. The useful information in the biomedical long texts can be preserved by the dynamic routing algorithm that is used in our two capsule layers. In the capsule network, the output is a vector instead of a scalar, this means that the output of the primary capsule layer will give a very clean and accurate signal to the appropriate subsequent capsule for the exact transmission of information. The capsule vector dimension is set to 32 in our experiments and the number of the dynamic routing iterations is set to 3 which can be seen from Fig. 6. In the dynamic routing algorithm, the vectors $(v_1, v_2)$ are the input which is transformed by the affine transformation to get the outputs $u_1, u_2$. The $s_1, s_2, s_3$ will be calculated as the sum of $u_1, u_2$. Then, $s_1, s_2, s_3$ will be used in formula to calculate the vector outputs $(a_1, a_2, a_3)$. The results of $a_1, a_2$ will be fed back to optimize the last result $a_3$.

### Classification and training

We used the *'Softmax'* function on the output layer to implement the detection and classification of the biomedical documents about PPI and PPIm in this paper. In the experiments, the programming language was Python, and the version was 3.5. The python libraries we used included numpy, scikit-learn, gensim, etc. Our model was implemented by Keras with the TensorFlow 1.13.1 backend. We employed the dropout layer mechanism before the word representation layer and output layer to prevent the overfitting of the neural network model. The hyperparameters used in our experiments are listed as follows. The dimension of our models in the word embedding was initialized to 200 to adapt to the time and space computational complexity. The number of hidden layer neurons was set to 100. The batch size was set to 128 and the number of epochs was set to 50 during the training. The size of each convolution kernel was 3, 4, and 5, which was set to 128 feature maps, and the dropout probability was set 0.5 and 0.8. The convolutional layer activation function was a ReLU function. In the hierarchical attention layer, the heads number of the first attention mechanism layer was set to 8, the vector dimension was set to 32, the second layer of attention mechanism had a headcount of 8, and a vector dimension of 32. In the capsule neural network, the dimension was set to 32, and the dynamic route iteration number was 3. The learning rate was set to 0.01. Then, all parameters of the models were optimized by using Adam [30] to minimize the categorical cross-entropy loss. The computing environment is Ubuntu 16.04.5 LTS and the hardware environment is GeForce GTX Titan Xp. Our source code is available at https://github.com/dqshuai/text_classification-of-biomedicine.

## Results and discussion

### Datasets and evaluation metrics

In our experiment, we used the corpus from the BioCreative II Interaction Article Sub-task, BioCreative III Article Classification Tasks and the BioCreative VI Precision Medicine Track to mine for protein interactions and mutations for a precision medicine task. The statistics of all corpus are presented in Table 1.

In the PM corpus experiments, 10% of the PM training sets were randomly selected as the development set to tune the models' hyperparameters and the remaining data were used to train our model. We evaluated the models on the open test set provided by the organizers of the BioCreative VI PM document triage task. We proved the effectiveness of our model by testing it on the other corpus, the test sets of IAS and ACT.

**Table 1** Dataset statistics

| Corpus | Positive | Negative | Total |
|---|---|---|---|
| PM Training set | 1729 | 2353 | 4028 |
| PM Test set | 704 | 723 | 1427 |
| IAS Training set | 3536 | 1959 | 5495 |
| IAS Test set | 338 | 339 | 667 |
| ACT Training set | 1140 | 1140 | 2280 |
| ACT Development set | 682 | 3318 | 4000 |
| ACT Test set | 910 | 5090 | 6000 |

The Precision (P), Recall (R) and F-score (F1) were used to measure our model's performance on the biomedical document triage task of PM corpus and IAS corpus, which were calculated by the official evaluation scripts . P, R, and F were calculated as follows:

$$P = TP/((TP + FP))$$

$$R = TP/((TP + FN))$$

$$F1 = 2PR/((P + R))$$

The F1 value is a comprehensive evaluation of the precision and the recall, which is an evaluation the overall performance. TP compares the actual positive examples to the positive examples that the model identified correctly, known as true positive examples, FP compares the actual negative examples to the positive examples that the model identifies incorrectly, known as false positive examples, and FN compares the actual positive examples to the negative examples, known as false negative examples. We can use the confusion matrix to express those clearly.

### Performance of our method

#### *Baseline methods*

Before inputting text into the model, some preprocessing of the original corpus was conducted. First, we extracted the texts and lables from the original corpora. Secondly, the text data was removed some invalid data points, such as those with empty text. Moreover, we use regular matching to keep only the letters and numbers. Then we began to process the sentence-level and word-level corpus to prepare for input into the model. For the input data of the attention mechanism, the extracted text needed to be split into sentences, then the sentences were split into words, and some noisy sentences were selectively removed. In order for the length of each text to be consistent when the model was trained, text padding was performed with $< PAD/ >$.

We compared our biomedical document triage method with some baseline methods, including CNN, capsule network, and self-attention, using the word em-bedding trained by Word2Vec tool as the model input.

CNN: This is a traditional method for text classification. In our work, this network consisted of the word representing layer, convolution layer, max-pooling layer, dropout layer, dense layer, and *Softmax* layer, successively. We used two convolution layers for the 128 feature maps, which were learned for each of two different filters size 3,8 and the step size is set to 2 in the max-pooling layer. The parameters of the dropout layer were set at 0.5 and 0.8.

Capsule network: This method was first used to recognize highly overlapping digits. In our work, we use the capsule network based on a CNN to process the text, which is for the BioNLP field. We used the neuron vectors to replace the neuron scalar nodes in the traditional deep neural network and used the dynamic routing protocol to replace the max-pooling layer in the CNN to train the new neural network. The capsule vector dimension was set to 32, and the dynamic route iteration was set to 3. Therefore, the capsule network was made up of the word representing layer, convolution layer, capsule layer, dropout layer, dense layer, and *Softmax* layer, successively.

Self-attention: Self-attention was first used in machine translation tasks to surpass and replace the recurrent neural network. In our baseline work, we combine the CNN with the self-attention to classify biomedical documents. We set the number of multi-head attention heads at 8 and the vector dimension of the attention mechanism at 64. This neural network was made up of a word embedding representation, self-attention layer, convolution layer, dropout layer, dense layer, and *Softmax* layer, successively.

The results of the baseline methods are shown in Table 2. From Table 2, we can make the followings observations. Firstly, CNN only used the language pretraining model trained by Word2Vec tool to get information from the text, which achieves an F-score of 0.664. Secondly, the capsule network effectively improved the F-score from 0.664 to 0.686. The results indicate that the dynamic routing algorithm of capsule networks was able to capture more features from the text information. Thirdly, integrating self-attention can significantly improve the performance of the CNN model (by an average improvement of 4.3% in F-score). The experimental results suggest that both capsule networks and attention mechanism are helpful in biomedical document triage task.

#### *Effects of hierarchical attention mechanism*

In Table 3, we evaluated the effect of a hierarchical attention mechanism on the PM corpus. From Table 3, we can see that the performance of the hierarchical attention mechanisms based on CNN was significantly higher than the CNN alone (an average improvement of 5.1% in F-score) and the self-attention based on CNN (an average improvement of 0.8% in F-score). The experimental results show that the hierarchical attention mechanisms greatly improved the precision value and captured more dependency features between words and sentences than the self-attention mechanism which only mines word-level information. The attention mechanism can reduce the problem of dependence information loss in the long biomedical document text.

#### *Effects of capsule network*

In Table 4, we evaluated the effect of capsule networks on the PM corpus. From Table 4, we can see that the single capsule network performed better than the single CNN (an average improvement of 2.2% in F-score). The capsule network based on hierarchical attention achieved better performance than and the CNN based on the hierarchical attention (an average improvement of 0.8% in F-score). The proposed hierarchical attention-based capsule network model achieved the best performance in all methods when we use the

**Table 2** The results of baseline methods

| Methods | P | R | F1 |
|---|---|---|---|
| CNN | 0.581 | 0.774 | 0.664 |
| Capsule network | 0.629 | 0.755 | 0.686 |
| Self-Attention | 0.584 | 0.895 | 0.707 |

**Table 3** The results of hierarchical attention

| Methods | P | R | F1 |
|---|---|---|---|
| CNN | 0.581 | 0.774 | 0.664 |
| CNN+self-attention | 0.584 | 0.895 | 0.707 |
| CNN+hierarchical attention | 0.623 | 0.840 | 0.715 |

capsule network based on hierarchical attention, which had an average improvement of 5.9% in F-score compared to the baseline of CNN. The precision and recall were both improved by this method. Hierarchical attention can cover the shortage of the self-attention in precision while the capsule network can further improve the hierarchical attention in recall. The capsule network can capture more feature information at lower levels even there are complicated sentences in biomedical document.

#### Performance comparison on PM corpus

In Table 5, we compared our method with other state-of-the-art methods on the PM corpus. It should be noted that the PrecMed Baseline [10] is the baseline method in the BioCreative VI PM document triage task and the PrecMed-best [10] is the method that got the highest F-score in the BioCreative VI PM document triage task challenge. "Team 418" got the highest precision and "Team 421" got the highest recall in the BioCreative VI PM document triage task challenge team competition. To the best of our knowledge, the ensemble model has had the best performance to date. Many deep learning models were used to improve the performance including five individual neural network models including LSTM (long-short term memory), CNN, LSTM-CNN (combine the LSTM and the CNN), recurrent CNN, and hierarchical LSTM. At last, they got an F-score of 0.710 by combining five models' results with three different alternatives.

Compared with other methods, our method has achieved the highest F-score (0.723) on the PM corpus. From Table 5, we can see our model's F-score is 2.8 percentage points higher than the best performance in the BioCreative VI PM challenge. In particular, our method achieved an improvement of 1.3% over the ensemble model (0.710 F-score). What's more, our model is relatively simple to implement unlike the ensemble model where high effort is required to construct the neural network models, which need much more time and energy to combine and integrate. To our knowledge, it is the first time to explore the complementarity of hierarchical attention and capsule network to classify long biomedical texts.

#### Performance of our model on ACT corpus and IAS corpus

To demonstrate the generalization of our methods, we added some comparison experiments on the other corpus. One is the ACT corpus from the BioCreative III Article

**Table 4** The results of capsule network

| Methods | P | R | F1 |
|---|---|---|---|
| CNN | 0.581 | 0.774 | 0.664 |
| Capsule network | 0.629 | 0.755 | 0.686 |
| CNN+hierarchical attention | 0.623 | 0.840 | 0.715 |
| CapsNet+hierarchical attention | 0.624 | 0.895 | 0.723 |

**Table 5** Performance compared with other methods on PM corpus

| Methods | P | R | F1 |
|---|---|---|---|
| PrecMed Baseline [10] | 0.610 | 0.636 | 0.622 |
| Team 418 | 0.629 | 0.766 | 0.691 |
| Team 421 | 0.570 | 0.874 | 0.690 |
| PrecMed-best [10] | 0.603 | 0.821 | 0.695 |
| Ensemble model | 0.629 | 0.815 | 0.710 |
| CapsNet+hierarchical attention | 0.624 | 0.895 | 0.723 |

Classification Task and the other is the IAS corpus from the BioCreative II Interaction Article Sub-task. The corpus statistics are shown in Table 1.

We evaluated our hierarchical attention-based capsule model on the ACT corpus. The results of our method compared with the top teams in the BioCreative III ACT challenge are shown in Table 6. The participating teams were provided with a training set of 2,280 abstracts and a development set of 4,000 abstracts, while the evaluation was carried out on a test set of 6,000 abstracts through comparison to manual labels generated by domain experts. They measured the performance of the ten participating teams in this task over a total of 52 runs. Table 6 gives the best result of each participating team in their many runs. The experimental results show that our method achieves an F-score of 0.618 on the ACT corpus, which outperforms the top teams in BioCreative III ACT challenge.

We also evaluated our model on the IAS corpus, shown in Table 7. Table 7 gives the best result achieved by each participating team. From Table 7, it can be seen that most of the teams achieved a high recall, but a modest precision. Our method also achieved the best F-score on the IAS corpus. All in all, our method is superior to the other state-of-the-art methods when applied to these three public corpora.

### Visualization of attention mechanisms

Considering that almost all the neural networks related to deep learning are used in the form of "black boxes", we visualize the attention mechanism in order to increase the interpretability of the model. An example sequence is used to visualize the attention mechanisms in Fig. 7. The line connecting the two words represents the association of the two words. The deeper the color is , the closer the relationship between the two words will be. As can be seen from the figure, the self-attention mechanism can learn the dependence of the words in-side the sentence. The blue lines represent the dependence

**Table 6** Performance compared with other methods on ACT corpus

| Methods | P | R | F1 |
|---|---|---|---|
| Team 65 | - | - | 0.598 |
| Team 70 | - | - | 0.549 |
| Team 73 | - | - | 0.614 |
| Team 81 | - | - | 0.311 |
| Team 88 | - | - | 0.344 |
| Team 89 | - | - | 0.608 |
| Team 90 | - | - | 0.596 |
| Team 92 | - | - | 0.572 |
| Team 100 | - | - | 0.594 |
| Team 104 | - | - | 0.539 |
| Our method | 0.570 | 0.676 | 0.618 |

**Table 7** Performance compared with other methods on IAS corpus

| Methods | P | R | F1 |
|---|---|---|---|
| Team 4 | 0.712 | 0.792 | 0.750 |
| Team 6 | 0.708 | 0.860 | 0.777 |
| Team 7 | 0.684 | 0.858 | 0.761 |
| Team 11 | 0.676 | 0.781 | 0.725 |
| Team 14 | 0.746 | 0.470 | 0.757 |
| Team 19 | 0.645 | 0.565 | 0.602 |
| Team 27 | 0.607 | 0.852 | 0.709 |
| Team 28 | 0.750 | 0.810 | 0.779 |
| Team 30 | 0.686 | 0.789 | 0.643 |
| Team 31 | 0.667 | 0.594 | 0.629 |
| Team 37 | 0.575 | 0.946 | 0.715 |
| Team 41 | 0.619 | 0.890 | 0.730 |
| Team 44 | 0.688 | 0.868 | 0.764 |
| Team 48 | 0.588 | 0.863 | 0.700 |
| Team 49 | 0.526 | 0.985 | 0.685 |
| Team 51 | 0.717 | 0.828 | 0.769 |
| Team 52 | 0.692 | 0.834 | 0.757 |
| Team 57 | 0.703 | 0.875 | 0.780 |
| Team 58 | 0.667 | 0.730 | 0.697 |
| Our method | 0.704 | 0.879 | 0.782 |

information of a single-head attention mechanism, from which we can find that the word 'hypersensitive' can learn the dependence relationship with the word 'role', 'has', 'development' and 'adaptation'. The purple lines represent the dependence information of another single-head attention mechanism, from which we can see that the word 'hypersensitive' learns some different dependence relationship with different word 'acid', 'aba', 'role', 'stress', 'adaptation', 'aba1', 'hab1', and 'a'. The multi-head attention mechanism in Fig. 7 combines the dependence information of the two single-head attention to extend the model's ability to focus on different locations and provide multiple "representative subspaces" for the attention layer. It is possible to objectively observe the ability of self-attention mechanism finds the dependencies between words in a sentence by visualization of self-attention, indicating that there is an important position in the field of BioNLP for the attention mechanism. In addition, the attention mechanism is also one of the few deep learning models that can be visualized and can be analyzed in detail.

### Error analysis

We manually analyzed why our hierarchical attention-based capsule network model failed to better classify biomedical documents in the PM corpus experiments. The prediction results confusion matrix is shown in Table 8, from which we can see the number of false positives and true negatives for the classification error.

**Table 8** The confusion matrix on PM corpus

| Actual | Predict | |
|---|---|---|
| | True | False |
| True | 865 | 142 |
| False | 522 | 98 |

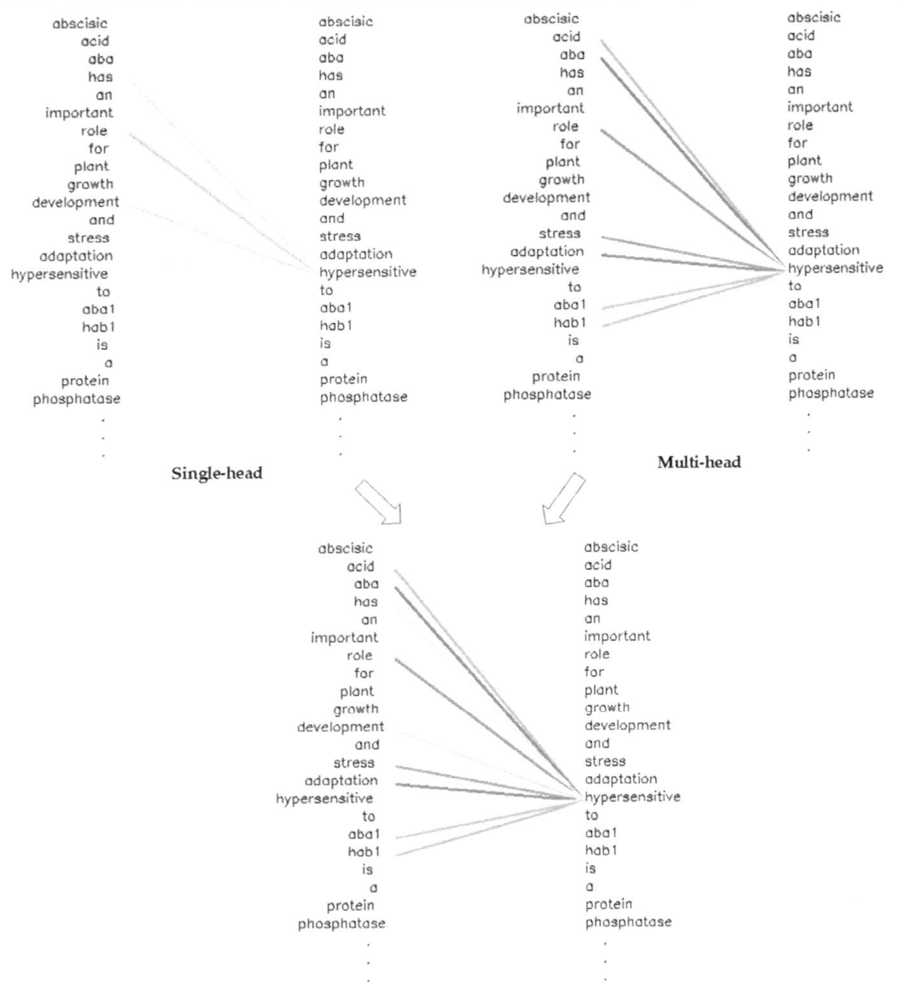


**Fig. 7** The visualization of the multi-head attention mechanism. The blue lines represent the dependence relationship between the words in the single-head self-attention mechanism and the purple lines represent the dependence relationship between the words in the multi-head self-attention mechanism

We found that the significant classification error was the identification of negative examples as positive examples by our model. We analyzed the reason as follows. When an article included some strong PPI indicators or some words similar to the examples, our model mistook it as a positive example that is actually negative. Figure 8 gives the false positive examples, the words causing the misunderstanding are in bold. We can see that the text of PMID (PubMed ID): 9685346 contains some words such as protein, interaction, and mutation, which are strong positive keywords in PPIm articles, however, it does not describe PPI influenced by genetic mutations. Similarly, our model misclassified the PMID: 9685346 as a positive example because it has similar expressions with positive articles while it is actually negative.

When we analyzed the reason that the actual positives are deemed negatives, we found that some strong positive keywords were missing, or that the positive indicators did not appear in the positive articles. It is difficult to accurately classify true positive PPIm articles when their appearance is rare (even zero) in the training set. For example, the article

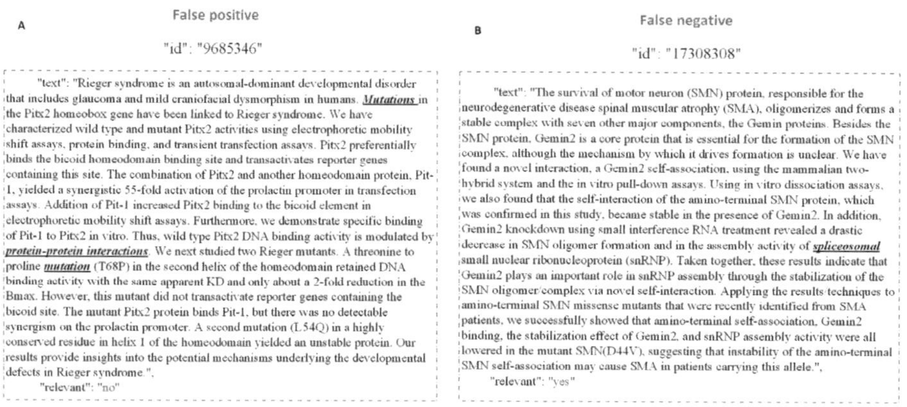

A

False positive

"id": "9685346"

"text": "Rieger syndrome is an autosomal-dominant developmental disorder that includes glaucoma and mild craniofacial dysmorphism in humans. ***Mutations*** in the Pitx2 homeobox gene have been linked to Rieger syndrome. We have characterized wild type and mutant Pitx2 activities using electrophoretic mobility shift assays, protein binding, and transient transfection assays. Pitx2 preferentially binds the bicoid homeodomain binding site and transactivates reporter genes containing this site. The combination of Pitx2 and another homeodomain protein, Pit-1, yielded a synergistic 55-fold activation of the prolactin promoter in transfection assays. Addition of Pit-1 increased Pitx2 binding to the bicoid element in electrophoretic mobility shift assays. Furthermore, we demonstrate specific binding of Pit-1 to Pitx2 in vitro. Thus, wild type Pitx2 DNA binding activity is modulated by ***protein-protein interactions***. We next studied two Rieger mutants. A threonine to proline ***mutation*** (T68P) in the second helix of the homeodomain retained DNA binding activity with the same apparent KD and only about a 2-fold reduction in the Bmax. However, this mutant did not transactivate reporter genes containing the bicoid site. The mutant Pitx2 protein binds Pit-1, but there was no detectable synergism on the prolactin promoter. A second mutation (L54Q) in a highly conserved residue in helix 1 of the homeodomain yielded an unstable protein. Our results provide insights into the potential mechanisms underlying the developmental defects in Rieger syndrome.",

"relevant": "no"

B

False negative

"id": "17308308"

"text": "The survival of motor neuron (SMN) protein, responsible for the neurodegenerative disease spinal muscular atrophy (SMA), oligomerizes and forms a stable complex with seven other major components, the Gemin proteins. Besides the SMN protein, Gemin2 is a core protein that is essential for the formation of the SMN complex, although the mechanism by which it drives formation is unclear. We have found a novel interaction, a Gemin2 self-association, using the mammalian two-hybrid system and the in vitro pull-down assays. Using in vitro dissociation assays, we also found that the self-interaction of the amino-terminal SMN protein, which was confirmed in this study, became stable in the presence of Gemin2. In addition, Gemin2 knockdown using small interference RNA treatment revealed a drastic decrease in SMN oligomer formation and in the assembly activity of ***spliceosomal*** small nuclear ribonucleoprotein (snRNP). Taken together, these results indicate that Gemin2 plays an important role in snRNP assembly through the stabilization of the SMN oligomer/complex via novel self-interaction. Applying the results/techniques to amino-terminal SMN missense mutants that were recently identified from SMA patients, we successfully showed that amino-terminal self-association, Gemin2 binding, the stabilization effect of Gemin2, and snRNP assembly activity were all lowered in the mutant SMN(D44V), suggesting that instability of the amino-terminal SMN self-association may cause SMA in patients carrying this allele.",

"relevant": "yes"

**Fig. 8** The false positive examples and false negative examples on PM corpus. **a** the false positive examples on PM corpus. **b** the false negative examples on PM corpus

with PMID: 17308308 actually describes PPI influenced by genetic mutations, but common positive keywords such as 'mutation' are replaced by the words like 'spliceosomal,' a word that rarely appears in the training set. Our model misclassified such positive instances as negative. In the future, a post-processing step could be helpful for these cases.

## Conclusion

Biomedical document triage is a crucial task in biomedical NLP, which is the first step in assisting literature curation workflows. Both attention mechanism and capsule networks are the recent advantages in neural networks. In this paper, we present a hierarchical attention-based capsule model for biomedical document triage. The proposed model employed the dynamic route algorithm and hierarchical attention mechanism to capture the important features across sentences. We evaluated our model on three BioCreative corpora. Experimental results showed that both hierarchical attention mechanism and capsule networks can improve performance in biomedical document triage. It is encouraging to see that our method achieved the state-of-the-art performance on all three corpora.

In future work, we will explore the effectiveness of pretrained deep contextualized word representations, such as Bert and ELMo, in biomedical document triage tasks. In addition, post-processing may further improve the performance of our model. The current state-of-the-art methods in biomedical document triage are primarily based on supervised machine learning and thus are highly dependent on sufficient labeled data. However, creating labeled datasets is prohibitively expensive and labor-intensive in the biomedical domain. Hence, reducing the dependency of methods on labeled training data is a key challenge in this domain. We will also plan on employing semi-supervised learning or transfer learning in biomedical document triage.

### Abbreviations

BioNLP: Biomedical Natural Language Processing; CNN: Convolutional Neural Networks; RNN: Recurrent Neural Network; PPIm: genetic mutations affecting Protein-Protein Interactions; PPI: Protein-Protein Interactions; PM: Precision Medicine; IAS: Interaction Article Sub-task; ACT: Article Classification Tasks; PMC: PubMed Central Open Access subset; LSTM: Long-Short Term Memory

### Acknowledgements

Not applicable.

**About this supplement**
This article has been published as part of *BMC Bioinformatics Volume 21 Supplement 13, 2020: Selected articles from the 18th Asia Pacific Bioinformatics Conference (APBC 2020): bioinformatics*. The full contents of the supplement are available online at https://bmcbioinformatics.biomedcentral.com/articles/supplements/volume-21-supplement-13.

**Authors' contributions**
ZYJ, WJ, LHF and YZH designed the study. DQS,ZYJ and LMY implemented the analysis. LMY wrote the manuscript. All author(s) have read and approved the final manuscript.

**Competing interests**
The content is that the authors declare that they have no competing interests.

**Funding**
This work has been supported by the grants from the Natural Science Foundation of China No. 61572098 and 61572102. Publication costs are funded by the Natural Science Foundation of China No. 61572098.

**Availability of data and materials**
PM, IAS and ACT data are all publicly available at https://biocreative.bioinformatics.udel.edu/resources/.

**Ethics approval and consent to participate**
No ethics approval was required for the study.

**Consent for publication**
Not applicable.

Published: 17 September 2020

**References**

1. Simmons M, Singhal A, Lu Z. Text mining for precision medicine: Bringing structure to ehrs and biomedical literature to understand genes and health. Adv Exp Med Biol. 2016;939:139–66.
2. Ayush S, Michael S, Zhiyong L, Andrey R. Text mining genotype-phenotype relationships from biomedical literature for database curation and precision medicine. PLoS Comput Biol. 2016;12(11):1005017.
3. Collins FS, Varmus H. A new initiative on precision medicine. N Engl J Med. 2015;372(9):793–5.
4. Rindflesch TC, Tanabe L, Weinstein JN, Hunter L. Edgar: extraction of drugs, genes and relations from the biomedical literature. In: Biocomputing 2000. Singapore: World Scientific; 1999. p. 517–28.
5. Krauthammer M, Nenadic G. Term identification in the biomedical literature. J Biomed Inform. 2004;37(6):512–26.
6. Cohen AM, Bhupatiraju RT, Hersh WR. Feature generation, feature selection, classifiers, and conceptual drift for biomedical document triage. In: TREC, vol. 500–261. America: National Institute of Standards and Technology (NIST); 2004.
7. Cohen AM. An effective general purpose approach for automated biomedical document classification. In: AMIA Annual Symposium Proceedings. vol. 2006. America: American Medical Informatics Association; 2006. p. 161.
8. Krallinger M, Leitner F, Rodriguez-Penagos C, Valencia A. Overview of the protein-protein interaction annotation extraction task of biocreative ii. Genome Biol. 2008;9(2):4.
9. Krallinger M, Vazquez M, Leitner F, Salgado D, Chatr-Aryamontri A, Winter A, Perfetto L, Briganti L, Licata L, Iannuccelli M, et al. The protein-protein interaction tasks of biocreative iii: classification/ranking of articles and linking bio-ontology concepts to full text. BMC bioinformatics. 2011;12(8):3.
10. Islamaj Doğan R, Kim S, Chatr-aryamontri A, Wei C-H, Comeau DC, Antunes R, Matos S, Chen Q, Elangovan A, Panyam NC, et al. Overview of the biocreative vi precision medicine track: mining protein interactions and mutations for precision medicine. Database. 2019;147(2019):.
11. Huang C-C, Lu Z. Community challenges in biomedical text mining over 10 years: success, failure and the future. Brief Bioinform. 2015;17(1):132–44.
12. Si L, Kanungo T. Thresholding strategies for text classifiers: Trec 2005 biomedical triage task experiments. In: TREC. America: National Institute of Standards and Technology (NIST); 2005.
13. Almeida H, Meurs M-J, Kosseim L, Butler G, Tsang A. Machine learning for biomedical literature triage. PLoS ONE. 2014;9(12):115892.
14. LeCun Y, Bengio Y, Hinton G. Deep learning. Nature. 2015;521(7553):436–44.
15. Kim Y. Convolutional neural networks for sentence classification. In: Proceedings of the 2014 Conference on Empirical Methods in Natural Language Processing, EMNLP. Doha: ACL; 2014. p. 1746–51.
16. Lai S, Xu L, Liu K, Zhao J. Recurrent convolutional neural networks for text classification. In: Twenty-ninth AAAI Conference on Artificial Intelligence. Austin: AAAI Press; 2015.
17. Shweta, Ekbal A, Saha S, Bhattacharyya P. A deep learning architecture for protein-protein interaction article identification. In: 2016 23rd International Conference on Pattern Recognition (ICPR). Mexico: IEEE; 2016.
18. Rios A, Kavuluru R. Convolutional neural networks for biomedical text classification:application in indexing biomedical articles. In: ACM Conference Bioinform. Atlanta: ACM; 2015.
19. Du J, Chen Q, Peng Y, Xiang Y, Tao C, Lu Z. Ml-net: multi-label classification of biomedical texts with deep neural networks. J Am Med Inform Assoc JAMIA. 2019;26(11):1279–85.
20. Zhang Y, Lin H, Yang Z, Wang J, Zhang S, Yuanyuan, Sun, Yang L. A hybrid model based on neural networks for biomedical relation extraction. J Biomed Inform. 2018;81:83–92.
21. Kumar SS, Ashish A. Drug-drug interaction extraction from biomedical texts using long short-term memory network. J Biomed Inform. 2018;86:15–24.

22. Pappas N, Popescu-Belis A. Multilingual hierarchical attention networks for document classification. In: Proceedings of the Eighth International Joint Conference on Natural Language Processing, IJCNLP, vol. 1. Taipei: Asian Federation of Natural Language Processing; 2017. p. 1015–25.
23. Sabour S, Frosst N, Hinton GE. Dynamic routing between capsules. In: AdNIPS'17: Proceedings of the 31st International Conference on Neural Information Processing Systems. Beach, CA; 2017. p. 3856–66.
24. Zhao W, Ye J, Yang M, Lei Z, Zhang S, Zhao Z. Investigating capsule networks with dynamic routing for text classification. In: Proceedings of the 2018 Conference on Empirical Methods in Natural Language Processing. Brussels: Association for Computational Linguistics; 2018. p. 3110–19.
25. Ramasinghe S, Athuralya CD, Khan S. A context-aware capsule network for multi-label classification. In: omputer Vision - ECCV 2018 Workshops Proceedings, Part III. Lecture Notes in Computer Science, vol. 11131. Munich: Springer; 2018. p. 546–54.
26. Mikolov T, Sutskever I, Chen K, Corrado G, Dean J. Distributed representations of words and phrases and their compositionality. AAdv Neural Inf Process Syst. 2013;26:3111–9.
27. Lai S, Liu K, He S, Zhao J. How to generate a good word embedding. IEEE Intell Syst. 2016;31(6):5–14.
28. Bahdanau D, Cho K, Bengio Y. Neural machine translation by jointly learning to align and translate. In: 3rd International Conference on Learning Representations, ICLR. San Diego: Conference Track Proceedings; 2014.
29. Kim Y, Denton C, Hoang L, Rush AM. Structured attention networks. In: 5th International Conference on Learning Representations, ICLR 2017. Toulon; 2017.
30. Kingma DP, Ba J. Adam: A method for stochastic optimization. In: 3rd International Conference on Learning Representations, ICLR (poster). San Diego; 2014.


## Publisher's Note

You *et al. BMC Bioinformatics* 2020, **21**(Suppl 13):383
https://doi.org/10.1186/s12859-020-03674-4

BMC Bioinformatics

RESEARCH Open Access

# Developing the novel bioinformatics algorithms to systematically investigate the connections among survival time, key genes and proteins for Glioblastoma multiforme

Yujie You[1], Xufang Ru[2], Wanjing Lei[1], Tingting Li[3], Ming Xiao[1], Huiru Zheng[4], Yujie Chen[2*] and Le Zhang[1*]

*From* The 18th Asia Pacific Bioinformatics Conference
Seoul, Korea. 18-20 August 2020

* Correspondence: yujiechen6886@foxmail.com; zhangle06@scu.edu.cn
[2]Department of Neurosurgery, Southwest Hospital, Third Military Medical University, Chongqing, P.R. China
[1]College of Computer Science, Sichuan University, Chengdu 610065, China
Full list of author information is available at the end of the article

## Abstract

**Background:** Glioblastoma multiforme (GBM) is one of the most common malignant brain tumors and its average survival time is less than 1 year after diagnosis.

**Results:** Firstly, this study aims to develop the novel survival analysis algorithms to explore the key genes and proteins related to GBM. Then, we explore the significant correlation between *AEBP1* upregulation and increased *EGFR* expression in primary glioma, and employ a glioma cell line LN229 to identify relevant proteins and molecular pathways through protein network analysis. Finally, we identify that *AEBP1* exerts its tumor-promoting effects by mainly activating mTOR pathway in Glioma.

**Conclusions:** We summarize the whole process of the experiment and discuss how to expand our experiment in the future.

**Keywords:** Glioblastoma multiforme (GBM), Survival analysis, Adipocyte enhancer-binding protein 1 (AEBP1), Data mining, mTOR pathway

## Background

Glioblastoma multiforme (GBM) is one of the most common malignant brain tumors and its average survival time is less than 1 year after diagnosis [1]. The occurrence of GBM is considered as a complicated biological phenomenon with multiple external simulating factors, genes and stages. The major challenge in the field is to translate the almost unique progress in deciphering the highly complex molecular genetic nature of GBM into advances that allow for better prognosis rate and survival of affected patients [2].

Currently, cancer researchers usually employ the survival analysis to explore genes that are closely related to GBM [2–5]. For example, Xia et al. [2] developed a CoxSisLasso survival analysis algorithm to identify gene signature for GBM by processing P> > N (the dimension of the factors P are less than the number of samples N) type of data [2]. However, since CoxSisLasso survival analysis algorithm does not consider the interactions among genes, its predictive accuracy is very limited. Therefore, we propose our first research question: Can we develop a more efficient survival analysis algorithm to explore the gene signature of GBM for P> > N type of data?

Secondly, current commonly used genomics [6] or proteomics [7] biotechnologies have limitations in quantitatively measuring gene expression and protein contents, which prevent us from developing the precise mathematical model. Thus, we propose our second research question: In order to explore the key proteins by using quantitative gene/protein data, can we employ cutting-edge biotechnologies, such as the CRIS PR(clustered regularly interspaced short palindromic repeat sequences) [7–9] to knock out the key gene and then use RPPA(reverse phase protein arrays) [10–15] to high-throughput screen the corresponding protein data?

Thirdly, although a few GBM studies ([2, 3, 16–22]) considered the multi-scale data from intracellular, cellular and tissue scales, they neither develop the high efficient computational biology algorithm nor use experimental data to validate their findings in genomic and proteomic level. Thus, we propose our third research question: can we build a precise multi-scale mathematical model that can be used to understand the origin of the GBM from systematic view?

This study aimed to address the above research questions and the following novel approaches have been taken: (1) Firstly, we improved previous additive survival analysis algorithm(CoxSisLasso) [2, 9, 23] to investigate the key genes for GBM survival by considering genes' interaction as well as used immune-chemistry experiments and TCGA data [24] to validate the findings. (2) Secondly, in order to study the effect of key gene on the expression of which proteins, we employed the CRISPR [7–9] and RPPA [10–15] biotechnologies to knock out the explored key gene and obtain the quantitative gene/protein data. (3) Thirdly, we used the aforementioned quantitative gene/protein data to develop a precise multi-scale mathematical model to find key proteins, which can be used to investigate the origin of the GBM from the genetic, protein and tissue level starting from explored key gene as well as use the related experiments to evaluate our findings.

In summary, this study has developed an efficient survival analysis algorithm to identify GBM related *Adipocyte enhancer-binding protein 1* (*AEBP1*) gene, and then used the related quantitative gene/protein data to explore the key proteins and molecular mechanism for GBM. Results show significant correlation between *AEBP1* upregulation and increased *EGFR* expression in primary glioma. A glioma cell line LN229 was used to identify major protein players and molecular pathways through RPPA analysis after *AEBP1* overexpression and AEBP1 knockdown. We reveal that *AEBP1* exerts its tumor-promoting effects by mainly activating mTOR pathway in Glioma.

## Methods

The pipeline of the study is illustrated in Fig. 1.

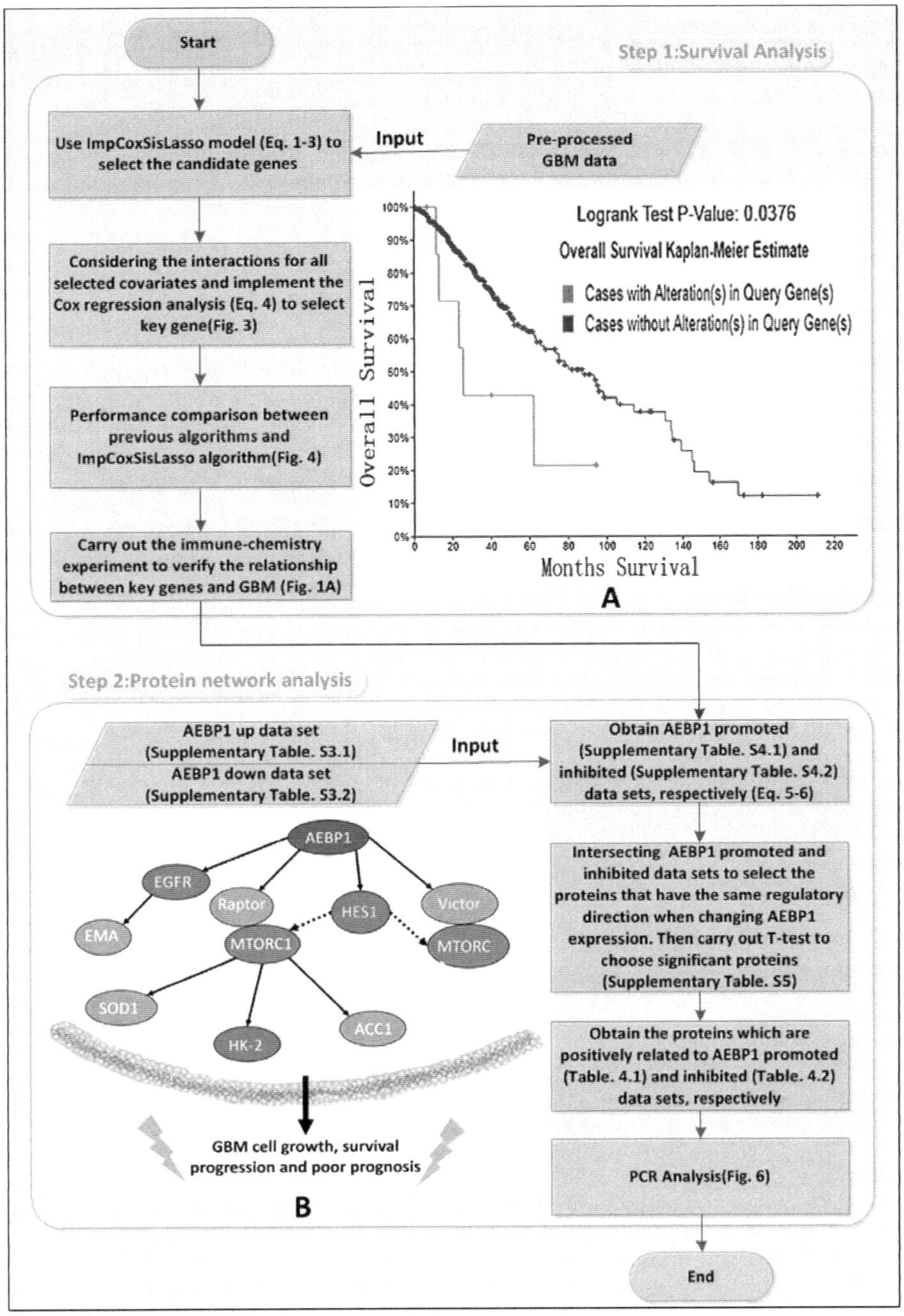


**Fig. 1** Pipeline of the study. **a** The survival curves for AEBP1 [24]. **b** Schematic diagram illustrating the protein network regulated by AEBP1

### Survival data

We use a multi-study microarray database of GBM expression profiles ($n = 227$) from the Georgetown Database of Cancer G-DOC [2], based on the Affymetrix U133 plus 2.0 GeneChip microarray platform. The original microarray datasets are normalized and preprocessed by R software package. After preprocessing step, there are 227 samples and 54,675 genes left in the data matrix. Next, the interquartile range (IQR) threshold is employed to screen out the genes with small variance value. After that, there are only 227 samples and 10,992 genes left in the GBM gene expression and survival time data matrix.

**Improved survival analysis algorithm (ImpCoxSisLasso)**

We extend our previously well-developed CoxSisLasso strategy [2] by considering the gene interactions to develop ImpCoxSisLasso algorithm(Eq. 1–4).

$$\hat{\beta}_{LASSO} = \arg\min\left\{ -\sum_{k\in D}\left[x_k^T\beta - \log\left(\sum\nolimits_{j\in R_k}\exp\left(x_j^T\beta\right)\right)\right] + \lambda\sum_{j=1}^{p}\left|\beta_j\right| \right\} \quad (1)$$

where β is an unknown p-dimensional regression coefficient vector and xi is a vector of potential predictors for the $i^{th}$ individual. Based on the samples, $\hat{\beta}_{LASSO}$ is the LASSO [2] estimator of the unknown parameter coefficients β. D is the set of indices of the events and $R_k$ denotes the set of indices of the individuals at risk in time $t_k$. k and i represent the index for the set D and set $R_k$, respectively. The tuning parameter λ is used to control the sparsity of the estimator.

$$\hat{\beta}_m = \arg\max_{\beta_m}\left\{ \sum_{k\in D}\left[x_{k,C_0}^T\beta_{C_0} + x_{k,m}\beta_m - \log\left(\sum\nolimits_{j\in R_k}\exp\left(x_{j,C_0}^T\beta_{C_0} + x_{j,m}\beta_m\right)\right)\right] \right\} \quad (2)$$

where the index of selected covariates for the nonzero components in $\hat{\beta}_{LASSO}$ is denoted by $C_0$ and each remaining covariate except $C_0$ is denoted by $x_m$, where m∈{1,2,...,p}. We choose the significant covariates with *P*-Value smaller than a threshold value, 1/p for example.

$$\min_{\beta_\Theta}\left\{ -\sum_{k\in D}\left[x_{k,\Theta}^T\beta_\Theta - \log\left(\sum\nolimits_{j\in R_k}\exp\left(x_{j,\Theta}^T\beta_\Theta\right)\right)\right] + \lambda\sum_{j\in\beta_\Theta}\left|\beta_j\right| \right\} \quad (3)$$

where Θ is the collection of the augmented selected predictors $C_0 \cup C_1$ with $C_0$ denoting the index of selected covariates with Lasso, and $C_1$ denoting the chosen covariates by Eq. 2.

$$\max_{\beta_C,\beta_{i_1,i_2}}\sum_{k\in D}\left[\begin{array}{c} x_{k,C}^T\beta_C + \sum\limits_{i_1\neq i_2, i_1, i_2\in C} x_{ki_1}x_{ki_2}\beta_{i_1,i_2} \\ -\log\left(\sum\nolimits_{j\in R_k}\exp\left(x_{j,C}^T\beta_C + \sum\limits_{i_1\neq i_2, i_1, i_2\in C} x_{ji_1}x_{ji_2}\beta_{i_1,i_2}\right)\right)\end{array}\right] \quad (4)$$

where C denotes the index of finally selected covariates by Eq. 3. and the product $x_{i_1}x_{i_2}$ with $i_1 \neq i_2$, $i_1$ and $i_2$∈C denote the interactions between the selected genes. Here, $i_1$ and $i_2$ represent the index for the interactions' terms. Then, Fig. 2 describes ImpCoxSisLasso algorithm as the following.

1: Input: gene data $G$ with $X$ vectors and $p$ dimensions
2: $C_0 \leftarrow$ Lasso( $G$)   Obtain the index of selected gene data by *Eq.1*
3: for each $X_m$ in $G$ and $m \notin C_0$ do by *Eq.2*
    $(Coef_m, P\text{-}Value_m) \leftarrow CoxRegression(X_m)$
  end for
  for each index $i$ in $m$
    if $P\text{-}Value_i \leqslant 1/p$
      $C_1 \leftarrow i$
    end if
  end for
4: $C \leftarrow$ Lasso( $X_{C_0 \vee C_1}$ )   Obtain the index of selected gene data by *Eq.3*
5: $(Coef_n, Pvalue_n) \leftarrow CoxRegression(X_C \vee (X_{i1} * X_{i2}))i_1 \neq i_2, i_1, i_2 \in C$ by *Eq.4*
  for each index $i$ in $n$
   if $Pvalue_i \leq 0.05$   Obtain the name of selected gene
    *ImpCoxSisLassoGene* $\leftarrow$ *GeneName(i)*
   end if
  end for
6: Output: *ImpCoxSisLassoGene*

**Fig. 2** The ImpCoxSisLasso algorithm

### Immune chemistry experiment

Immunofluorescence staining was performed on GBM tissue sections as previously described. The specimens were rapidly isolated and post fixed in 4% PFA for 24 h and then soaked in 40% sucrose for 1 day. GBM tissue sections (10 μm) were obtained using a cryostat (Leica CM3050S-3-1-1, Bannockburn, IL, USA) and permeabilized with 0.3% Triton X-100 in PBS for 30 min. Sections were blocked with 5% donkey serum for 1 h and slices were incubated with a solution of 0.6 mg/ml diaminobenzidine and 0.05% $H_2O_2$ for 2 min. After that, incubation was terminated with three 10-min phosphate-buffered saline washes. Finally, slices were mounted onto gelatin-coated slides and dried overnight before placing a coverslip on them. Representative sections for each GBM tissue were then photographed.

### Cell lines, plasmids

The LN229 cell line was purchased from ATCC [25]. The *pcDNA-AEBP1* Plasmid was provided by Professor Ro from Dalhousie University, Canada [26]. Plasmid for *sgRNA-AEBP1* was constructed by inserting two guide RNA from exon 12 of *AEBP1* into pPB-Cas9-puro at the cloning site. We named the construct with insert as *pPB-Cas9-puro-AEBP1-f1/r4.*

### *AEBP1* overexpression experiment

Cell line LN229 was seeded at 4 × 105 cells/well in a 60 mm plate and incubated in standard culture medium at 370 °C overnight. Culture plates designated for pcDNA vector or pcDNA-AEBP1 transfection were done in triplicates. Transfection was done the next day using Fugene 6 transfection reagent from Roche. We used Reagent to DNA ratio of 3:1 during the transfection. Cells were harvested 48 h after transfection. Growing Cells in culture plates were trypsinized, spun down and washed with PBS for two times. Cell pellet was stored at -80 °C for RPPA analysis. Cells from one plate of vector transfection and another plate of pcDNA-AEBP1 transfection were lysed for Western blotting analysis to confirm the expression of *AEBP1* at the same time of RPPA cell pellet preparation.

### AEBP1 CRISPR-cas9 knockdown experiment

Cell line LN229 was seeded at 4 × 105 cells/well in a 60 mm plate as above. Culture plates designated for vector or sgRNA-AEBP1 transfection were done in triplicates. Transfection was performed the next day after seeding using Fugene 6 transfection reagent from Roche and followed the same procedure as above. 48 h after transfection, puromycin was added to each culture dish at a final concentration of 1.5μg/ml. Transfected LN229 cells survived puromycin selection and grew well. LN229 cell pellet was prepared in the same way as above and stored at -800C for RPPA analysis. RPPA processing of frozen cell pellets: Frozen cell pellets were submitted to the RPPA core facility at M.D. Anderson Cancer Center. At the RPPA core, protein was extracted by applying RPPA lysis buffer to the cell pellets. Protein lysates were serially-diluted in lysis buffer and printed on nitrocellulose-coated RPPA slides. Slides were incubated with around 300 validated primary antibodies followed by binding with corresponding Biotinylated secondary antibodies and Avidin-Biotinylated Peroxidase (Vectastain Elite ABC kit, Vector Lab). Signals were detected by DAB colorimetric reaction. Signals on the slides were scanned and quantified as per protocol at the core facility.

### Polymerase chain reaction (PCR)

Total RNA was extracted from cultured cells using Trizol reagent (Invitrogen, Camarillo, CA, USA). Isolated RNA was reverse-transcribed into cDNA using a cDNA synthesis kit (Vazyme, Jiangsu, China), in accordance with the manufacturer's protocols. qPCR was performed using synthetic primers and SYBR Green (Thermo, Rockford, IL, USA) with an IQ5 Detection System. After incubating at 50 °C for 2 min and 95 °C for 10 min, the samples were subjected to 40 cycles of 95 °C for 15 s and 60 °C for 1 min. The sequences of the primers specific for target genes are listed in Table 1.

## Results

### Key genes selection

By integrating the genes selected by ImpCosSisLasso and previous studies [2], Fig. 3 demonstrates that AEBP1 is the only mutual explored gene for ImpCoxSisLasso, Cox Lasso, Coxsis and CoxSisLasso, which implies that AEBP1 is very potential for the survival time of GBM. The key genes selected by ImpCoxSisLasso are listed in Table 2 and

**Table 1** The sequences of the primers specific for target genes

| Target gene | The sequences of the primers specific |
|---|---|
| HES1-F | 5'GGCGGCTAAGGTGTTTGGAGG3' |
| HES1-R | 5'GGGCCGCTGGAAGGTGACAC3' |
| HK2-F | 5'AGGGGACTTTGATATCGACATTG3' |
| HK2-R | 5'GCCCCCCACTCCATATTGATAC3' |

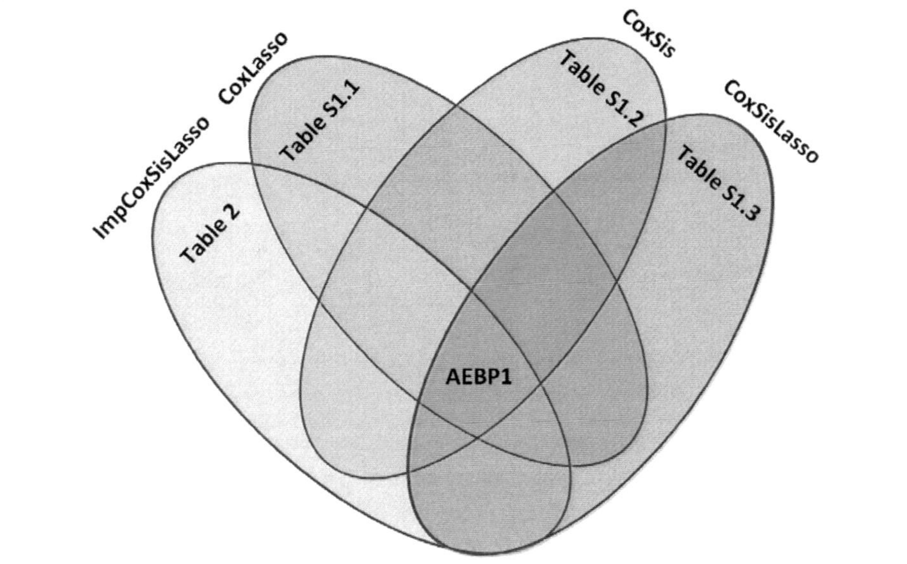


**Fig. 3** Venn plot for the key gene

Supplementary Tables S1.1, 1.2, 1.3. The genes selected by previous algorithms [2] are shown in Supplementary Table S1.4.

### Performance comparison for different survival analysis algorithms

Here, we use Receiver operating Characteristic (ROC) and Area Under Curve (AUC) [21] curves to compare the performance of the aforementioned algorithms [2]. Both ROC (Fig. 4a) and AUC (Fig. 4b) curves show that ImpCoxSisLasso outperforms the CoxLasso, CoxSis and CoxSisLasso algorithms [2].

### Immune-chemistry experiment

Figure 5a, b and c demonstrate that *AEBP1* is positively detectable in GBM. Figure 5d, e and f also show that *EGFR* is positively expressed in GBM. Next, Table 3 shows that there is a strong correlation between *AEBP1* and *EGFR*, since the $p$ value of Chi square test [16] is less than 0.05. Lastly, Supplementary Table S2 lists the protein expressions for *AEBP1* and *EGFR* in immune-chemistry experiment, as well as we use Fig. 1a to demonstrate the impact of *AEBP1* on the survival time for GBM patients.

### Protein network analysis

Firstly, we increase and keep the *AEBP1* expression for LN229 cell lines in the experimental group and control group, respectively. Then, we carry out RPPA experiment twice for these two groups. Finally, the *AEBP1* Up data set (Supplementary Table S3.1) shows 287 related proteins' expression.

**Table 2** Selected genes by ImpCoxSisLasso strategy

| Key gene |
| --- |
| AEBP1, CDCA7L, SNTB1, TELO2, SLC35D1, FOXG1, ARIH2_INTS1, ZNF786_AEBP1, ZNF786_INTS1, ZNF786_SGCD, ZNF786_EIF3A, ZNF786_CDCA7L, ZNF786_TELO2, AEBP1_IL17RC, INTS1_SGCD, INTS1_IL17RC, INTS1_SLC35D1, GDNF_IL17RC, SGCD_TELO2, IL17RC_TELO2, CBLN1_TELO2, SLC35D1_TELO2 |

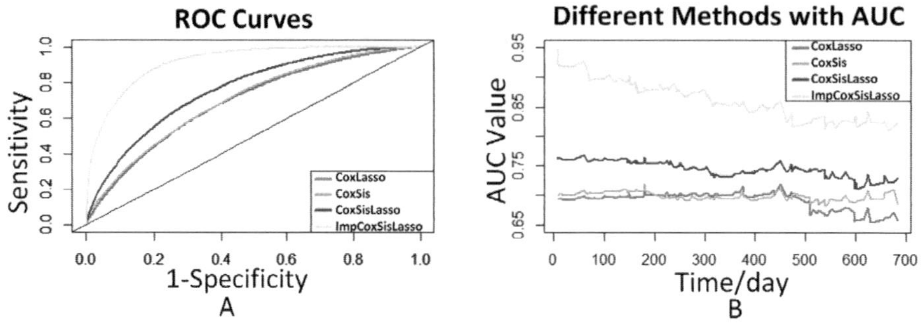


**Fig. 4** Algorithm performance comparison (**a**) ROC and (**b**) AUC

Secondly, we knock out the *AEBP1* by CRISPR-cas9 and keep the *AEBP1* expression for LN229 cell lines in the experimental group and control group, respectively. Then, we carry out RPPA experiment twice for these two groups. Finally, the *AEBP1* Down data set (Supplementary Table S3.2) shows 302 related proteins' expression.

Thirdly, we use Eq. 5. to compute the fold change between experimental and control group for both AEBP1 Up and Down data set, respectively.

$$FC_i = \frac{\sum_{j=1}^{m} ExP_{\mathrm{ij}}}{\sum_{j=1}^{m} CoP_{\mathrm{ij}}} \tag{5}$$

Here, $FC_i$ is the fold change between experimental ($ExP_{ij}$) and control group ($CoP_{ij}$). i represents the index of the proteins. j represents the index for the experiment, and m represents the replicates number. $ExP_{ij}$ and $CoP_{ij}$ are the proteins' expression measured by RPPA experiment.

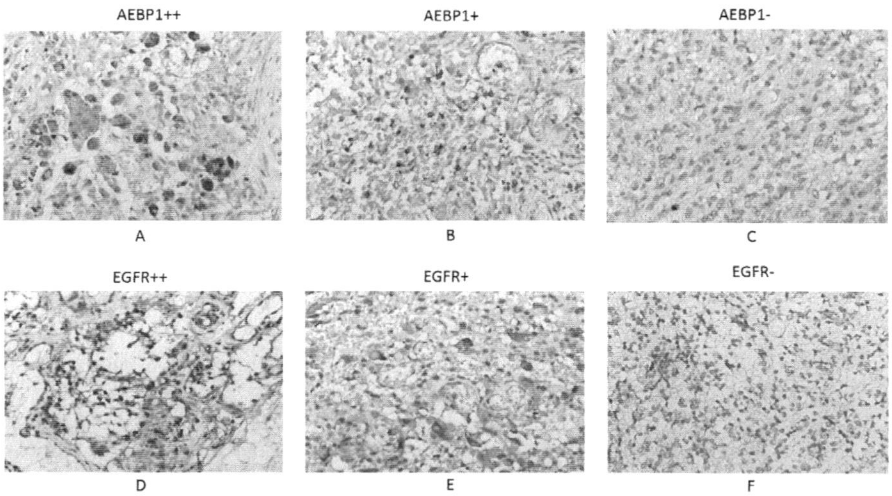


**Fig. 5** Immunohistochemistry and survival analysis results. A, B and C show high (++), low (+) and negative (−) AEBP1 expression in GBM, respectively. And D, E and F show high (++), low (+) and negative (−) EGFR expression, respectively

**Table 3** Statistical analysis of IHC data for AEBP1 and EGFR protein in GBM

| | AEBP1 positive (++&+) | AEBP1 negative (–) | Row Total |
|---|---|---|---|
| EGFR+ | 18 | 11 | 29 |
| EGFR- | 2 | 7 | 9 |
| Column total | 20 | 18 | 38 |
| *P*-Value | 0.036485093 | | |

$$\text{Judge}_i(\text{FC}_i) = \begin{cases} \text{UP} & \text{FC}_i > 1 \\ \text{DOWN} & \text{FC}_i \leq 1 \end{cases} \tag{6}$$

Next, we use Eq. 6. to determine which proteins are promoted or inhibited by AEBP1. For *AEBP1* Up data set, the promoted proteins are listed in Supplementary Table S4.1 as AEBP1 promoted data set. For AEBP1 Down data set, the inhibited proteins are listed in Supplementary Table S4.2 as AEBP1 inhibited data set.

Lastly, we compare the experimental group (ExPij) and control group (CoPij) for both AEBP1 Promoted and Inhibited data sets by T test [3]. The null hypothesis is that the average expression level of a protein under experimental condition is equal to the level under control conditions. After we compute the *P*-value for both *AEBP1* promoted data set and *AEBP1* inhibited data set, 7 proteins' expression are statistically significant in AEBP1 promoted data set (Table 4) and AEBP1 inhibited data set (Table 5), respectively. The detail data is listed in Supplementary Table S5.

Since manually reviewed experimental evidences already turn out that *ACC1-R-C* (*ACC1*) [27], *EMA-M-C* (*EMA*) [28], Raptor-R-V (Raptor) [29], *Rictor_pT1135-R-V* (*Rictor*) [30] and *SOD1-M-V* (*SOD1*) [31] are related to *AEBP1*, we then carry out PCR experiment for *HES1-R-V* (*HES1*) and *Hexokinase-II-R-V* (*HK2*) in Cell Line LN-18, LN-229, and U251 to confirm the correlation between the expression of these two proteins and *AEBP1*, respectively.

### PCR verification

Figure 6 and Supplementary Table S6 indicate that the expressions of both *HES1* and *HK2* are positively related to the expression of *AEBP1*. Therefore, we conclude that all the proteins listed in Tables 4 and 5 are related to the expression of *AEBP1*.

## Discussion

This study aims to investigate the origin of the GBM from systematic view. Three research questions were proposed and addressed.

Since previous additive survival analysis algorithms (CoxLasso, CoxSis and CoxSisLasso [2]) do not consider the gene's interaction, they neither agree with the nature of biology [32–35] nor have the high predictive accuracy. Here, after we integrate interaction item into Eq. 4 of ImpCoxSisLasso algorithm, we not only can explore the gene

**Table 4** AEBP1 promoted dataset

| Protein Name | $FC_i$ | *P*-value |
|---|---|---|
| ACC1-R-C | 1.06701864 | 0.01841297 |
| EMA-M-C | 1.36047171 | 0.02989766 |
| HES1-R-V | 1.15596 | 0.00029086 |

**Table 5** AEBP1 inhibited dataset

| Protein Name | $FC_i$ | *P*-value |
|---|---|---|
| Hexokinase-II-R-V | 0.98022947 | 0.00639911 |
| Raptor-R-V | 0.9132965 | 0.03564656 |
| Rictor_pT1135-R-V | 0.95940456 | 0.02726222 |
| SOD1-M-V | 0.97722104 | 0.03054562 |

signature of GBM for P> > N type of data (Table 2), but also Fig. 4 turns out that ImpCoxSisLasso algorithm is much better than the previous [2]. Moreover, Fig. 3 shows that both ImpCoxSisLasso algorithm and our previous survival algorithms (Cox-Lasso, CoxSis and CoxSisLasso [2]) mutually find *AEBP1* as the key gene related to the survival time of GBM patients. Lastly, we use TCGA data [24] (Figs. 1a) and 5 to cross validate the impact of *AEBP1* gene on GBM.

In this report, Table 3 and Fig. 5 demonstrated that there is a significant correlation between the expression of *AEBP1* and *EGFR* in patient samples of GBM. Through over-expression or knockdown of *AEBP1* gene in glioma cell line LN229 and RPPA analysis, we can show that a panel of proteins (Tables 4 and 5) are significantly affected by the manipulation of AEBP1 expression. By a stringent cutoff of $p < 0.05$, *HK2*(Fig. 6), *SOD1* [36], *Raptor* [37], *Rictor* [38] were markedly downregulated when *AEBP1* expression was knocked down. Using the same cutoff standard, *ACC1* [39], *HES1*(Fig. 6) and *EMA* [28] were significantly upregulated when *AEBP1* was overexpressed.

Through extensive literature research [36, 37, 40, 41], we found that affected proteins by either *AEBP1* overexpression or knockdown converge to the regulation of mainly one signaling pathway: the MTOR pathway. MTOR forms two different complexes, MTORC1 and *MTORC2. HK2* is a downstream effector of MTORC1 through the transcriptional regulation of transcription factor *HIF1* [41]. It integrates energy metabolism with cancer cell survival. MTORC1 directly regulates *Superoxide dismutase 1* (*SOD1*) through phosphorylation [36]. *SOD1* can in turn act as a nuclear transcription factor to

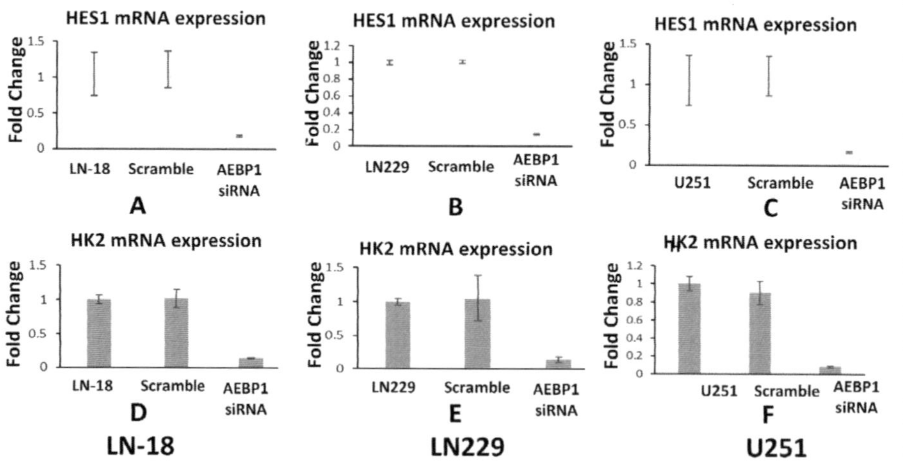


**Fig. 6** Gene expression of HES1 and HK2 in different cell lines with AEBP1 siRNA knockdown. Here, X-axis represents the cell line, Scramble siRNA and AEBP1 siRNA treatment. Y-axis represents the fold changes of different genes in mRNA levels as a result of AEBP1 knockdown. **a**, **b** and **c** show the HES1 mRNA expression in Cell Line LN-18, LN-229, and U251 respectively. **d**, **e** and **f** show the HK2 mRNA expression in Cell Line LN-18, LN-229, and U251 respectively. Data presented were from qRT-PCR experiment

fend off oxidative stress [14] in cancer cells [31]. Raptor is an interacting partner of MTORC1 [40] and phosphorylation of Raptor by MTORC1 is essential for the kinase activity of MTORC1 on its substrates [37]. Dissociation of Raptor from MTORC1 resulted in the inhibition of MTOR signaling activity [29].

On the other hand, *Rictor* is an essential partner of MTORC2 [42, 43]. Blocked association of Rictor from MTORC2 led to the inhibition of MTORC2 activity and cancer cell death in glioma cell line in vitro [30]. In addition, overexpression of Rictor was associated with increased MTORC2 activity and tumor growth in glioma patients [44].

*ACC1* is a *SREBP1* target gene [39]. MTORC1 signaling controls the expression of *SREBP1* and consequently regulates *SREBP1* target gene *ACC1* expression [27]. HES1 (Hairy and Enhancer of Split 1) is a downstream target of Notch signaling pathway [45, 46]. *HES1* antagonizes PTEN-induced inhibition of MTOR signaling activity [47]. *EMA* (Epithelial Membrane Antigen) is a glycosylated protein encoded by *MUC1* gene [28]. Although the cytoplasmic tail of EMA has roles in signal transduction, the relationship between *EMA* and MTOR signaling pathway is not clear at this moment. But *EGFR* signaling activation increased *MUC1* gene expression in some cancer cell lines [48].

Taken together, our results indicate that *AEBP1* regulates the expression of *Raptor* and *Rictor*, key binding partners of MTORC1 and MTORC2. *AEBP1* also has an important role in regulating the expression of *HES1* which promotes MTOR activity by inhibiting PTEN function [47, 49]. Enhanced MTOR signaling by *AEBP1* stimulates downstream expression of MTOR targets *HK2* and *ACC1* or increases phosphorylation of MTOR target *SOD1*. The net effect of *AEBP1* overexpression in glioma is the activation and promotion of MTOR signaling pathway in cancer cells. We illustrate the effects of *AEBP1* with different molecular players on MTOR pathway in a schematic model (Fig. 1b).

*AEBP1* overexpression in glioma is associated with higher tumor grade and worse prognosis [50]. *AEBP1* as therapeutic target for cancer treatment has not drawn much attention. However, clinical trials are under way to test the effectiveness of MTOR inhibitors in the treatment of different types of cancers [51]. Indeed, it has been reported that MTOR inhibitors are effective in a subset of patients with *Rictor* gene amplification in lung cancer [52] and gastric cancer [38]. MTOR inhibitors are actively investigated in preclinical cancer cell line and mouse models of GBM with various success [51, 53]. Experimental results from these preclinical studies place MTOR as the pivotal target in treating GBM [53]. A search of clinical trials in the USA revealed 34 trials using MTOR inhibitors in glioma patients (www.clinicaltrials.gov last accessed on July 20th, 2019). Most of these clinical trials are still under way. Completed trials showed limited successes. In terms of our findings, we may need to take into consideration the expression levels of *AEBP1* in patients' tumor samples when planning clinical trials of MTOR inhibitors on GBM or evaluating clinical responses of MTOR inhibitors in the treatment of GBM.

## Conclusions

In conclusion, in step one, our improved survival analysis model not only finds that the key gene, but also has better performance and efficiency. In step two, through CRISPR and RPPA techniques, we get reliable protein data, and then search for the seven key proteins by a precise multi-scale mathematical model. And last, we analyze the relationship between AEBP1 and key proteins to identify that AEBP1 exerts its tumor-promoting effects by mainly activating mTOR pathway in Glioma.

Although the results show that we can find the mechanism of AEBP1 in mTOR, we can still improve it. For instance, in protein network, we can consider adding machine learning algorithm to find the key protein. In the future, we can even improve the survival time of patients through the explored pathway mechanism.

## Supplementary information

**Supplementary information** accompanies this paper at https://doi.org/10.1186/s12859-020-03674-4.

**Additional file 1.** This file contains all the supplementary tables and figures.

### Abbreviations

GBM: Glioblastoma multiforme; AEBP1: Adipocyte enhancer-binding protein 1; CRISPR: Clustered regularly interspaced short palindromic repeat sequences; RPPA: Reverse phase protein arrays; IQR: Interquartile range; PCR: Polymerase Chain Reaction; ROC: Receiver operating Characteristic; AUC: Area Under Curve

### Acknowledgements

Not applicable.

### About this supplement

This article has been published as part of *BMC Bioinformatics Volume 21 Supplement 13, 2020: Selected articles from the 18th Asia Pacific Bioinformatics Conference (APBC 2020): bioinformatics*. The full contents of the supplement are available online at https://bmcbioinformatics.biomedcentral.com/articles/supplements/volume-21-supplement-13.

### Authors' contributions

LZ participated in the design of the study. YJY and WJL collected and analyzed the data. XFR did the experiment. YJY, WJL, TTL, MX, HRZ and LZ wrote and edited the manuscript. All authors read and approved the final manuscript.

### Funding

Publication costs are funded by the National Natural Science Foundation of China [Nos.61372138 and Nos.11701469] and the National Science and Technology Major Project [2018ZX10201002]. Top-notch Talent Cultivation Plan of Southwest Hospital (SWH2018BJKJ-05).

### Availability of data and materials

The datasets supporting the conclusions of this article are included within the article and the additional file.

### Ethics approval and consent to participate

the experimental protocol and the retrospective analysis of GBM specimens were approved by the Ethical Committee of Southwest Hospital.

### Consent for publication

Not applicable.

### Competing interests

The authors declare that they have no competing interests.

### Author details

[1]College of Computer Science, Sichuan University, Chengdu 610065, China. [2]Department of Neurosurgery, Southwest Hospital, Third Military Medical University, Chongqing, P.R. China. [3]College of Mathematics and Statistics, Southwest University, Chongqing 400715, P.R. China. [4]School of Computing, Ulster University, Coleraine, Londonderry, Northern Ireland, UK.

Published: 17 September 2020

### References

1. Ahmadloo N, et al. Treatment outcome and prognostic factors of adult glioblastoma multiforme. J Egypt Natl Canc Inst. 2013;25(1):21–30.
2. Xia Y, et al. Exploring the key genes and signaling transduction pathways related to the survival time of glioblastoma multiforme patients by a novel survival analysis model. BMC Genomics. 2017;18(Suppl 1):950.
3. Zhang L, et al. EZH2-, CHD4-, and IDH-linked epigenetic perturbation and its association with survival in glioma patients. J Mol Cell Biol. 2017;9(6):477–88.
4. Ladha J, et al. Identification of genomic targets of transcription factor Aebp1 and its role in survival of Glioma cells. Mol Cancer Res. 2012;10(8):1039–51.
5. Wangaryattawanich P, et al. Survival analysis of pre-operative GBM patients by using quantitative image features. In: International Conference on Control; 2014.

6. Pudelko L, et al. Glioblastoma and glioblastoma stem cells are dependent on functional MTH1. Oncotarget. 2017;8(49): 84671–84.
7. Wei N, et al. Efficient gene knockout in goats using CRISPR/Cas9 system. PLoS One. 2014;9(9):e106718.
8. Walters BJ, et al. Advanced In vivo Use of CRISPR/Cas9 and Anti-sense DNA Inhibition for Gene Manipulation in the Brain. Front Genet. 2016;6:362.
9. Xie Y, et al. An episomal vector-based CRISPR/Cas9 system for highly efficient gene knockout in human pluripotent stem cells. Sci Rep. 2017;7(1):2320.
10. Nishizuka SS, Mills GB. New era of integrated cancer biomarker discovery using reverse-phase proteinarrays ☆. Drug Metab Pharmacokinet. 2016;31(1):35–45.
11. Pushparaj PN, et al. siRNA, miRNA, and shRNA: in vivo applications. J Dent Res. 2008;87(11):992–1003.
12. Shull AY, et al. RPPA-based protein profiling reveals eIF4G overexpression and 4E-BP1 serine 65 phos-phorylation as molecular events that correspond with a pro-survival phenotype in chronic lymphocytic leukemia. Oncotarget. 2015; 6(16):14632–45.
13. Ludovini V, et al. Reverse phase protein array (RPPA) combined with computational analysis to unravel relevant prognostic factors in non- small cell lung cancer (NSCLC): a pilot study. Oncotarget. 2017;8(47): 83343–53.
14. Ummanni R, et al. Evaluation of reverse phase protein array (RPPA)-based pathway-activation profiling in 84 non-small cell lung cancer (NSCLC) cell lines as platform for cancer proteomics and biomarker discovery ☆. Biochim Biophys Acta. 2014;1844(5):950–9.
15. O., N., BerEP4 and AE1/3 are Reliable Markers of Epithelial Content for Biomarker Discovery Using Reverse Phase Protein Arrays (RPPA). 2016.
16. Zhang L, et al. Building up a robust risk mathematical platform to predict colorectal Cancer. Complexity. 2017;2017:14.
17. Zhang L, Zhang S. Using game theory to investigate the epigenetic control mechanisms of embryo development: comment on: "epigenetic game theory: how to compute the epigenetic control of maternal-to-zygotic transition" by Qian Wang et al. Phys Life Rev. 2017;20:140–2.
18. Zhang L, et al. Lineage-associated underrepresented permutations (LAUPs) of mammalian genomic sequences based on a jellyfish-based LAUPs analysis application (JBLA). Bioinformatics. 2018;34(21):3624–30.
19. Zhang L, et al. Investigation of mechanism of bone regeneration in a porous biodegradable calcium phosphate (CaP) scaffold by a combination of a multi-scale agent-based model and experimental optimization/validation. Nanoscale. 2016;8(31):14877–87.
20. Zhang L, et al. Revealing dynamic regulations and the related key proteins of myeloma-initiating cells by integrating experimental data into a systems biological model. Bioinformatics. 2019;26:btz542. https://doi.org/10.1093/bioinformatics/btz542.
21. Zhang L, et al. Computed tomography angiography-based analysis of high-risk intracerebral haemorrhage patients by employing a mathematical model. BMC Bioinformatics. 2019;20(Suppl 7):193.
22. Zhang L, et al. Comprehensively benchmarking applications for detecting copy number variation. PLoS Comput Biol. 2019;15(5):e1007069.
23. Tong X, et al. Development of an agent-based model (ABM) to simulate the immune system and integration of a regression method to estimate the key ABM parameters by fitting the experimental data. PLoS One. 2015;10(11): e0141295.
24. Katarzyna T, Patrycja C, Maciej W. The Cancer genome atlas (TCGA): an immeasurable source of knowledge. Contemp Oncol. 2015;19(1A):68–77.
25. Maglott DR, Nierman WC. Clone and genomic repositories at the American type culture collection. Genomics. 1990;8(3): 601–5.
26. Majdalawieh A, Zhang L, Ro H-S. Adipocyte enhancer-binding protein-1 promotes macrophage inflammatory responsiveness by up-regulating NF-kappaB via IkappaBalpha negative regulation. Mol Biol Cell. 2007;18(3):930–42.
27. Jeon T, Osborne TF. SREBPs: metabolic integrators in physiology and metabolism. Trends Endocrinol Metab. 2012;23(2): 65–72.
28. C, Y., M. JL, and I. NK. MUC1 and cancer immunotherapy, vol. 1; 2017.
29. Oshiro N, et al. Dissociation of raptor from mTOR is a mechanism of rapamycin-induced inhibition of mTOR function. Genes Cells. 2004;9(4):359–66.
30. Angelica BS, et al. Specific blockade of Rictor-mTOR association inhibits mTORC2 activity and is cytotoxic in glioblastoma. PLoS One. 2017;12(4):e0176599.
31. Tsang CK, et al. Superoxide dismutase 1 acts as a nuclear transcription factor to regulate oxidative stress resistance. Nat Commun. 2014;5(1):3446.
32. Van Dijk D, Sharma R, Nainys J, et al. Recovering Gene Interactions from Single-Cell Data Using Data Diffusion. Cell. 2018;174(3):716–729.e27.
33. Van Dijk D, et al. MAGIC: A diffusion-based imputation method reveals gene-gene interactions in single-cell RNA-sequencing data. bioRxiv. 2017:111591.
34. Howard TD, et al. Gene-gene interaction in asthma: IL4RA and IL13 in a Dutch population with asthma. Am J Hum Genet. 2002;70(1):230–6.
35. Upstillgoddard R, et al. Machine learning approaches for the discovery of gene–gene interactions in disease data. Brief Bioinform. 2013;14(2):251–60.
36. Tsang CK, et al. SOD1 Phosphorylation by mTORC1 Couples Nutrient Sensing and Redox Regulation. Mol Cell. 2018; 70(3):502–15.
37. Foster KG, et al. Regulation of mTOR complex 1 (mTORC1) by raptor Ser863 and multisite phosphorylation. J Biol Chem. 2010;285(1):80–94.
38. Kim ST, et al. Rapamycin-insensitive companion of mTOR (RICTOR) amplification defines a subset of advanced gastric Cancer and is sensitive to AZD2014-mediated mTORC1/2 inhibition. Ann Oncol. 2016;28(3):547–54.
39. Yin G, et al. mTOR complex 1 signaling regulates the balance between lipid synthesis and oxidation in hypoxic lymphocytes. Biosci Rep. 2017;37(1):BSR20160479.

40. Kim D, et al. mTOR interacts with raptor to form a nutrient-sensitive complex that signals to the cell growth machinery. Cell. 2002;110(2):163–75.
41. Roberts DJ, Miyamoto S. Hexokinase II integrates energy metabolism and cellular protection: Akting on mitochondria and TORCing to autophagy. Cell Death Differ. 2015;22(2):248–57.
42. Kocalis H, et al. Rictor/mTORC2 facilitates central regulation of energy and glucose homeostasis. Mol Metab. 2014;3(4): 394–407.
43. Sarbassov DD, et al. Rictor, a novel binding partner of mTOR, defines a rapamycin-insensitive and raptor-independent pathway that regulates the cytoskeleton. Curr Biol. 2004;14(14):1296–302.
44. Masri J, et al. mTORC2 activity is elevated in Gliomas and promotes growth and cell motility via overexpression of Rictor. Cancer Res. 2007;67(24):11712–20.
45. Schnell SA, et al. Therapeutic targeting of HES1 transcriptional programs in T-ALL. Blood. 2015;125(18):2806–14.
46. Schreck KC, et al. The notch target Hes1 directly modulates Gli1 expression and hedgehog signaling: a potential mechanism of therapeutic resistance. Clin Cancer Res. 2010;16(24):6060–70.
47. Wong GW, et al. HES1 opposes a PTEN-dependent check on survival, differentiation, and proliferation of TCR -selected mouse thymocytes. Blood. 2012;120(7):1439–48.
48. Dharmaraj N, Engel BJ, Carson DD. Activated EGFR stimulates MUC1 expression in human uterine and pancreatic cancer cell lines. J Cell Biochem. 2013;114(10):2314–22.
49. Bhattacharya K, Maiti S, Mandal C. PTEN negatively regulates mTORC2 formation and signaling in grade IV glioma via Rictor hyperphosphorylation at Thr1135 and direct the mode of action of an mTORC1/2 inhibitor. Oncogenesis. 2016; 5(5):e227.
50. Kohsaka S, et al. Epiregulin enhances tumorigenicity by activating the ERK/MAPK pathway in glioblastoma. Neuro-oncology. 2014;16(7):960.
51. Fan Q, Nicolaides T, Weiss WA. Inhibiting 4EBP1 in Glioblastoma. Clin Cancer Res. 2018;24(1):14–21.
52. Cheng H, et al. RICTOR amplification defines a novel subset of patients with lung cancer who may benefit from treatment with mTORC1/2 inhibitors. Cancer Discov. 2015;5(12):1262–70.
53. Fan Q, et al. A kinase inhibitor targeted to mTORC1 drives regression in Glioblastoma. Cancer Cell. 2017;31(3):424–35.

## Publisher's Note

Pan et al. BMC Bioinformatics 2020, 21(Suppl 13):384
https://doi.org/10.1186/s12859-020-03675-3

BMC Bioinformatics

RESEARCH Open Access

# Computationally identifying hot spots in protein-DNA binding interfaces using an ensemble approach

Yuliang Pan[1], Shuigeng Zhou[2] and Jihong Guan[1*]

*From* The 18th Asia Pacific Bioinformatics Conference
Seoul, Korea. 18-20 August 2020

*Correspondence:
jhguan@tongji.edu.cn
[1] Department of Computer Science and Technology, Tongji University, No. 4800 Caoan Road, 201804 Shanghai, China
Full list of author information is available at the end of the article

## Abstract

**Background:** Protein-DNA interaction governs a large number of cellular processes, and it can be altered by a small fraction of interface residues, i.e., the so-called *hot spots*, which account for most of the interface binding free energy. Accurate prediction of hot spots is critical to understand the principle of protein-DNA interactions. There are already some computational methods that can accurately and efficiently predict a large number of hot residues. However, the insufficiency of experimentally validated hot-spot residues in protein-DNA complexes and the low diversity of the employed features limit the performance of existing methods.

**Results:** Here, we report a new computational method for effectively predicting hot spots in protein-DNA binding interfaces. This method, called *PreHots* (the abbreviation of *Pre*dicting *Hot* spots), adopts an ensemble stacking classifier that integrates different machine learning classifiers to generate a robust model with 19 features selected by a sequential backward feature selection algorithm. To this end, we constructed two new and reliable datasets (one benchmark for model training and one independent dataset for validation), which totally consist of 123 hot spots and 137 non-hot spots from 89 protein-DNA complexes. The data were manually collected from the literature and existing databases with a strict process of redundancy removal. Our method achieves a sensitivity of 0.813 and an AUC score of 0.868 in 10-fold cross-validation on the benchmark dataset, and a sensitivity of 0.818 and an AUC score of 0.820 on the independent test dataset. The results show that our approach outperforms the existing ones.

**Conclusions:** *PreHots*, which is based on stack ensemble of boosting algorithms, can reliably predict hot spots at the protein-DNA binding interface on a large scale.
(Continued on next page)

(Continued from previous page)
Compared with the existing methods, *PreHots* can achieve better prediction performance. Both the webserver of *PreHots* and the datasets are freely available at: http://dmb.tongji.edu.cn/tools/PreHots/.

**Keywords:** Protein-DNA complexes, Hot spots, Ensemble stacking classifier, Feature selection

## Background

With the rapid development of structural biology technologies such as X-ray crystallography, NMR spectroscopy, and cryo-electron microscopy, a large number of tertiary structures of biological macromolecules have been generated [1]. However, the interpretation of these structures and the recognition of critical residues lie far behind the step of structure generation. Proteins and DNA are two kinds of most important biological macromolecules of life compounds. The interactions of proteins and DNA are essential for many crucial cellular processes, including gene expression and regulation, DNA replication and repair. For example, genes are regulated by the DNA-binding proteins that bind to some specific DNA sequences [2, 3]. Although DNA-protein binding interfaces contain a large number of residues, the associations between DNA and proteins are governed by a small fraction of residues with high binding affinity, which are also called *hot spots*. Hot spots are considered the most crucial residues for the formation and stabilization of protein complexes. Hence, accurate identification of hot spots is important to understand molecular regulation mechanisms and provide solutions to disease diagnosis and treatment [4].

At present, many experimental techniques have been used to measure protein-DNA binding free energy by site-directed mutagenesis, such as surface plasmon resonance (SPR) [5], isothermal titration calorimetry (ITC) [6] and fluorescence resonance energy transfer (FRET) [7]. However, these experimental techniques are not only inefficient and laborious, but also not suitable for dealing with the vast amounts of residues. Therefore, efficient and effective computational methods for identifying protein-DNA binding hot spots are greatly desirable and urgently needed.

Computational approaches can complement the experimental methods and make large-scale predictions efficiently. Molecular dynamics simulations and feature-based approaches are effective ways to identify hot spots. Two molecular dynamics simulation methods, SAMPDI [8] and PremPDI [9], were proposed to predict the change of protein-DNA binding free energy. SAMPDI utilizes the modified Molecular Mechanics Poisson-Boltzmann Surface Area (MM/PBSA) approach [10] along with additional knowledge-based features to predict binding affinity changes upon single mutation, while PremPDI relies on molecular mechanics force fields and fast side-chain optimization algorithms to evaluate the effects of single mutations on protein-DNA interactions. As for feature based approaches, a method called mCSN-NA [11] was developed, which uses graph-based signatures to predict the impact of a single mutation on protein-nucleic acid binding. Another feature-based approach PrPDH [12] was developed to predict protein-DNA binding hot spots. Although substantial advances have been made, there is still much space to explore for accurately identifying DNA-binding hot spots.

In this work, we develop a novel computational approach *PreHots* (the abbreviation of *Pre*dicting *Hot* spots), which is based on stack ensemble of boosting algorithms, for effectively predicting hot spots in protein-DNA binding interfaces. To this end, a dataset was constructed, which contains 260 samples from 89 protein-DNA complexes. More than half of the data are manually collected from the literature by ourselves, and the rest data are from the databases of ProNIT [13] and dbAMEPNI [14]. We totally calculated 157 features for fully representing hot spots, including not only the properties of the target residue but also target residues' network information. From these features, a set of 19 informative features are selected by using a sequential backward selection algorithm. Extensive experiments were conducted on the benchmark dataset and the independent dataset to evaluate the proposed method, with comparison to existing methods. The experimental results show that our method can significantly boost the performance of DNA-binding hot spots prediction.

## Methods

Figure 1 shows the workflow of the proposed method *PreHots*. First, a new reference dataset that consists of 123 hot spots and 137 non-hot spots from 89 protein-DNA complexes is constructed. The data are manually collecting from the literature and databases with a strict process of redundancy removal. Then, four types of features are encoded to characterize the target residues, including network features, exposure features, sequence features and structural features. Next, the informative features are selected by using sequential backward selection method. Following that, three boost classifiers, including categorical boosting (Catboost) [15], extreme gradient boosting (XGBoost) [16] and gradient tree boosting (GTB) [17] classifiers, are taken as the base models to form an ensemble stacking classifier (ESC), by a meta-model that adopts logistic regression (LR) [18] classifier. Finally, prediction results are output by the ESC model based on the selected feature set.

### Datasets

We constructed an initial dataset, containing experimentally measured binding free energy changes of 660 mutations from 162 protein-DNA complexes, which were obtained

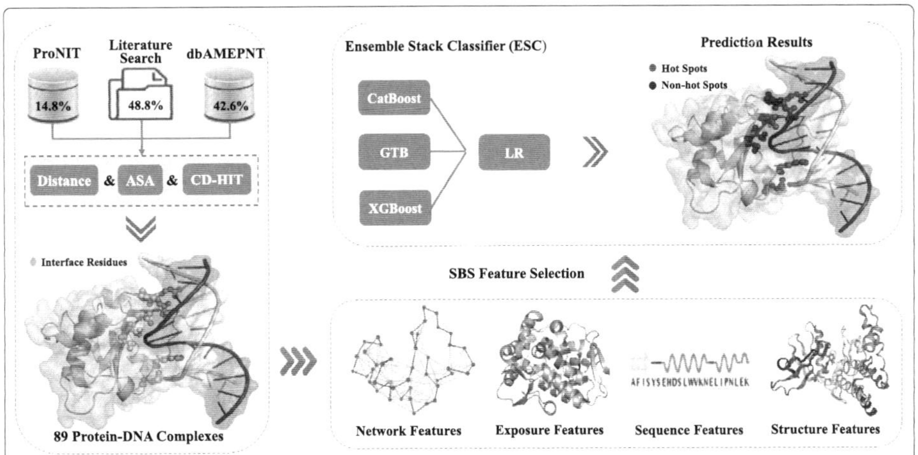


**Fig. 1** The workflow of *PreHots*

by combining two databases and manually searching the literature. Among them, 79 protein-DNA crystal structures were obtained from the database of ProNIT [13] and dbAMEPNI [14], and the other 83 protein-DNA crystal structures were manually collected from the literature.

To build high quality protein-DNA binding hot spots dataset, we used two methods to determine the interface residues. Solvent accessibility area (SAS) is widely used to identify interfacial residues, which can be obtained by calculating the difference of absolute solvent accessibility ($\Delta$ASA $>$1Å) and the ratio of relative solvent accessibility (RASA $>$5%). And to make the results more accurate and stable, the ASA and RASA values of residues are calculated from protein structures by using Naccess [19]. Another method is to calculate the distance between the target residue and the DNA strand. If the distance is less than 5Å, the target residue can be considered as the interface residue. Moreover, we removed redundant homology sequences, where the similarity of protein sequences is more than 40% by using CD-HIT [20]. In this study, we define hot spots as the interface residues with the change in binding free energy ($\Delta\Delta$G) $\geq$1.0 kcal/mol, and the others are defined as non-hot spots. Finally, the constructed dataset consists of 123 hot spots and 137 non-hot spots from 89 complexes. In order to construct a balanced dataset to reduce the potential bias of the machine learning method, 64 protein-DNA complexes were randomly selected to form the benchmark dataset, which contains 90 hot spots and 90 non-hot spots. The rest of 25 protein-DNA complexes constitute the independent dataset, including 33 hot spots and 47 non-hot spots. To the best of our knowledge, our dataset is the largest one for predicting protein-DNA binding hot spots.

### Performance measures

We do performance evaluation by 10-fold cross-validation. The benchmark dataset is randomly divided into 10 subsets, each of which contains approximately the same number of samples. For each round, nine subsets are merged as the training set, while the remaining one subset is used for testing.

For comprehensively assessing the performance of our model, we adopted seven widely used evaluation metrics, including accuracy (ACC), sensitivity (SEN/Recall), specificity (SPE), precision (PRE), F1-score (F1), Matthew's correlation coefficient (MCC) and the area under the ROC curve (AUC). ACC, SEN, SPE, PRE, F1 and MCC are defined as follows:

$$SEN = \frac{TP}{TP + FN} \tag{1}$$

$$SPE = \frac{TN}{TN + FP} \tag{2}$$

$$PRE = \frac{TP}{TP + FP} \tag{3}$$

$$ACC = \frac{TP + TN}{TP + TN + FP + FN} \tag{4}$$

$$F1 = \frac{2 \times SEN \times Precision}{SEN + Precision} \tag{5}$$

$$MCC = \frac{TP \times TN - FP \times FN}{\sqrt{(TP + FP)(TP + FN)(TN + FP)(TN + FN)}} \tag{6}$$

Above, *TP* is the number of true positives, *FP* is the number of false positives, *TN* is the number of true negatives and *FN* is the number of false negatives, respectively.

## Feature description

In order to explore informative features that play important role in the prediction of protein-DNA binding hot spots, we collected a comprehensive feature set that consists of 157 features, which can be roughly divided into four groups: residue interaction network features, solvent exposure features, and traditional features based on protein sequence and structure. More details about these feature are given below.

### *Residue interaction network features*

As a representative kind of protein structures, residue interaction networks (RINs) have been widely and successfully used for revealing the effect of residues mutation, functional region and protein folding [21]. The traditional way to build RINs is to calculate the distance between $C_\alpha$ atoms of two residues within a certain threshold, which ranges from 5 to 9Å [22, 23]. But in fact, the interaction of protein-DNA depends on several intermolecular factors such as hydrogen bonds, van der Waals contacts, ionic bond and several other factors [24, 25]. The stability of protein-DNA interaction is maintained by forming hydrogen bonds between amino acid side chain residues of protein and DNA bases [26]. Therefore, the construction of RINs based on whether there is an intermolecular interaction between any two nodes, including residue and DNA, in the protein-DNA complexes by using RING [27]. In this study, five intermolecular interactions are considered: hydrogen bond, Van der Waals, disulfide bond, salt bridge, $\pi$-$\pi$ stacking and $\pi$-cation.

To make the network contain more knowledge, each edge weight is assigned with the distance between two corresponding nodes. We calculate 10 RINs features that represent the importance of the target residue in the RINs, including node degree, clustering, closeness, betweenness, eigenvector, eccentricity, average neighbor degree, flow closeness, square clustering and Katz centrality.

### *Solvent exposure features*

Solvent exposure of amino acid is crucial for exploring and predicting protein interaction and function. Solvent exposure features consist of several types of features, including half-sphere exposure (HSE), contact number (CN), residue depth (RD), accessible surface area (ASA) and relative accessible surface area (RASA). The solvent accessible has been extensively and successfully utilized to predict protein-protein interaction hot spots [28–31]. The limitation of solvent accessible is that it cannot provide any information about completely buried residues. Compared with traditional solvent accessible, half-sphere exposure (HSE) can describe the local environment of the target residue better from another perspective [32]. RD represents the average atom depth of target residue atoms, while CN is the number of residues in the sphere within a specific distance [33].

In this study, we calculated the characteristics of half-sphere exposure, contact number and residue depth, which could complement the solvent exposure information of interface residues. Based on protein sequence, a series of computing tools have been developed for predicting HSE, CN. We choose the method of HSEpred [32] and SPOT-1D [34] to calculate these features. For protein structure, we use hsexpo [33] to calculate the above three types of features, including HSE, CN and RD.

### *Structure-based features*

Based on the three-dimensional structures of proteins, structure-based features were calculated, including hydrogen bonds, consensus scores, secondary structures, fluctuation score and solvent accessible surface area.

**1. Hydrogen bonds (Hbond).** The stability of protein-DNA interaction is maintained by forming hydrogen bonds between amino acid side chain residues of protein and DNA bases [26]. The hydrogen bond of protein-DNA complexes were calculated by using HBPLUS [35].

**2. Consensus scores.** Consensus score is a linear combination of residue interface propensity score, residue energy score and residue conservation score. Here, we used ENDES [36] to calculate consensus score, while the side chain energy score and relative solvent accessibility can also be obtained.

**3. Secondary structure (SS).** As an important feature, the secondary structural characteristics of residues were obtained from both sequences and structures of proteins. The definition of secondary structures of proteins (DSSP) [37] defines the secondary structure according to atomic coordinates in the protein data bank (PDB) [1]. In addition, several tools can predict the secondary structure of residue from protein sequence, including SPOT-1D [34], NetSurfp2 [38] and SPIDER3 [39].

**4. Fluctuation score.** The study of protein fluctuation is helpful to understand protein structures. FlexPred was used to predict the value of residue fluctuations [40]. Meanwhile, B-factor, represents the dynamic motion of atoms in a protein, was extracted from the PDB file.

**5. Solvent accessible surface area.** Solvent accessible surface area, including available surface area (ASA) and relatively accessible surface area (RASA), which has a strong correlation with hot spot prediction [12]. We applied Naccess [19] to calculate the ASA and RASA of residues from protein-DNA complexes.

### *Sequence-based features*

Based on previous studies, we calculated many features of protein-DNA binding residues from protein sequences.

**1. Position-specific scoring matrix (PSSM).** It is well known that PSSM is an essential feature for predicting hot spots [4, 28, 31]. PSSM score represents the relationship between the frequency of amino acid substitutions and that expected by chance. Negative numbers indicate less frequent substitutions than expected by chance, while positive numbers mean more frequent substitutions than expected.

**2. Conservation score.** Conservative analysis of residues is extensively used to identify functionally important residues in protein sequences. The conservation score of residues can be calculated by using Jensen-Shannon divergence [41].

**3. Solvent accessible surface area.** Apart from deriving solvent accessibility from protein structure, we also used SPIDER3 [39] and NetSurfp2 [38] to calculating ASA and RASA from protein sequence.

**4. Physicochemical features.** Amino acid indices database (AAindex) collects various biochemical and physicochemical characteristics of amino acids [42]. In this work, protein-DNA binding hot spots are described by eight physicochemical characteristics: propensities, polarity, hydrophilicity, average accessible surface area, atom-based hydrophobic moment, flexibility parameter for no rigid neighbors, hydrophobicity and polarizability.

**5. Blocks substitution matrix (BLOSUM).** BLOSUM62 [43] means that sequence similarity is more than 62% in terms of sequence alignment. We calculated BLOSUM62, the most widely used amino acid scoring matrix, whose scores indicate the similarity between two types of amino acids.

**6. Local structural entropy (LSE).** Previous research found that local structural entropy is related to the stability of protein, and it was successfully used for predicting protein-protein interaction hot spots. In this work, we calculated the LSE [44] value of each residue within a protein sequence.

**7. Disordered regions (DISO).** Recognizing protein disorder regions contributes to the understanding of protein function and protein fold pathway. SPOT-Disorder [45] and RaporX-Property [46] were used to predict disorder regions of protein-DNA binding residues.

### Feature selection

For high-dimensional datasets, feature selection can effectively remove some irrelevant features, which contributes to lifting the efficiency of learning tasks and making the model easier to be understood. We used a sequential backward selection (SBS) algorithm to select a subset of informative features that are highly relevant to protein-DNA binding hot spots from the initial set of 157 features. Sequential backward selection (SBS), which is a heuristic search algorithm, removes one feature each time till an optimal feature subset is generated. Here, each resulting feature set is evaluated by using 10-fold cross-validation with the ESC classifier. Such 10-fold cross-validation procedure is repeated 30 times and the average performance over 30 trials is taken as the result. Besides, we combine the independent dataset and each cross-validation test dataset as the test dataset, which is used to evaluate features and obtain the evaluation score at each 10-fold cross-validation. The evaluation metric of feature selection is represented by $E_c$, calculated as follows:

$$E_c = \frac{1}{R}\sum_{R=1}^{R}\left\{\frac{1}{n}\sum_{n=1}^{n}(ACC_i + SEN_i + SPE_i + MCC_i + AUC_i)\right\} \quad (7)$$

where $R$ is the number of cross-validation; $n$ is the number of iterations of 10-fold cross-validation; $ACC_i$, $SEN_i$, $SPE_i$, $MCC_i$, and $AUC_i$ indicate the values of accuracy, sensitivity, specificity, Matthew's correlation coefficient and AUC score of the $i$-th 10-fold cross-validation, respectively.

In the SBS method, features are iteratively removed one by one from the initial feature set. In the first round, each feature is deleted once (resulting in 157 subsets of 156 features). If the ESC classifier based on a certain feature subset achieves the higher $E_c$, this feature subset is left for the next round of feature selection. Such a feature selection process would continues till $E_c$ does not increase any more.

### Ensemble stacking classifier

Stacking, also called super learning [47], is an ensemble machine learning method that constructs the base-level models and meta-model by combining different machine learning classifiers. The construction of base-level models is based on the benchmark dataset, and the meta-model is trained on the outputs of the base-level models. The ensemble stacking classifier (ESC) can overcome the disadvantage of single classifier and make the prediction more robust than a single model. In this study, we choose three boost classifiers as the base-level models, which are categorical boosting (Catboost) [15], extreme gradient boosting (XGBoost) [16] and gradient tree boosting (GTB) [17] classifiers, and the meta-model adopts logistic regression (LR) [18] classifier.

## Results and discussion

### Performance of the ensemble stacking classifier

Ensemble stacking classifier (ESC) is an ensemble technique that the output of the first-level (base) classifiers is taken as the input of the second-level classifier by constructing a two-level model. In this study, the first-level classifiers consist of categorical boosting (Catboost) [15], extreme gradient boosting (XGBoost) [16] and gradient tree boosting (GTB) [17] models, and the second-level classifier is a logistic regression (LR) [18] model. To check whether ESC is suitable for predicting hot spots in the complexes, we compared ESC with ensemble vote classifier (EVC) and some popular machine learning models, including random forests (RF) [48], GTB, support vector machine (SVM) [49], Catboost and XGBoost. Among them, the ensemble vote classifier (EVC) is another ensemble technique, which integrates different machine learning algorithms and predicts hot spots by using the average predictive probability of all algorithms. To avoid the randomness of cross-validation results, we do 10-fold cross-validation 30 times and the averaged result of all 30 cross-validation trials is taken as the final result. Table 1 shows the results of ESC and the compared methods. We can see that the ensemble techniques are generally superior to the other machine learning methods. And, ECS outperforms EVC and can significantly improves the performance of hot- spots prediction.

**Table 1** Performance comparison between ESC and five existing classifiers

| Method | ACC | SEN | SPE | PRE | F1 | MCC | AUC |
|---|---|---|---|---|---|---|---|
| RF | 0.683 | 0.696 | 0.684 | 0.687 | 0.669 | 0.374 | 0.758 |
| SVM | 0.685 | 0.673 | 0.695 | 0.670 | 0.665 | 0.366 | 0.793 |
| CatBoost | 0.722 | 0.731 | 0.726 | 0.734 | 0.721 | 0.455 | 0.806 |
| GTB | 0.711 | 0.743 | 0.733 | 0.718 | 0.705 | 0.468 | 0.816 |
| EVC | 0.725 | 0.741 | 0.721 | 0.699 | 0.694 | 0.446 | 0.826 |
| ECS | 0.783 | 0.795 | 0.753 | 0.784 | 0.782 | 0.562 | 0.833 |

### Performance of feature selection

Feature selection is crucial for building accurate classification models, which aims to select a small number of informative features. In this study, our initial feature set consists of 157 candidate features, which can be divided into four groups: residue contact network features (network), solvent exposure features (exposure), sequence features and structural features. We used a sequential backward selection (SBS) method to choose relevant and informative features from the initial feature set. SBS uses a stepwise feature selection scheme, which iteratively removes features one by one from the feature set. The evaluation criterion ($E_c$) represents the average prediction performance of ESC when selecting features. Figure 2 shows how $E_c$ changes during the process of stepwise feature selection. $E_c$ reaches the maximum when the number of selected features is 19. Consequently, these 19 features form our optimal feature set.

To assess the advantage of the SBS method, we compared it with four existing methods, including random forest (RF), recursive feature elimination (RFE) [50], maximum relevance minimum redundancy (mRMR) [51] and the block Hilbert-Schmidt independence criterion (HSIC) Lasso [52]. The commonly used methods are RF, RFE and mRMR, which use the mean decrease Gini index (MDGI), SVM-based recursive feature elimination and max relevance and min redundancy criteria to evaluate the importance of features, respectively. The block HSIC Lasso (HSIC Lasso) is a relatively novel method, which adopts an effective nonlinear feature selection algorithm based on HSIC Lasso to select informative biological features. To obtain reliable results, we ran 30 times of 10-fold cross-validation and took the average performance as final result. Table 2 shows the performance of the five feature selection methods on the benchmark dataset. We can see that SBS can select better features, which are helpful to predict protein-DNA binding hot spots. And the ESC classifier with SBS achieves the best prediction performance, with a 0.535 MCC and a 0.853 AUC.

### Significance of selected features

By using the SBS feature selection method, we obtain an optimal feature set, which contains 19 features as shown in Table 3. The ranking of these selected features is based

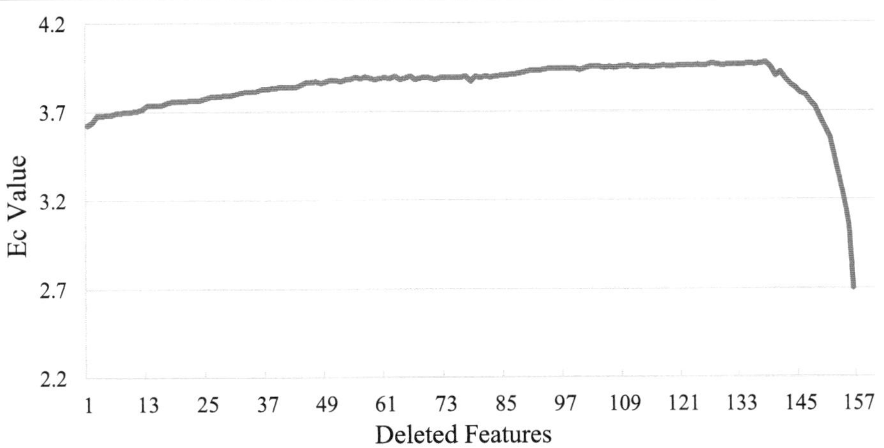


**Fig. 2** The change of $E_c$ value in the process of stepwise feature selection

**Table 2** Performance comparison between SBS and four existing feature selection methods

| Method | ACC | SEN | SPE | PRE | F1 | MCC | AUC |
|---|---|---|---|---|---|---|---|
| RF (28) | 0.744 | 0.739 | 0.749 | 0.716 | 0.715 | 0.483 | 0.823 |
| RFE (20) | 0.739 | 0.723 | 0.730 | 0.719 | 0.718 | 0.452 | 0.830 |
| mRMR (25) | 0.755 | 0.787 | 0.746 | 0.766 | 0.761 | 0.531 | 0.835 |
| HSIC Lasso (30) | 0.740 | 0.777 | 0.727 | 0.746 | 0.744 | 0.500 | 0.841 |
| SBS (19) | 0.767 | 0.784 | 0.766 | 0.776 | 0.741 | 0.535 | 0.853 |

on F-score, which is to measure the distinguishing ability of features between hot and non-hot spots. The most important features include PSSM, hydrogen bonds, secondary structure and RINs features. Two exposure features (as novel features) are selected into the optimal feature set, which indicates that they are important features for identifying DNA-binding hot spots. Fig. 3 shows more details about the distribution of selected features in different feature categories. Six secondary structural features are selected. In previous works, secondary structural has been considered as a fundamental and essential features to improve prediction performance. In this work, we derived secondary structural features from two levels of protein structures and sequences, which can provide a more comprehensive description of secondary structural characteristics of target residues. Besides, ASA, exposure features and consensus score also contribute significantly to the prediction of hot spot residues. These results suggest that the ten categories of 19 optimal features can complement each other and accurately describe the hot spot residues, thus collectively improve the prediction performance.

#### Performance comparison with state-of-the-art methods

Here, we further compare our method with four existing protein-DNA binding hot spots prediction methods, including PrPDH [12], PremPDI [9], mCSM-NA [11] and SAMPDI [8], on the benchmark dataset and the independent test dataset. PrPDH uses a classification model to identify hot-spots from various interface residues, while PremPDI, mCSM-NA and SAMPDI use regression models to predict the change of Protein-DNA binding free energy.

Table 4 presents the results on the benchmark dataset, where the prediction results of existing methods are from their websites. In general, our method performs better than

**Table 3** The rankings of the 19 selected features

| Rank | Feature name | Type | Rank | Feature name | Type |
|---|---|---|---|---|---|
| 1 | PSSM(R) | Sequence | 11 | Lse score | Sequence |
| 2 | H-Bond in HBPLUS | Structure | 12 | phi in SPOT-1D | Sequence |
| 3 | ALPHA in Xssp | Structure | 13 | COMBINED2 score in ENDES | Structure |
| 4 | Current_flow_closeness_centrality | Network | 14 | ACC in Xssp | Structure |
| 5 | Q3_prob_3 in NetSurfp2 | Sequence | 15 | RSA in NetSurfp2 | Sequence |
| 6 | HSEa-u in SPOT-1D | Exposure | 16 | P(8-G) in SPOT-1D | Sequence |
| 7 | COMBINED1 score in ENDES | Structure | 17 | CN in hsexpo | Exposure |
| 8 | SIDESCORE score in ENDES | Structure | 18 | Conservation score | Sequence |
| 9 | Q8_prob_1 in NetSurfp2 | Sequence | 19 | Blosum(E) | Sequence |
| 10 | P(8-I) in SPOT-1D | Sequence | | | |

These features fall intro four types, i.e., network features, exposure features, structure features and sequence features

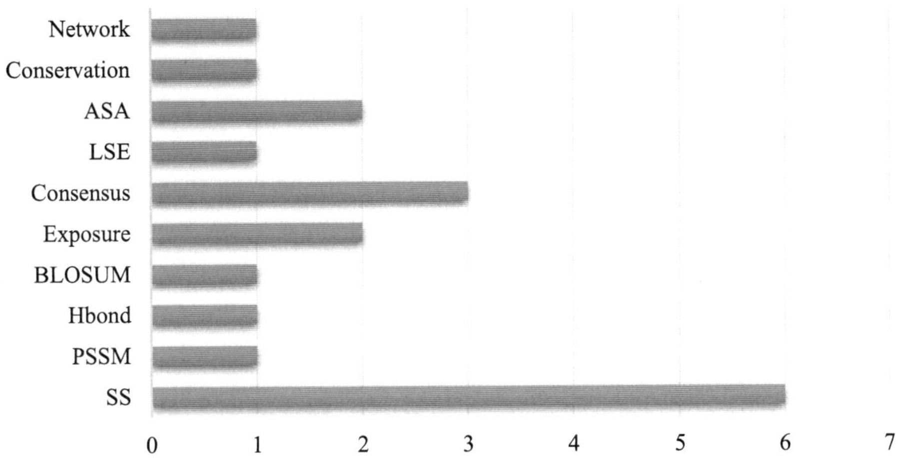


**Fig. 3** The number of each class features on the optimal feature set

the other methods in terms of six of the seven metrics (ACC, SEN, SPE, FRE, F1, MCC and AUC). Only our SPE is smaller than that of the mCSM-NA method.

Table 5 gives the results on the independent test dataset. Compared with the existing methods, our method significantly improves the prediction performance. Concretely, 81.8% of the true hot spots are correctly predicted (SEN = 0.818) and 76.6% of the non-hot spots are correctly predicted (SPE = 0.766). Except for SPE, our method achieves the highest values of the other metrics, especially for the comprehensive indexes MCC (0.576) and AUC (0.82). These results show that our method is superior to the existing methods in identifying protein-DNA binding hot spots.

## Case study

### *The λ exonuclease (λexo) and DNA complex.*

λexo is an ATP-independent enzyme that binds double-stranded DNA (dsDNA) to form the λexo-DNA complex (PDB ID: 3SM4, chain: A) [53]. Four mutated interfacial residues of the λexo-DNA complex have experimentally been identified and shown in Fig. 4. The hot spots residues ($\Delta\Delta G$) $\geq$1.0 kcal/mol are K49_A and R137_A, and the rest are non-hot spots (K76_A and M53_A). Our approach successfully identified all the hot spots, while only a non-hot spot (K76_A) was wrongly identified. In addition, PremPDI, PrPDH and SAMPDI only correctly predicted two non-hot spots (K76_A and M53_A), while the two hot spots were wrongly predicted. mCSM-NA only correctly predicted one non-hot spots (M53_A). This example shows that our method can effectively identify hot spots from protein-DNA complexes than the major existing methods.

**Table 4** Performance comparison between our method with four existing methods on the benchmark dataset

| Method | ACC | SEN | SPE | PRE | F1 | MCC | AUC |
|---|---|---|---|---|---|---|---|
| PreHots | 0.789 | 0.813 | 0.801 | 0.785 | 0.784 | 0.597 | 0.868 |
| PrPDH | 0.683 | 0.667 | 0.700 | 0.690 | 0.678 | 0.367 | 0.779 |
| PremPDI | 0.756 | 0.711 | 0.800 | 0.780 | 0.744 | 0.513 | 0.790 |
| mCSM-NA | 0.461 | 0.056 | 0.867 | 0.284 | 0.093 | -0.133 | 0.314 |
| SAMPDI | 0.544 | 0.444 | 0.644 | 0.556 | 0.494 | 0.091 | 0.522 |

**Table 5** Performance comparison between our method with four existing methods on the independent dataset

| Method | ACC | SEN | SPE | PRE | F1 | MCC | AUC |
|---|---|---|---|---|---|---|---|
| PreHots | 0.788 | 0.818 | 0.766 | 0.711 | 0.761 | 0.576 | 0.820 |
| PrPDH | 0.600 | 0.545 | 0.638 | 0.514 | 0.529 | 0.182 | 0.628 |
| PremPDI | 0.463 | 0.333 | 0.553 | 0.344 | 0.338 | -0.114 | 0.411 |
| mCSM-NA | 0.563 | 0.121 | 0.872 | 0.400 | 0.186 | -0.010 | 0.472 |
| SAMPDI | 0.545 | 0.272 | 0.727 | 0.400 | 0.324 | 0.000 | 0.525 |

***DNA-bound SUP-12$_{28-121}$ complex.***

The structure of DNA-bound SUP-12$_{28-121}$ (PDB ID:4CH1, chain: A) can provide accurate clue to the mechanism of DNA recognition [54]. The defined hot spot residues are K36_M, Y44_A, E63_K, R103_M, A110_T and G113_E, and the remaining three residues (Y78_F, N106_A and N108_A) are non-hot spots (see Fig. 5). For these nine mutated residues, PremPDI identified three of the six hot spots (K36_M, R103_M and A110_T) and one non-hot spot (Y78_F). PrPDH predicted two residues as hot spots (R103_M and A110_T) and the others as non-hot spots. SAMPDI identified one residue as hot spot (Y44_A) and the others non-hot spots, while mCSM-NA predicted all residues as non-hot spots. On the contrary, except for a hot spot (K36_M), our method predicted correctly all the other residues. This suggests that our method has the highest accuracy, which is desirable for many biological applications.

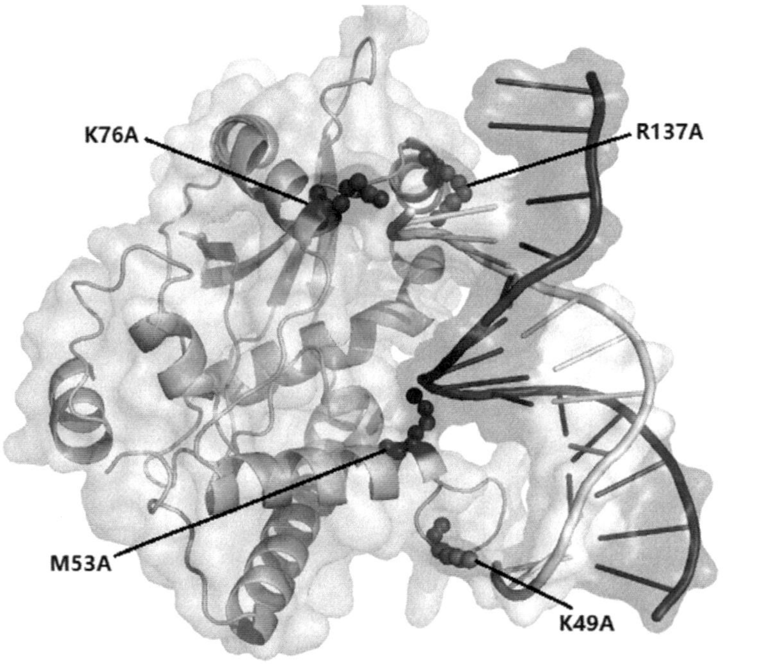


**Fig. 4** The hot spot residues of λexo-DNA complex (PDB ID: 3SM4) identified by experiments. The green surface denotes the protein chain (chain A) while the purple and yellow surfaces represent the DNA chains (purple: chain E and yellow: chain D). The red color represents experimentally identified hot spot residues and the blue color represents experimentally determined non-hot spot residues

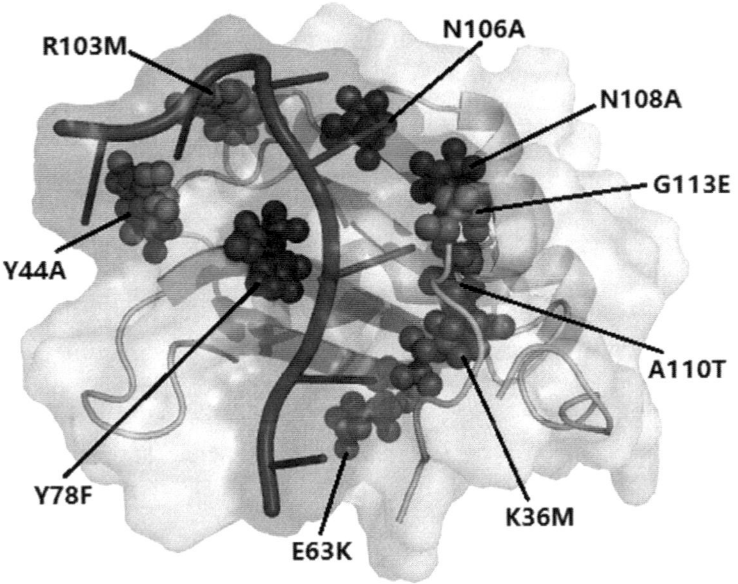


**Fig. 5** The hot spot residues of DNA-bound SUP-$12_{28-121}$ complex (PDB ID: 4CH1) identified by experiments. The green surface denotes the protein chain (chain A) while the purple surface represents the DNA chain (chain B). The red color represents experimentally identified hot spot residues and the blue color represents experimentally determined non-hot spot residues

### Webserver

A user-friendly webserver of *PreHots* has been implemented, which is available at: http://dmb.tongji.edu.cn/tools/PreHots/. The input to *PreHots* should be the PDB file, which contains at least one protein chain and one DNA strand. The user can select the chain of protein and DNA, and submit the job to the server. Then, *PreHots* will return a list of residues, which are predicted to be potential DNA-binding hot spots based on the ensemble classifier with optimally features. Interface residues are colored according to the predicted confidence score. For visual display, users can use the 3D viewer to display prediction results and download the results. Multiple PDB files can be submitted simultaneously, and the jobs are executed in parallel on a cluster server with multiple computing nodes to lift prediction efficiency.

## Conclusion

Computational approaches can effectively and efficiently distinguish hot spots and non-hot spots from protein-DNA complexes on a large scale. In this work, we present a new computational method named *PreHots* for predicting hot spots in protein-DNA complexes. Compared with the existing methods, *PreHots* uses a high-quality dataset manually curated from literature and databases and with a strict process of redundancy removal. A large number of related features (network, exposure, sequence and structure) were calculated to characterize the residues from various aspects. To improve prediction performance, we used the SBS feature selection method to get the optimal feature set and constructed the classification model by the ESC method that integrates four well-performing models. Our method overcomes the drawbacks of single classifiers and makes the prediction more robust. We conducted extensive experiments to evaluate the proposed method, and compared it with existing methods on both a benchmark dataset and an independent test dataset. Experimental results show that our approach achieves higher overall performance than the existing methods. We believe that our method is an invaluable tool of identifying hot spot residues in protein-DNA complexes and can provide insights for the characterization of protein-DNA binding sites.

#### Abbreviations
PreHots: The abbreviation of predicting hot spots; SPR: Surface plasmon resonance; ITC: Isothermal titration calorimetry; MM/PBSA: Molecular mechanics poisson-boltzmann surface area; Catboost: Categorical boosting; XGBoost: Extreme gradient boosting; GTB: Gradient tree boosting; ESC: Ensemble stacking classifier; LR: Logistic regression; SAS: Solvent accessibility area; RASA: Relative solvent accessibility area; ACC: Accuracy; SEN: Sensitivity; SPE: Specificity; PRE: Precision; F1: F1-score; MCC: Matthews correlation coefficient; AUC: Area under curve; HSE: Half-sphere exposure; CN: Contact number; RD: Residue depth; Hbond: Hydrogen bonds; SS: Secondary structure; PDB: Protein data bank; PSSM: Position-specific scoring matrix; AAindex: Amino acid indices database; BLOSUM: Blocks substitution matrix; LSE: Local structural entropy; DISO: Disordered regions; SBS: Sequential backward selection; ESC: Ensemble stacking classifier; EVC: Ensemble vote classifier; RF: Random forests; SVM: Support vector machine; RFE: Recursive feature elimination; mRMR: Maximum relevance minimum redundancy; HSIC: Hilbert Schmidt independence criterion; MDGI: The mean decrease Gini index; FRET: Fluorescence resonance energy transfer

#### Acknowledgements
Not applicable.

#### About this supplement
This article has been published as part of *BMC Bioinformatics Volume 21 Supplement 13, 2020: Selected articles from the 18th Asia Pacific Bioinformatics Conference (APBC 2020): bioinformatics*. The full contents of the supplement are available online at https://bmcbioinformatics.biomedcentral.com/articles/supplements/volume-21-supplement-13.

#### Authors' contributions
YL and JH conceived this work and designed the experiments. YL carried out the experiments, drafted the manuscript and developed the webserver. YL and JH collected the data and analyzed the results. SG participated technical discussions and revised the manuscript. All authors have read and approved the final manuscript.

#### Funding
This work was supported by the National Natural Science Foundation of China (61772367, 61972100) and the National Key Research and Development Program of China (grant No. 2016YFC0901704). Publication costs were funded by the National Natural Science Foundation of China (61772367, 61972100).

#### Availability of data and materials
*PreHots* is free available at http://dmb.tongji.edu.cn/tools/PreHots/.

#### Ethics approval and consent to participate
Not applicable.

#### Consent for publication
Not applicable.

#### Competing interests
The authors declare that they have no competing interests.

#### Author details
[1]Department of Computer Science and Technology, Tongji University, No. 4800 Caoan Road, 201804 Shanghai, China. [2]Shanghai Key Laboratory of Intelligent Information Processing, and School of Computer Science, Fudan University, No. 220 Handan Road, 200433 Shanghai, China.

Published: 17 September 2020

#### References
1. Berman MH. The protein data bank. Nucleic Acids Res. 28(1):235–42.
2. Orphanides G, Reinberg D. A unified theory of gene expression. Cell. 2002;108(4):439–51.
3. Roeder R. Role of general and gene-specific cofactors in the regulation of eukaryotic transcription. In: Cold Spring Harbor Symposia on Quantitative Biology, vol. 63. Cold Spring Harbor Symposia on Quantitative Biology; 1998. p. 201–18.
4. Pan Y, Wang Z, Zhan W, Deng L. Computational identification of binding energy hot spots in protein-RNA complexes using an ensemble approach. Bioinformatics. 2017;34(9):1473–80.
5. Teh HF, Peh WY, Su X, Thomsen JS. Characterization of protein-DNA interactions using surface plasmon resonance spectroscopy with various assay schemes. Biochemistry. 2007;46(8):2127–35.
6. Freire E, Mayorga OL, Straume M. Isothermal titration calorimetry. Anal Chem. 1990;62(18):950–9.

7. Hillisch A, Lorenz M, Diekmann S. Recent advances in fret: distance determination in protein-DNA complexes. Curr Opin Struct Biol. 2001;11(2):201–7.
8. Peng Y, Sun L, Jia Z, Li L, Alexov E. Predicting protein-DNA binding free energy change upon missense mutations using modified MM/PBSA approach: SAMPDI webserver. Bioinformatics. 2017;34(5):779–86.
9. Zhang N, Chen Y, Zhao F, Yang Q, Simonetti FL, Li M. PremPDI estimates and interprets the effects of missense mutations on protein-DNA interactions. PLoS Comput Biol. 2018;14(12):1006615.
10. Hou T, Wang J, Li Y, Wang W. Assessing the performance of the MM/PBSA and MM/GBSA methods. 1. the accuracy of binding free energy calculations based on molecular dynamics simulations. J Chem Inf Model. 2010;51(1):69–82.
11. Pires DE, Ascher DB. mCSM-NA: predicting the effects of mutations on protein–nucleic acids interactions. Nucleic Acids Res. 2017;45(W1):241–6.
12. Zhang S, Zhao L, Zheng C-H, Xia J. A feature-based approach to predict hot spots in protein-DNA binding interfaces. Brief Bioinform. 2019;21(3):1038–46.
13. Kumar MS, Bava KA, Gromiha MM, Prabakaran P, Kitajima K, Uedaira H, Sarai A. Protherm and pronit: thermodynamic databases for proteins and protein–nucleic acid interactions. Nucleic Acids Res. 2006;34(suppl_1): 204–6.
14. Liu L, Xiong Y, Gao H, Wei D-Q, Mitchell JC, Zhu X. dbAMEPNI: a database of alanine mutagenic effects for protein–nucleic acid interactions. Database. 2018;2018:. https://doi.org/10.1093/database/bay034.
15. Dorogush AV, Ershov V, Gulin A. Catboost: gradient boosting with categorical features support. 2018. arXiv preprint arXiv:1810.11363.
16. Chen T, Guestrin C. Xgboost: A scalable tree boosting system. In: In 22nd SIGKDD Conference on Knowledge Discovery and Data Mining. ACM; 2016. p. 785–94.
17. Friedman JH. Stochastic gradient boosting. Comput Stat Data Anal. 2002;38(4):367–78.
18. Wright RE. Logistic regression. Reading & Understanding Multivariate Stats. 1995;68(3):497–07.
19. Hubbard SJ, Thornton JM. Naccess. Computer Program, Department of Biochemistry and Molecular Biology, University College London. 1993;2(1):.
20. Li W, Godzik A. Cd-hit: a fast program for clustering and comparing large sets of protein or nucleotide sequences. Bioinformatics. 2006;22(13):1658–9.
21. Yan W, Zhou J, Sun M, Chen J, Hu G, Shen B. The construction of an amino acid network for understanding protein structure and function. Amino Acids. 2014;46(6):1419–39.
22. Chakrabarty B, Parekh N. NAPS: Network analysis of protein structures. Nucleic Acids Res. 2016;44(W1):375–82.
23. Pan Y, Liu D, Deng L. Accurate prediction of functional effects for variants by combining gradient tree boosting with optimal neighborhood properties. PLoS ONE. 2017;12(6):0179314.
24. Hogan M, Austin RH. Importance of DNA stiffness in protein-DNA binding specificity. Nature. 1987;329(6136):263.
25. Luscombe NM, Laskowski RA, Thornton JM. Amino acid–base interactions: a three-dimensional analysis of protein-DNA interactions at an atomic level. Nucleic Acids Res. 2001;29(13):2860–74.
26. Rohs R, West SM, Sosinsky A, Liu P, Mann RS, Honig B. The role of dna shape in protein-DNA recognition. Nature. 2009;461(7268):1248.
27. Piovesan D, Minervini G, Tosatto SC. The RING 2.0 web server for high quality residue interaction networks. Nucleic Acids Res. 2016;44(W1):367–74.
28. Deng L, Guan J, Wei X, Yi Y, Zhang QC, Zhou S. Boosting prediction performance of protein-protein interaction hot spots by using structural neighborhood properties. J Comput Biol. 2013;20(11):878–91.
29. Deng L, Zhang QC, Chen Z, Meng Y, Guan J, Zhou S. PredHS: a web server for predicting protein–protein interaction hot spots by using structural neighborhood properties. Nucleic Acids Res. 2014;42(Webserver-Issue):290–5.
30. Tuncbag N, Gursoy A, Keskin O. Identification of computational hot spots in protein interfaces: combining solvent accessibility and inter-residue potentials improves the accuracy. Bioinformatics. 2009;25(12):1513–20.
31. Deng L, Guan J, Dong Q, Zhou S. Prediction of protein-protein interaction sites using an ensemble method. BMC Bioinformatics. 2009;10(1):426.
32. Song J, Tan H, Takemoto K, Akutsu T. HSEpred: predict half-sphere exposure from protein sequences. Bioinformatics. 2008;24(13):1489–97.
33. Hamelryck T. An amino acid has two sides: a new 2D measure provides a different view of solvent exposure. Proteins Struct Funct Bioinforma. 2005;59(1):38–48.
34. Hanson J, Paliwal K, Litfin T, Yang Y, Zhou Y. Improving prediction of protein secondary structure, backbone angles, solvent accessibility and contact numbers by using predicted contact maps and an ensemble of recurrent and residual convolutional neural networks. Bioinformatics. 2018;10:2403–10.
35. McDonald IK, Thornton JM. Satisfying hydrogen bonding potential in proteins. J Mol Biol. 1994;238(5):777–93.
36. Liang S, Meroueh SO, Wang G, Qiu C, Zhou Y. Consensus scoring for enriching near-native structures from protein–protein docking decoys. Proteins Struct Funct Bioinforma. 2009;75(2):397–403.
37. Kabsch W, Sander C. Dictionary of protein secondary structure: pattern recognition of hydrogen-bonded and geometrical features. Biopolymers Orig Res Biomol. 1983;22(12):2577–637.
38. Klausen MS, Jespersen MC, Nielsen H, Jensen KK, Jurtz VI, Soenderby CK, Sommer MOA, Winther O, Nielsen M, Petersen B, et al. Netsurfp-2.0: Improved prediction of protein structural features by integrated deep learning. Proteins Struct Funct Bioinforma. 2019;87(6):520–7.
39. Heffernan R, Yang Y, Paliwal K, Zhou Y. Capturing non-local interactions by long short-term memory bidirectional recurrent neural networks for improving prediction of protein secondary structure, backbone angles, contact numbers and solvent accessibility. Bioinformatics. 2017;33(18):2842–9.
40. Jamroz M, Kolinski A, Kihara D. Structural features that predict real-value fluctuations of globular proteins. Proteins Struct Funct Bioinforma. 2012;80(5):1425–35.
41. Capra JA, Singh M. Predicting functionally important residues from sequence conservation. Bioinformatics. 2007;23(15):1875–82.
42. Kawashima S, Kanehisa M. AAindex: amino acid index database. Nucleic Acids Res. 2000;28(1):374.
43. Henikoff S, Henikoff JG. Amino acid substitution matrices from protein blocks. Proc Natl Acad Sci. 1992;89(22): 10915–9.

44. Chan C-H, Liang H-K, Hsiao N-W, Ko M-T, Lyu P-C, Hwang J-K. Relationship between local structural entropy and protein thermostabilty. Proteins Struct Funct Bioinforma. 2004;57(4):684–91.
45. Hanson J, Yang Y, Paliwal K, Zhou Y. Improving protein disorder prediction by deep bidirectional long short-term memory recurrent neural networks. Bioinformatics. 2016;33(5):685–692.
46. Wang S, Li W, Liu S, Xu J. Raptorx-property: a web server for protein structure property prediction. Nucleic Acids Res. 2016;44(W1):430–5.
47. Van der Laan MJ, Polley EC, Hubbard AE. Super learner. Stat Appl Genet Mol Biol. 2007;6(1):.
48. Breiman L. Random forests. Mach Learn. 2001;45(1):5–32.
49. Chang C-C, Lin C-J. Libsvm: A library for support vector machines. ACM Trans Intell Syst Technol (TIST). 2011;2(3):27.
50. Guyon I, Weston J, Barnhill S, Vapnik V. Gene selection for cancer classification using support vector machines. Mach Learn. 2002;46(1-3):389–422.
51. Peng H, Long F, Ding C. Feature selection based on mutual information: criteria of max-dependency, max-relevance, and min-redundancy. IEEE Trans Pattern Anal Mach Intell. 2005;8:1226–38. https://doi.org/10.1109/tpami.2005.159.
52. Climente-González H, Azencott C-A, Kaski S, Yamada M. Block hsic lasso: model-free biomarker detection for ultra-high dimensional data. bioRxiv. 2019532192. https://doi.org/10.1093/bioinformatics/btz333.
53. Pan X, Smith CE, Zhang J, McCabe KA, Fu J, Bell CE. A structure–activity analysis for probing the mechanism of processive double-stranded DNA digestion by λ exonuclease trimers. Biochemistry. 2015;54(39):6139–48.
54. Amrane S, Rebora K, Zniber I, Dupuy D, Mackereth CD. Backbone-independent nucleic acid binding by splicing factor sup-12 reveals key aspects of molecular recognition. Nat Commun. 2014;5:4595.

## Publisher's Note

Liu *et al. BMC Bioinformatics* 2020, **21**(Suppl 13):379
https://doi.org/10.1186/s12859-020-03676-2

BMC Bioinformatics

RESEARCH Open Access

# Explicit representation of protein activity states significantly improves causal discovery of protein phosphorylation networks

Jinling Liu[1,2], Xiaojun Ma[1], Gregory F. Cooper[1] and Xinghua Lu[1*]

*From* The 18th Asia Pacific Bioinformatics Conference
Seoul, Korea. 18-20 August 2020

* Correspondence: xinghua@pitt.edu
[1]Department of Biomedical Informatics, University of Pittsburgh, 5607 Baum Blvd, Suite 500, Pittsburgh, PA 15206, USA
Full list of author information is available at the end of the article

## Abstract

**Background:** Protein phosphorylation networks play an important role in cell signaling. In these networks, phosphorylation of a protein kinase usually leads to its activation, which in turn will phosphorylate its downstream target proteins. A phosphorylation network is essentially a causal network, which can be learned by causal inference algorithms. Prior efforts have applied such algorithms to data measuring protein phosphorylation levels, assuming that the phosphorylation levels represent protein activity states. However, the phosphorylation status of a kinase does not always reflect its activity state, because interventions such as inhibitors or mutations can directly affect its activity state without changing its phosphorylation status. Thus, when cellular systems are subjected to extensive perturbations, the statistical relationships between phosphorylation states of proteins may be disrupted, making it difficult to reconstruct the true protein phosphorylation network. Here, we describe a novel framework to address this challenge.

**Results:** We have developed a causal discovery framework that explicitly represents the activity state of each protein kinase as an unmeasured variable and developed a novel algorithm called "InferA" to infer the protein activity states, which allows us to incorporate the protein phosphorylation level, pharmacological interventions and prior knowledge. We applied our framework to simulated datasets and to a real-world dataset. The simulation experiments demonstrated that explicit representation of activity states of protein kinases allows one to effectively represent the impact of interventions and thus enabled our framework to accurately recover the ground-truth causal network. Results from the real-world dataset showed that the explicit representation of protein activity states allowed an effective and data-driven integration of the prior knowledge by InferA, which further leads to the recovery of a phosphorylation network that is more consistent with experiment results.

(Continued on next page)

(Continued from previous page)
**Conclusions:** Explicit representation of the protein activity states by our novel framework significantly enhances causal discovery of protein phosphorylation networks.

**Keywords:** Causal inference, Protein kinase activity state, Protein phosphorylation networks, Cancer signaling pathways

## Background

Protein phosphorylation networks play a significant role in cellular signaling under physiological and pathological conditions [1]. Many drugs are specifically designed to block signals originating from aberrantly activated protein kinases [2, 3]. A better understanding of protein phosphorylation networks will enhance our understanding of cellular signaling systems and facilitate design of new drugs.

In essence, a protein phosphorylation network is a causal network. In such networks, phosphorylation of a protein kinase usually leads to its activation, and the activated kinase will in turn phosphorylate downstream target proteins [4]. It is the kinase activity state, rather than its phosphorylation state, that drives the causal chain among different protein kinases. To discover such causal networks, researchers often systematically apply genetic or pharmacological perturbations to a cellular system to resolve the causal relationships among the proteins of interest [5, 6]. There has been a rapid accumulation of phosphorylation level data in which the levels of the phosphorylated forms of proteins are measured [7]. Researchers commonly apply causal inference approaches to such data to infer protein phosphorylation networks. However, direct use of phosphorylation level data may be insufficient or even misleading, because phosphorylation levels do not necessarily reflect the activity states of protein kinases. Particularly when a cellular system is subjected to genetic or pharmacological perturbations (e.g., activating mutations of kinases in cancers or pharmacological agents blocking activities of kinases) the phosphorylation status of a kinase and its target protein can be decoupled. Under such circumstances, conventional approaches of attempting to learn causal relationships among phosphoproteins solely based on protein phosphorylation status may fail.

To address such challenges, we have designed a novel causal discovery framework in which the phosphorylation status of each protein kinase is represented as a measured variable and its activation status is explicitly represented as a latent variable; the statistical relationships between the phosphorylation status of a protein kinase and its activity is represented as a stochastic relationship. The impact of interventions on kinase activity can thereby be explicitly represented. Moreover, prior knowledge can be integrated so that inference of kinase activity states is both knowledge-guided and data-driven. We designed a learning framework that iteratively infers the activity states of protein kinases. Through a series of simulation experiments and an application to a real-world data set, we found that explicitly representing the activity states of protein kinases can help identify a more accurate protein phosphorylation network.

## Methods

### The novel causal discovery framework

As an example, we use a simple phosphorylation network containing three proteins to illustrate our framework. In this system, there are three types of variables (nodes)

representing quantities of interest. A "*P*" node in Fig. 1 represents the phosphorylation state of a kinase, which can be measured by its phosphorylation level. If protein2 is phosphorylated as indicated by its phosphorylation level, protein2 likely becomes active, and it will further phosphorylate protein3 to likely activate protein3 (Fig. 1a). However, when protein2's kinase activity is inhibited by an inhibitor that does not affect protein2's phosphorylation state, protein2 is inactive even though it is phosphorylated, and as such it cannot further phosphorylate protein3 (Fig. 1b). In this case, the phosphorylation states of protein2 and protein3 are decoupled. If we represented each protein only with its phosphorylation state (a *P* node), it would be difficult to learn the true causal relationship between protein2 and its downstream protein3, when such perturbation events occur. We propose a novel causal discovery framework where each protein is represented by both its phosphorylation state (a *P* node) and its activity state (an *A* node), an unmeasured latent variable. Figure 1c illustrates this representation for the example. A default causal arc from the *P* node to the *A* node for each protein is required, since the protein phosphorylation state can affect its activity state (Fig. 1c). If a protein is intervened upon, an intervention (*V*) node representing the intervention action will be included in the model, and a causal arc from this *V* node to the protein's activity node will be added (Fig. 1c) since the intervention can directly affect the

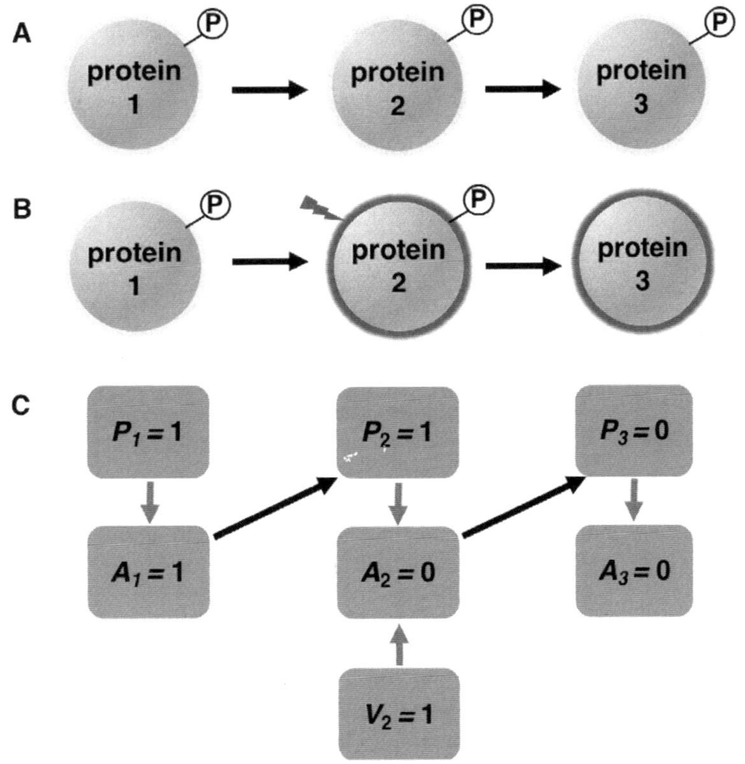


**Fig. 1** The novel causal discovery framework. **a** A phosphorylation network without interventions. A yellow glow indicates the protein is active while 'p' in the circle shows the protein is phosphorylated. **b** A phosphorylation network with an intervention. A red glow suggests the protein is inactive. **c** The novel causal discovery framework. *P* nodes represent protein phosphorylation states; *A* nodes represent protein activity states; *V* nodes represent the intervention states. Gray arcs are required default arcs and black arcs represent the protein causal relationships to be inferred

protein activity state. Finally, the activity state of a kinase causally influences the phosphorylation states of its downstream targets (Fig. 1c). This causal discovery framework models more accurately the underlying biological process, and it allows us to effectively represent the impact of interventions on protein activity states, as well as model the stochastic relationships between protein phosphorylation and activity states using causal Bayesian network (CBN) discovery methods.

#### Overview of CBN methods

A CBN is a probabilistic graphical model that contains a graphical structure and a set of parameters. The graphical structure is represented by a directed acyclic graph (DAG) which contains nodes and directed edges. Nodes represent domain variables and directed edges represent the causal relationships between variables. A set of parameters contains the conditional probability of each node given its parents. The joint probability of all the nodes in a CBN can be written as a product of each node's conditional probability, based on the causal Markov assumption [8]. A CBN is a flexible framework for modeling a complex domain by representing the joint probability distribution of variables, which allows unobserved variables and stochastic elements during network inference. A learned CBN with directed edges can reveal the causal relationships among variables [9]. CBNs are typically learned by two approaches: constraint-based and score-based methods. Constraint-based methods, such as the PC algorithm, use statistical tests of conditional independence to learn a CBN structure [10], which is relatively efficient, but sensitive to error propagation. Score-based algorithms use a score, such as a Bayesian posterior probability, to search for causal models with the highest score [9]. As compared to traditional constraint-based algorithms, score-based ones can assign a relative confidence score to a model, enabling selection among candidate models. In this study, we used a state-of-the-art score-based method, the FGES algorithm, which can efficiently analyze data sets containing tens of thousands of variables and performs well on simulated data sets [11].

#### Simulated data

To illustrate the potential impact of systematic perturbations in learning protein phosphorylation networks, we performed a series of simulation experiments. Based on prior knowledge of a protein phosphorylation network of 35 protein kinases that came from literature and also applied to the real-data to be introduced below [6], we created a "ground-truth" causal network of 16 kinases, which consists of all three types of nodes (Fig. 2). In this network, each kinase is represented by two types of nodes: protein phosphorylation states ($P$ nodes) and kinase activity states ($A$ nodes). For each kinase, there is a causal arc from its $P$ node to its $A$ node, and some or no causal arcs between its $A$ node and the $P$ nodes of other proteins, depending on the prior knowledge. In this simulated experiment, we set $P(A = 1 \mid P = 1) = 1$, as well as $P(A = 1 \mid P = 0) = 0$ if there is no intervention. We further simulated the impact of interventions on 9 proteins by adding a set of $V$ nodes reflecting the states of inventions and by directing these $V$ nodes to the corresponding $A$ nodes; when a $V$ node is set to "1", it directly inhibits the activity state of the corresponding $A$ node (i.e., set the activity of the kinases to "inactive" or "0" state), without changing the protein's phosphorylation

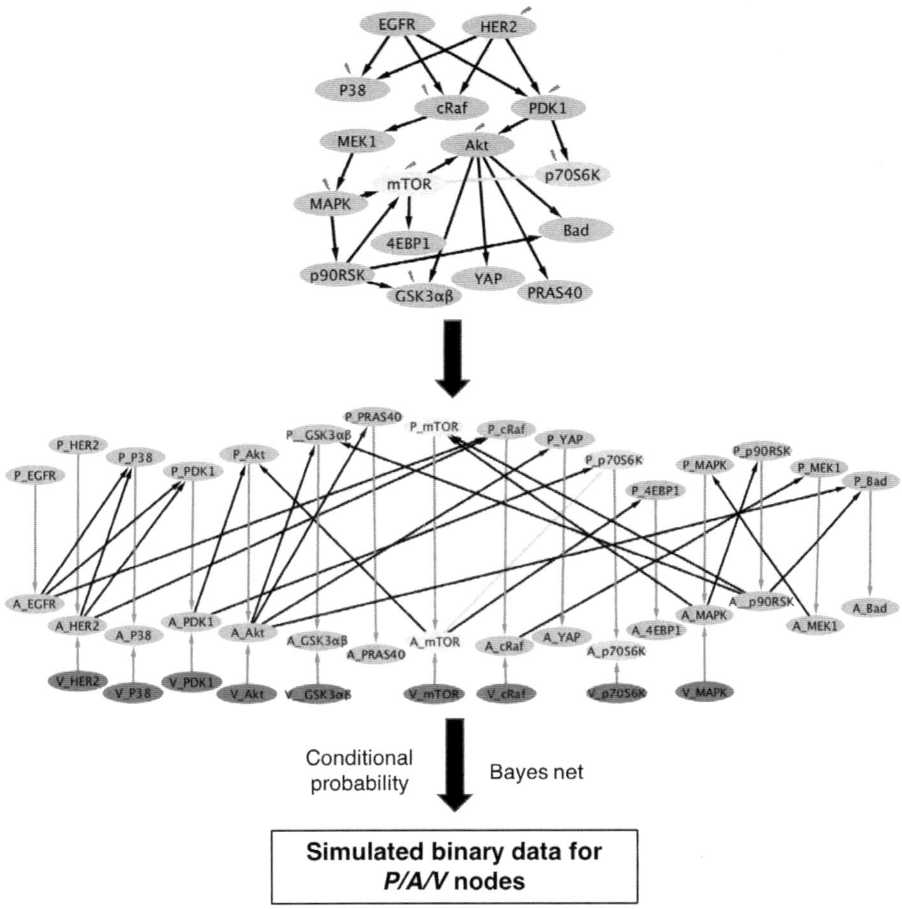


**Fig. 2** Outline of the simulation approach. The top graph is the ground-truth network where red lighting represents the applied interventions. The middle graph is the ground-truth network represented by our framework, where gray arcs are the required default arcs, black arcs come from the ground-truth network above and red nodes represent intervention nodes. Nodes for mTOR are highlighted in fluorescent yellow, nodes for p70S6K are highlighted in green, and the arc between them is highlighted in cyan. These highlights were for better visualization of the transformation from the ground-truth network to the middle network represented by our causal discovery framework

state (Fig. 2). Under these conditions, an *A* node integrates both information from its *P* node and *V* node: it will be a copy of *P* node if there is no intervention, and it will be set to "0" if there is an intervention. With this CBN, we used the Tetrad causal modeling and discovery system (https://www.ccd.pitt.edu/tools/) to instantiate the parameters of conditional probabilities and to further simulate binary values for all the *P* nodes, *V* nodes, and *A* nodes for 500 instances (Fig. 2).

#### Real data: protein phosphorylation data and preprocessing

The proteomic data set used in this study involved the perturbation of four different human breast cancer cell lines and measurement of the abundance of phosphoproteins [6]. The breast cancer cell lines included BT20(Basal), BT549(Claudin-low), MCF7(Luminal) and UACC812(HER2+), which were pre-treated using five different inhibitors and DMSO (as a control) for 2 hours and then stimulated by eight different stimuli. The interventions as well as their potential targets are summarized in Additional file 1. Reverse phase protein array (RPPA) assays were performed to measure the abundance of 35 phosphoproteins harvested at seven different time points within a four-hour period. The amounts of phosphorylated proteins (phosphorylation level), rather than the total protein abundance, were measured in the data set used here.

We log-transformed and discretized the continuous data for the abundance of 35 phosphoproteins that was downloaded from [6]. For each cell line, we extracted all the cases treated with DMSO and collected at time 0, a condition where cells are not perturbed by inhibitors nor stimuli. These cases served as controls. We compared all the other cases to the control group to examine whether there was any significant difference in phosphorylation levels. If the value of a protein at a non-control condition is significantly different from those derived under the control condition ($p$-value $< 0.05$), we set the phosphorylation state of the protein under the condition to 1; otherwise, a discretized value of 0 was used, indicating no changes of the protein phosphorylation states compared to the control condition. After discretizing all the cases for each cell line, we combined all the discretized values from all four cell lines to form a larger dataset for further analysis. If a protein was intervened upon by pharmacological agents, its intervention was represented as a binary variable and its value was set to 1.

### Inference of protein activity states

Since the activity state of each kinase of interest is not measured (latent) under the experimental conditions in which the data were collected [6], we designed a Markov chain Monte Carlo (MCMC) algorithm [12] to infer the protein activity states; we call it the InferA algorithm. Following our causal discovery framework (Fig. 1c), we instantiate the phosphorylation state ($P$ node) of each protein based on the observed phosphorylation level data. If a protein is targeted by a drug in the experiments, we also created a $V$ node and instantiated its value based on the experimental conditions. An initial network was constructed as the structure to connect the $A$ nodes and $P$ nodes for the 35 proteins of interest (described below in more details). If there was an arc from Protein1 to Protein2 in the initial network, we added a causal arc from the $A$ node of Protein1 to the $P$ node of Protein2.

We inferred the conditional probability of the A node for each protein based on its Markov blanket in the initial network, as shown in the pseudo code below. The Markov blanket of a node includes the node's parents, children, and the parents of its children, which represent all the information needed for inferring the value of this node [13]. We then used Gibbs sampling, a MCMC algorithm, to assign the value of 1 or 0 to each protein's A node based on its conditional probability. We sampled a random number from (0,1), and compared this number with an A node's conditional probability given its Markov blanket. If an A node's conditional probability is smaller than this random number, then the value of this A node was set as 0; otherwise the value of this A node was set as 1. We calculated the logarithm of the joint probability for all the observed variables ($P$ nodes and $V$ nodes) in all the cases as a convergence criterion. We repeated sampling (E-step) and updating the conditional probability Table (M-step) until a predefined maximum number of iterations when the logarithm of joint probability converged with a small-range oscillation. Upon convergence, the resulting values for all the $A$ nodes were outputted and used in the subsequent analysis. The InferA algorithm is coded in C++ and is very efficient. For example, it takes 48 s to output $A$ node values for 35 $P$ nodes with about 1000 instances after 1000 iterations on a server of 2 CPUs with 6 cores per CPU. The pseudo code is showed by Fig. 3a.

### Construction of initial networks

Inferring the values of the $A$ nodes requires an initial structure of a CBN. We adopted two approaches to define network structures to be used in the InferA algorithm. First, a network describing the known causal relationships between the proteins of interest was curated from the literature [6], which included cycles in regulation relationships. We turned this network into an acyclic causal graph by removing a minimal number of arcs to break all cycles, and we refer to this network as the knowledge-based network (K_Net). Second, we created an initial network based on the observed data from [6] using a generalized linear regression model, and we refer to this network as data-derived network (D_Net). More specifically, we first instantiated $A$ node values with the same value of the corresponding $P$ node values if there was no intervention. If there was intervention for a protein, then its $A$ value was set as 1. Based on the assumption that the phosphorylation state of a protein is affected by the activity states of other kinases, we performed a series of regression analyses, where a $P$ node of a protein is treated as a dependent variable and the $A$ nodes of other proteins as independent variables. We applied elastic network regression using the R package "glmnet." We then constructed an initial causal network by adding causal arcs from the top $A$ nodes (the absolute values of their coefficients in the regression model $\geq 0.5$) to a $P$ node. Cycles in the D_Net were eliminated by removing a minimal number of arcs. Finally, using the function of Random Graph provided by the Tetrad program, we generated a DAG of random network of the same number of nodes and arcs as K_Net by randomly adding causal arcs between $A$ nodes and $P$ nodes. We refer this network as rand_Net, which serves as a control of a network without any biological information.

### Network learning using FGES

Causal-cmd is a command-line tool (https://www.ccd.pitt.edu/) that allows a user to call many of the causal discovery algorithms implemented in Tetrad, including FGES. FGES is an optimized and parallelized version of the greedy equivalence search (GES) algorithm [14], which can efficiently learn CBN structure on thousands of variables and samples [11].

After inferring the state of $A$ nodes, we combine observed $P$ node data with inferred $A$ nodes as input to FGES to search for causal arcs from $A$ nodes to non-self $P$ nodes (Fig. 3b). FGES allows a user to set constraints on search scope. Based on our causal discovery framework (Fig. 1c), we require FGES to instantiate a default arc from the $P$ node to the $A$ node for each protein as well as a default arc from the $V$ node to the $A$ node, if there is an intervention; we prohibit FGES from adding direct edges among $P$ nodes; we also prohibit FGES from adding direct edges among $A$ nodes; we only allow FGES to search for edges from $A$ nodes to non-self $P$ nodes (Fig. 3b). To estimate the

**A**

**Algorithm InferA**

1. Initialize $A$ nodes
2. Infer values for A nodes

   While not yet converged:

   for i in (1, N_Instances)

   for j in (1, N_Proteins)

   $$P(A_{ij} = 1 \mid All_{\setminus ij}) = P(A_{ij} \mid MB(A_{ij}))$$
   $$\propto P(A_{ij} = 1 \mid pa(A_{ij})) * \prod P(A_{ij}{}^{C} \mid pa(A_{ij}{}^{C}))$$

   Use Gibbs sampling to update the value of $A_{ij}$ node

   Update the conditional probability table for $A_j$

   Calculate, $Log\ L(data \mid initial\_Net)$
3. Output all $A_{ij}$ node values upon convergence

**B**

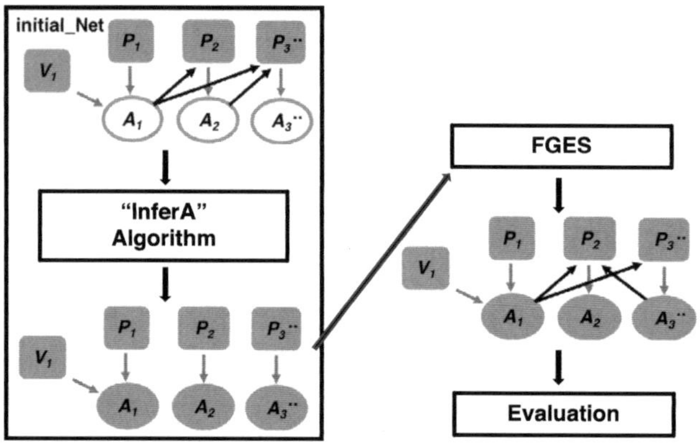


**Fig. 3** Overview of the InferA algorithm and its use with FGES. **a** The pseudo code of algorithm InferA. "$All_{\setminus ij}$" refers to all the nodes except $A_{ij}$, "pa($A_{ij}$)" refers to the parents of $A_{ij}$, and "$A_{ij}{}^{C}$" represents a child node of the $A_{ij}$ node. "initial_Net" refers to the initial network, which is discussed in the subsection of "construction of initial networks" within the methods section. **b** The pipeline of InferA-FGES for predicting the causal network among phosphoproteins. Within the three networks shown in the pipeline, $P$ nodes represent protein phosphorylation states; $A$ nodes represent protein activity states; the gray edges are the default edges between each protein's $P$ node to its $A$ node while the black edges represent the causal relationships among proteins; the black edges in the initial_Net were provided to InferA for inferring the $A$ node values while the black edges in the predicted network outputted by FGES were learned by the pipeline. The unfilled ovals represent $A$ nodes with unknown values while the filled ovals or squares denote nodes with known values

posterior probability that FGES assigns an edge between two nodes, we employed a bootstrap approach, where FGES was applied to 100 different datasets sampled from the original dataset with replacement. The reliability of each edge is estimated based on the frequency of discovering this edge over the 100 bootstrap samples.

### Evaluation of the learned protein causal networks

We have used two different approaches to evaluate the networks learned by the FGES algorithm. For the first approach, we compared the learned network to a ground-truth network. In this method, we only included edges with an estimated posterior probability $\geq 0.5$. We assessed the ability of FGES to discover the ground-truth causal network structure from three perspectives: 1) *"arc"*, which assesses how well the directed edges in the ground-truth network are recovered; 2) *"adjacency"*, which assesses how well true neighbor nodes are connected; and 3) *"path"*, which assesses how well directed paths connecting pairs of nodes are recovered. For each of the above measures, we derived the precision rate, recall rate, and their harmonic average (F1 score) for the learned network.

For the second approach, we used the evaluation criterion published by Hill et al. (2017). In their study, the mTOR inhibitor was applied to cell lines; those proteins that exhibited significant changes in phosphorylation status were identified, and they were regarded as the gold-standard descendants of mTOR. For a learned network, if there was a directed path from mTOR to a protein, we treated the protein as a descendant of mTOR. We compared the descendants of mTOR discovered by our approach to the gold-standard descendants to acquire precision and recall rates. We also calculated the area under the ROC curve (AUROC) by varying the thresholds of the learned arc probability.

## Results

We applied our causal discovery framework to both simulated and real-world data sets. We investigated whether our framework allowed a better illustration of intervention effects as well as a good usage of prior knowledge, and whether these helped obtain a more accurate protein causal network compared to the current model of using the phosphorylation levels alone.

### Simulation experiments

To investigate the impact of perturbations on disrupting statistical relationships between protein phosphorylation data and the performance of our causal discovery framework in learning protein phosphorylation networks, we performed a series of simulation experiments using CBNs with *P/A/V* nodes where 9 out of 16 protein kinases were intervened (Fig. 2). We generated samples in which the values of all nodes (*P/A/V*) were instantiated based on the joint probability defined by the structure and parameters of the CBN. Please note in the simulation experiments, the A node values were simulated based on predefined structure and parameters, not inferred by InferA. We further conducted the following experiments. First, to evaluate the impacts of perturbations on the statistical relationships among *P* nodes, we applied FGES model to infer causal network structures using just the data corresponding to *P* nodes, which is referred as the P-only experiment. Second, in previous studies [5, 15], in order to represent the impact of interventions on the systems, many researchers changed the protein phosphorylation state to a predefined value if the protein was intervened upon. While this approach helped to maintain the statistical relationships of phosphorylation states between a perturbed protein and its downstream proteins, it would disrupt the statistical relationships of its phosphorylation status with respect to its upstream proteins. To examine the impact of such approach, we produced a dataset, in which a *P*-node value was set to 0 if it was intervened upon in the sample, and we call this experiment as the **I**ntervened on ***P*** (I-P) experiment. Finally, we applied FGES to both P nodes and A nodes to search for causal arcs from *A* nodes to non-self *P* nodes, and we call this experiment as the A-P experiment.

In the A-P experiment from simulated data, the FGES algorithm completely recovered the ground truth network, with 100% recall and 100% precision for *arc* and *adjacency* (Fig. 4a,b). By contrast, the causal networks learned from only P nodes showed significantly lower recall (0.64) and precision (0.69) for adjacency, and much lower recall (0.32) and precision (0.35) for arcs. The I-P experiment acquired a high precision for adjacency (0.98) that is similar to the A_P experiment (Fig. 4b). But the I-P group had a worse recall for adjacency (0.8) and much lower recall (0.38) and precision (0.45) for arc, compared to the A-P experiment (Fig. 4a,b). The results provide support that under systematic perturbations, learning CBNs only based on P nodes results in relatively poor overall discovery performance. While the brute force representation of effects of interventions by discarding observed protein phosphorylation data and replacing it with a predefined value based on the interventions can partially restore the statistical relationships of phosphorylation states between an intervened protein and its downstream target proteins, the approach could not fully recover the ground truth network. Finally, inclusion of A nodes can accurately reflect the statistical relationships among the proteins and interventions, and thus achieved the complete reconstruction of the ground truth network.

We further examined the structures of the CBNs recovered by FGES from the different experiments discussed above. Compared with the ground truth network of 22 arcs, the P_only experiment included five false arcs, missed eight true arcs, failed to correctly

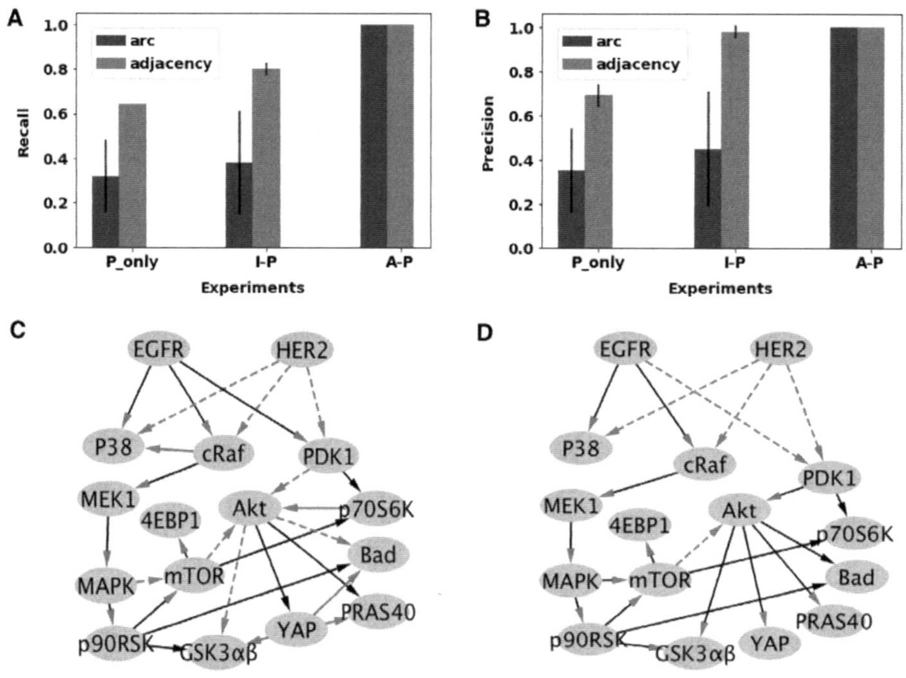


**Fig. 4** Explicit representation of protein activity states helped acquire a more accurate network with simulated datasets. **a** Recall relative to the ground-truth network. Error bars in this figure represent standard deviations from three simulation experiments. **b** Precision relative to the ground-truth network. **c** The comparison between the learned network by P_only experiment and the ground-truth network. **d** The comparison between the estimated network by I-P experiment and the ground-truth network. Dashed red arcs represent missed true arcs, solid red arcs show false arcs added, black edges with red arrowheads represent the misoriented edges, and black edges with black arrowheads show the true arcs discovered by the P_only or I-P group

orient eight edges and only obtained five true arcs (Fig. 4c); the I-P experiment did not add any false edges, but it removed five true arcs and failed to correctly orient thirteen edges (Fig. 4d); the A-P experiment had the same causal structure as the ground-truth network. The simulation experiments support our hypothesis that the explicit representation of protein activity states together with the protein phosphorylation states can improve the causal discovery of protein phosphorylation networks.

#### InferA enables an effective integration of prior knowledge in a data-driven manner in the real-data experiments

We set out to test if our framework can accurately learn causal networks based on the real-world data from [6] and evaluate whether we can recover network structures that conform with prior knowledge. Given experimental protein phosphorylation data and an initial network of K_Net that originates from a prior network curated from literature, we used our InferA algorithm to infer the *A* node values based on the observed *P/V* nodes. Within the algorithm of InferA, the *A* node values were either initialized randomly (initRand) or initialized with the corresponding *P* node values (initCopy). The logarithm of the joint probability of all the observed nodes (i.e., the *P* nodes and *V* nodes) increased as the algorithm iterated and it converged quickly for both initRand and initCopy (Fig. 5a). We outputted *A* node values at various iteration numbers before convergence. We applied FGES to the resulting *A* node values along with the observed *P/V* nodes to learn a protein-signaling causal network. Interestingly, the F1 scores for different network properties (arc, adjacency and path) exhibited positive correlations with the joint probabilities of the data under both initRand (Fig. 5b) and initCopy (Fig. 5c) situations. This positive correlation supported the convergence of the InferA algorithm to the initial network. Since initRand and initCopy worked similarly, we only show the initCopy results in the following figures.

We further compared the recall, precision, and F1 score of retrieving the K_Net for the different causal networks under different experiments (Table 1). For the P_only experiment, we only used the protein phosphorylation states represented by *P* nodes to learn the protein causal network. For the other experiments, we used both *P* nodes and *A* nodes as inputs to FGES, as illustrated in Table 1. The P_only experiment had a very low recall, precision, and F1 score for *arc*, *adjacency* and *path*. The A_as_P experiment obtained similar results to the P_only experiment in discovering K_Net, indicating that simply setting A nodes to the same values as P nodes does not enhance learning. For the K_Net experiment that uses K_Net as the initial network, the recall, precision, and F1 score for *arc*, *adjacency* and *path* were all greatly improved compared to the P_only experiment (Fig. 6a, b, c). Here, the higher recall suggests that the InferA algorithm effectively integrated the prior knowledge of the protein causal interactions from K_Net. We noted that since we used K_Net as input, and compared the resulting network to K_Net, this was the best-case scenario for achieving good performance. But it was indeed an effective way of integrating the prior knowledge of K_Net. To demonstrate that the learned causal networks are data-driven (not solely determined by the prior network structure), we permuted the data values of *P* nodes and *V* nodes, and we repeated the procedures of InferA (using K_Net) and FGES network search. Indeed, the recall for *arc*, *adjacency* and *path* decreased significantly (Fig. 6a). This indicated that

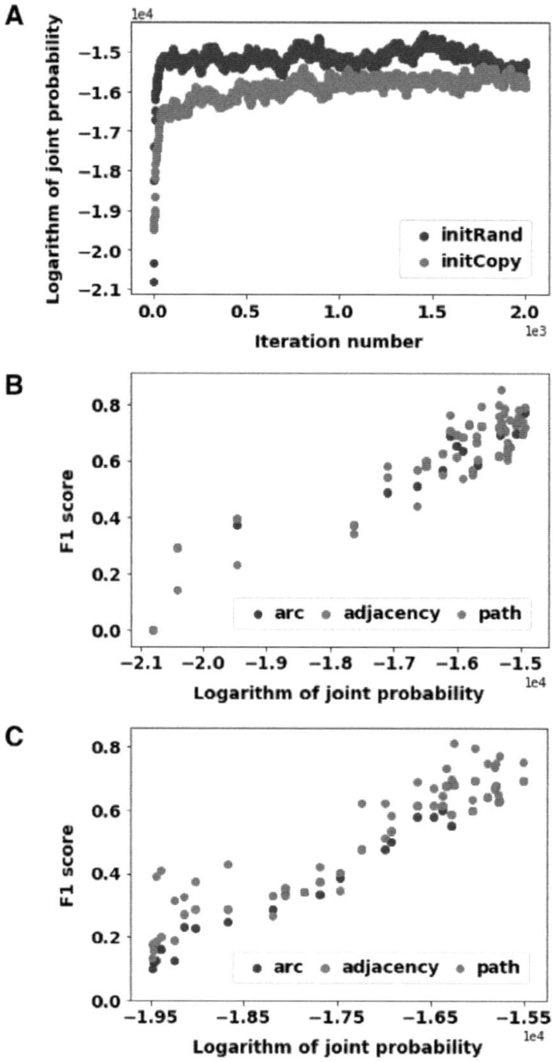


**Fig. 5** The InferA algorithm converged well for the real-world data analysis. **a** The changes of the logarithm of joint probability for observed *P* nodes and *V* nodes with iteration numbers when initializing *A* nodes with random 0/1(initRand) or initializing *A* nodes with the corresponding *P* node values (initCopy). Values on both axes are in scientific formats. **b** The F1 scores of *arc*, *adjacency* and *path* increase with the logarithm of joint probability for initRand. X-axis values are represented by scientific notations. **c** The F1 scores of *arc*, *adjacency* and *path* increase with the logarithm of joint probability for initCopy. X-axis values are shown in scientific formats

**Table 1** Different experiments for the real-world data analysis

| Names | Input to FGES | *A* values | Input to InferA |
|---|---|---|---|
| P_only | *P* | N/A | N/A |
| A_as_P | *P* + *A* | Copied *P* values | N/A |
| K_Net | *P* + *A* | Inferred by InferA | K_Net + *P/V* |
| D_Net | *P* + *A* | Inferred by InferA | D_Net + *P/V* |
| rand_Net | *P* + *A* | Inferred by InferA | rand_Net + *P/V* |
| P-permuted | permuted *P* + *A* | Inferred by InferA | K_Net + permuted *P/V* |

the effective integration of the prior network of K_Net by the InferA algorithm is data-driven. Without sensible data, it is unlikely to recover most of the causal arcs in K_Net. With disrupted statistical relationships between protein variables resulting from data permutation, FGES only reported few significant causal edges. With fewer causal edges recovered from K_Net, the precision for the P-permuted experiment became very high (Fig. 6b). The F1 scores of *arc*, *adjacency* and *path* in the P-permuted experiment were significantly lower than that in the K_Net experiment (Fig. 6c).

We examined further the structure of a network learned by our framework (Fig. 6d). The learned network based on the inferred A node values from K_Net recovered many edges in K_Net. Interestingly, our methods also learned some new edges in comparison to K_Net. The biological validity of these edges remains to be examined through experiments.

### Evaluating the learned networks with respect to held-out experiment data

We then evaluated the learned causal networks against the ground truth acquired from biology experiments (i.e. gold-standard descendants of mTOR) using the same metrics and methods as described by [6]. The causal networks learned from only *P* nodes have a low AUROC of 0.43 (Fig. 7). Also, the A_as_P experiment acquired similar AUROC to the P_only experiment. The experiments using either K_Net or D_Net as an initial network for inferring the *A* node values had significantly better AUROCs (0.75 for K_Net and 0.73 for D_Net), compared to the P_only experiment. This result suggests that the explicit representation of protein activity states with *A* node values inferred from a sensible initial network helped FGES find a more accurate protein causal network. It is rather exciting to see the D_Net experiment performed very well without using any prior knowledge. With a randomly-generated initial network to infer the *A* node values, using the resulting *A* values led to a learned causal graph with a low AUROC (0.45), which supports the utility of a good initial network. If using K_Net as an initial network, but at the same time permuting *P/V* node values, FGES failed to recover true causal relationships, with an AUROC of 0.31 that is greatly decreased compared to the K_Net group (Fig. 7). This result suggested an initial prior network (i.e. K_Net) itself is not sufficient, and InferA needs both a good prior network and good data in order to derive truthful values of *A* nodes which in turn allow the discovery of a more accurate causal network of the phosphoproteins.

## Discussion

Discovering protein phosphorylation networks in specific cell types remains an important task in understanding cellular signaling systems and mechanisms of certain diseases, particularly cancer. For example, the cancer genome atlas (TCGA) project has systematically collected RPPA data of many phosphoproteins [7], and understanding how these proteins causally interact with each other under the perturbations of diverse genomic and epigenomic alterations in cancer cells could shed light on disease mechanisms of cancers. However, as illustrated in our study, attempts to infer causal phosphorylation networks solely based on the abundance of phosphoproteins are insufficient in recovering the true causal networks. We conjecture that this problem will likely be exacerbated when inferring large causal networks of many phosphoproteins

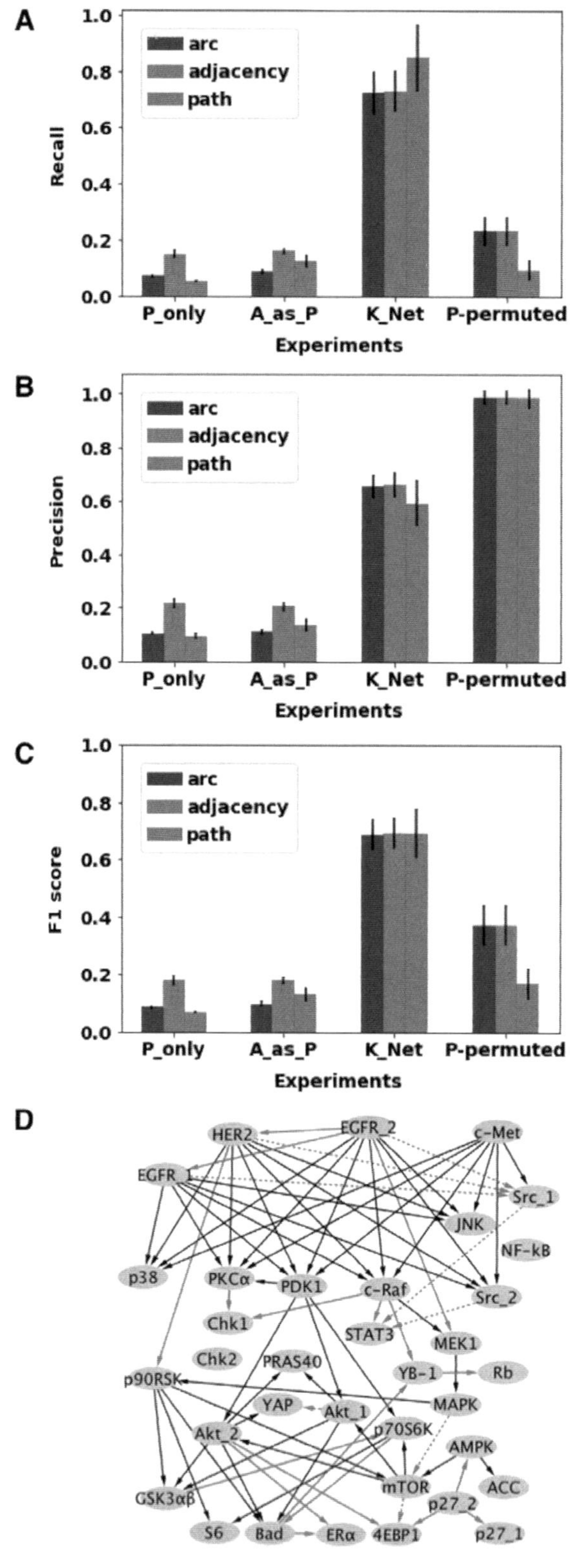


**Fig. 6** (See legend on next page.)

(See figure on previous page.)
**Fig. 6** The InferA algorithm effectively integrates the prior knowledge in the real-world data analysis. Summarizes the recall **(a)**, precision **(b)** and F1 score **(c)** of *arc*, *adjacency* and *path* for different experiments compared to K_Net. Error bars in this figure represent standard deviations among 50 experiments. **d** A learned protein phosphorylation network compared to K_Net. Solid black arcs showed the recovered true arcs from K_Net; dashed red arcs were the ones not recovered from K_Net; solid red arcs represent the new arcs learned by FGES

under the conditions of systematic perturbations, such as in cancers where many kinases are aberrantly activated by mutations without changing their phosphorylation states. In this study, we showed that in addition to *P* nodes representing the protein phosphorylation status, including *A* nodes to explicitly represent the activity states of kinases enables one to represent the impact of interventions on protein function in a biologically sensible way, and thereby enhances the capability of CBN discovery algorithms to recover biological sensible results.

Our study illustrated important factors that may influence the discovery of causal networks. First, constructing a biologically sensible initial network, either based on prior knowledge or constructed from data, will influence the accuracy of inferring the activity states of phosphoproteins and thereby the quality of the final causal network. We showed that both K_Net and D_Net tested in our study were able to lead to sensible discoveries of the protein causal networks, whereas random initial networks did not lead to accurate causal discovery. Second, in addition to a good initial network, data are also important to achieve accurate performance, which reflects the data-driven nature of our approach. The InferA algorithm was able to effectively integrate prior knowledge of known causal relationships in a data-driven manner, which further helped FGES acquire a final causal network that is consistent with experiment results. This result reflected a strength of this Bayesian learning approach, which combines prior knowledge (in the form of an initial network structure) and data. Our learned network postulates some new causal relationships beyond the prior knowledge, which serve as hypotheses for further investigation.

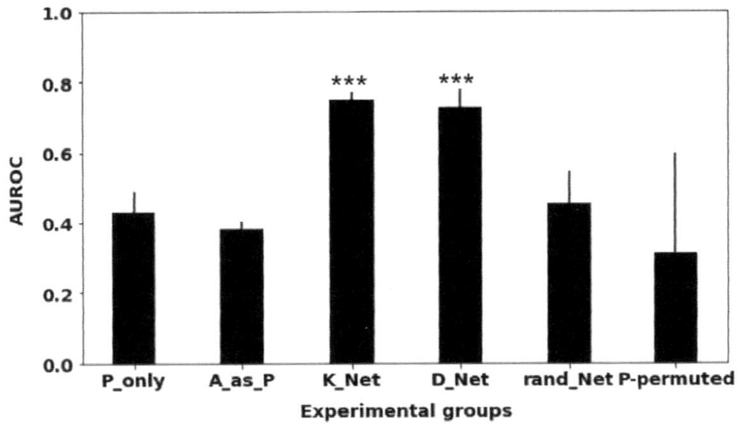


**Fig. 7** Network evaluation with respect to the held-out experiment data. AUROC results when comparing the mTOR descendants from the learned causal networks to the gold-standard ones for different experiments. Error bars in this figure represent standard deviations over 50 experiments. Three asterisks indicate significant differences ($p$-value < 0.001) compared with the P_only experiment

## Conclusions

In summary, the results support the hypothesis that explicit representation of protein activity states together with their phosphorylation states can help causal discovery algorithms more accurately learn protein phosphorylation networks. Our InferA algorithm can estimate the protein activity states by effectively and efficiently integrating various information including the observed phosphorylation level data, interventions applied in the data, and prior knowledge of protein causal relationships.

## Supplementary information

**Supplementary information** accompanies this paper at https://doi.org/10.1186/s12859-020-03676-2.

**Additional file 1.** Interventions (stimuli or inhibitors) and their most-direct target proteins among the 35 proteins of interest in the RPPA dataset. N/A means there is either no target protein among the proteins of interest or the target proteins are not clear.

### Abbreviations

TCGA: The cancer genome atlas; FGES: Fast greedy equivalence search; GES: Greedy equivalence search; CBN: Causal Bayesian Network; RPPA: Reverse phase protein array; K_Net: Knowledge-based network; D_Net: Data-derived network

### Acknowledgements

The authors would like to thank Dr. Harry Hochheiser for his feedback on the manuscript.

### About this supplement

This article has been published as part of *BMC Bioinformatics Volume 21 Supplement 13, 2020: Selected articles from the 18th Asia Pacific Bioinformatics Conference (APBC 2020): bioinformatics.* The full contents of the supplement are available online at https://bmcbioinformatics.biomedcentral.com/articles/supplements/volume-21-supplement-13

### Authors' contributions

XL and JL conceived and designed the experiments; JL carried out the data collection, modeling and analyses; XM wrote the code for the InferA algorithm; XL and JL wrote the paper; GC provided general advice on the causal discovery methodology and editing of the manuscript. The authors read and approved the final manuscript.

### Funding

This research was supported by the National Library of Medicine, training grant 5T15LM007059–32, by the National Human Genome Research Institute, grant U54HG008540 via the trans-NIH Big Data to Knowledge (BD2K) Initiative, and by the Pennsylvania Department of Health, grant 4100070287. Publication costs are funded by U54HG008540, R01LM012011, and NLM training grant 5T15LM007059–32.

### Availability of data and materials

Not applicable.

### Ethics approval and consent to participate

Not applicable.

### Consent for publication

Not applicable.

### Competing interests

The authors declared no competing interests.

### Author details

[1]Department of Biomedical Informatics, University of Pittsburgh, 5607 Baum Blvd, Suite 500, Pittsburgh, PA 15206, USA. [2]Present address: Department of Engineering Management and Systems Engineering and Department of Biological Sciences, Missouri University of Science and Technology, Rolla, MO 65409, USA.

Published: 17 September 2020

### References

1. Ardito F, Giuliani M, Perrone D, Troiano G, Lo ML. The crucial role of protein phosphorylation in cell signalingand its use as targeted therapy (review). Int J Mol Med. 2017;40:271–80.
2. Cohen P. Protein kinases--the major drug targets of the twenty-first century? Nat Rev Drug Discov. 2002;1:309–15.
3. Knight ZA, Lin H, Shokat KM. Targeting the cancer kinome through polypharmacology. Nat Rev Cancer. 2010;10:130–7.
4. Alberts B, Johnson A, Lewis J, Morgan D, Raff M, Roberts K, et al. In: Anderson M, Granum S, editors. Molecular biology of the cell. 5th ed. New York: Garland Science; 2007. p. 175. 2017.
5. Hill SM, Heiser LM, Cokelaer T, Linger M, Nesser NK, Carlin DE, et al. Inferring causal molecular networks: empirical assessment through a community-based effort. Nat Methods. 2016;13:310–22.

6. Hill SM, Nesser NK, Johnson-Camacho K, Jeffress M, Johnson A, Boniface C, et al. Context specificity in causal signaling networks revealed by phosphoprotein profiling. Cell Syst. 2017;4:73–83.e10.
7. Li J, Lu Y, Akbani R, Ju Z, Roebuck PL, Liu W, et al. TCPA: a resource for cancer functional proteomics data. Nat Methods. 2013;10:1046–47.
8. Hausman DM, Woodward J. Independence, invariance and the causal Markov condition. Br J Philos Sci. 1999;50:521–83.
9. Heckerman D, Meek C, Cooper G. A {B}ayesian approach to causal discovery. In: Computation, causation, & discovery; 1999. p. 141–65.
10. Spirtes P, Glymour C, Scheines R, Heckerman D, Meek C, Cooper GF, et al. Causation, prediction, and search. 2nd edition. Cambridge: MIT Press; 2000.
11. Ramsey J, Glymour M, Sanchez-Romero R, Glymour C. A million variables and more: the fast greedy equivalence search algorithm for learning high-dimensional graphical causal models, with an application to functional magnetic resonance images. Int J Data Sci Anal. 2017;3:121–29.
12. Andrieu C. An introduction to MCMC for machine learning; 2003. p. 5–43.
13. Pearl J. Probabilistic reasoning in intelligent systems: : networks of plausible inference (Morgan kaufmann series in representation and reasoning); 1988.
14. Chickering DM. Optimal structure identification with greedy search. J Mach Learn Res. 2002;3:507–54.
15. Spencer SEF, Hill SM, Mukherjee S. Inferring network structure from interventional time-course experiments. Ann Appl Stat. 2015;9:507–24.

## Publisher's Note

Peng et al. BMC Bioinformatics 2020, 21(Suppl 13):394
https://doi.org/10.1186/s12859-020-03677-1

BMC Bioinformatics

RESEARCH

Open Access

# A learning-based method for drug-target interaction prediction based on feature representation learning and deep neural network

Jiajie Peng[1,2], Jingyi Li[1,2] and Xuequn Shang[1,2*]

*From* The 18th Asia Pacific Bioinformatics Conference
Seoul, Korea. 18-20 August 2020

*Correspondence:
shang@nwpu.edu.cn
[1]The School of Computer Science, Northwestern Polytechnical University, 710072 Xian, China
[2]The Key Laboratory of Big Data Storage an Management, Northwestern Polytechnical Universitythe, Ministry of Industry and Information Technology, 710072 Xian, China

## Abstract

**Background:** Drug-target interaction prediction is of great significance for narrowing down the scope of candidate medications, and thus is a vital step in drug discovery. Because of the particularity of biochemical experiments, the development of new drugs is not only costly, but also time-consuming. Therefore, the computational prediction of drug target interactions has become an essential way in the process of drug discovery, aiming to greatly reducing the experimental cost and time.

**Results:** We propose a learning-based method based on feature representation learning and deep neural network named DTI-CNN to predict the drug-target interactions. We first extract the relevant features of drugs and proteins from heterogeneous networks by using the Jaccard similarity coefficient and restart random walk model. Then, we adopt a denoising autoencoder model to reduce the dimension and identify the essential features. Third, based on the features obtained from last step, we constructed a convolutional neural network model to predict the interaction between drugs and proteins . The evaluation results show that the average AUROC score and AUPR score of DTI-CNN were 0.9416 and 0.9499, which obtains better performance than the other three existing state-of-the-art methods.

**Conclusions:** All the experimental results show that the performance of DTI-CNN is better than that of the three existing methods and the proposed method is appropriately designed.

**Keywords:** DTIs prediction, Convolutional neural network, Feature representation learning

## Background

Drug targets are special molecules that can bind to drugs and produce effects in cells, the main molecular targets for drugs are proteins [1]. Drug-target interactions (DTIs) prediction is of great significance for drug repositioning [2], drug discovery [3], side-effect prediction [4] and drug resistance [5]. However, identifying the drug-target interactions via biochemical and chemical biological experiments is costly and time-consuming [6]. Recently, as genomic, chemical, and pharmacological data become more and more complete, new opportunities for identifying drug target interactions have been emerged [2]. Therefore, many researchers have attempted to predict DTIs by using silico or computational approaches to guide in vivo validation in recent years, and thus significantly reduce the cost and time for identifying the drug-target interactions [2].

The traditional computational DTIs prediction approaches are mainly categorized into docking-based approaches [7] [8] and ligand-based approaches [9]. However, the docking is difficult to play a good performance when the three-dimensional structures of the target protein are unknown [10]. The ligand-based approaches are very effective in DTIs prediction, but it often requires a large number of known binding data and thus the prediction results are poor with only a small amount of known data [11].

In recent years, network-based approaches have demonstrated great advantages compared to docking-based and ligand-based methods [12] [13]. First, network-based approaches have a good prediction performance even without the three-dimensional structure of the target. Secondly, they are simple and fast for only by performing simple physical processes [13]. In the past decade, DTIs prediction approaches based on machine learning have also been widely studied. A key idea of these approaches is the assumption that similar drugs may share similar targets [14].Thus predicting DTIs is often regarded as a binary classification issue by using chemical structures of drugs and targets as input features and considering known DTIs as labels [2]. However, most existing prediction methods are limited to homogeneous networks, which ignore a rich variety of topological information and the complex interaction relationship of heterogeneous data [3].

In recent years, a variety of computing methods based on heterogeneous data sources have been developed to predict DTI. *Wang et.al* used a heterogeneous network data to obtain the diffusion feature and directly use the obtained diffusion distributions to derive the prediction scores of DTIs [3]. However, the direct use of diffusion state as a feature or prediction score is vulnerable to the deviation caused by noise and high dimension of heterogeneous network data, resulting in inaccurate DTI prediction [2]. *Luo et.al* provided the DTINet as a new prediction method, which extracted the low-dimensional characteristic information from heterogeneous data sources and used the inductive matrix completion (IMC) approach to predict the drug-target interaction fraction [2]. The inductive matrix completion approach predicts the new DTI by using a small amount of known drug-protein interaction information, combined with the extracted drug and protein characteristic information. This method has been proved to be superior to the commonly used Laplacian regularization least square approach [15], heterogeneous network model [16], cooperation matrix factorization (CMF) [17] and Bipartite local model by learning from local information and neighbors [18].

In this paper, we improve the prediction method by learning low-dimensional vector representations of features from heterogeneous networks, and adopting convolution neural networks (CNN) as classification model. An efficient DTI prediction method is presented, DTI-CNN, which can be used to identify the drug-target interactions, guide biochemical experiments and reduce the cost of research. Here are four major contributions:

- We propose a learning-based method for drug-target interaction prediction that contains three components, named as heterogeneous-network-based feature extractor, denoising-autoencoder-based feature selector and CNN-based interaction predictor.
- Based on random walk with restart (RWR) and denoising autoencoder (DAE) model, DTI-CNN can cope with the noisy, incomplete and high-dimensional features from heterogeneous data sources, including drug, proteins, side-effects and diseases information.
- Based on a deep CNN model, DTI-CNN can handle the low dimensional feature vectors and predict the probability of interaction between each pair of drugs and proteins.
- Based on our DTI prediction task, the results indicate that DTI-CNN is better than the other three state-of-the-art methods and is appropriately designed.

## Methods

We propose an learning-based method called DTI-CNN to predict drug-target interactions. The workflow of DTI-CNN is shown in Fig. 1. First, the heterogeneous network was constructed by integrating a variety of drug and protein related information sources, and the initial drug feature vector and protein feature vector were obtained by RWR model.

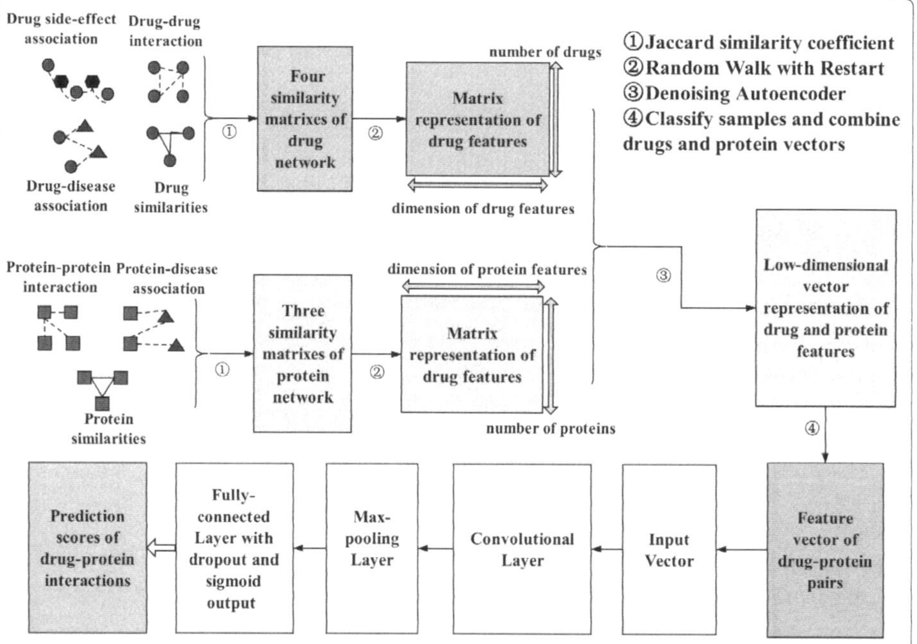


**Fig. 1** The flowchart of DTI-CNN pipeline. The DTI-CNN contains heterogeneous-network-based feature extractor, denoising-autoencoder-based feature selector and CNN-based interaction predictor. First, the features are extracted from seven networks of drug and protein by the Jaccard similarity coefficient and RWR algorithm, then we get the low-dimensional representation of drug and protein features by adopting the DAE model. Third, a deep CNN model is constructed to predict the interaction of each pair of drugs and proteins

In the second step, the high-dimensional features of drugs and proteins are reduced by adopting the DAE model, and the low-dimensional representations of them are obtained respectively. Finally, according to the known drug-protein interactions, the samples are divided into positive samples and negative samples. Combining the feature vector of drug-protein pairs, CNN was adopted to predict the association between each pair of drugs and proteins.

### Heterogeneous-network-based feature extractor

The heterogeneous networks is constructed based on two types of networks as follows. One is the drug-related networks, including drug-drug interactions, drug-disease associations, drug-side-effect associations, drug similarities (based on chemical structures of drugs) [2]. The other is protein-related networks, including protein-disease associations, protein-protein interactions, protein similarities (based on primary sequences of proteins) [2]. Firstly, the Jaccard similarity algorithm [19] is executed on each association and interaction matrix respectively, so we can generate a similarity matrix for each network.

Taking the drug-disease interactions matrix as an example, two sets $A$ and $B$ are given as two rows in the adjacent matrix, which represent the interactions between two different drugs and all diseases. The Jaccard similarity coefficient[20] is an indicator of the similarity between two sets, defined as follows:

$$Sim(A,B) = \frac{|A \cap B|}{|A \cup B|}. \tag{1}$$

The original adjacent matrix is a description of the relationship between a single row and column node, and the Jaccard similarity coefficient calculation is based on the adjacent two row vectors of the original adjacent matrix. Thus the similarity matrices $S$ represent the similarity between each drug or protein node and all features of the column nodes. The element $S_{i,j}$ represents similarity of row $i$ and row $j$ in the original adjacenct matrix.

After all the original data is transformed into similarity matrices, the RWR algorithm [21] is applied to each similarity matrix, which represents a weighted network. The diffusion state of each drug or protein is obtained on each network, which includes the topological structure relation of each drug or protein with all the other nodes in the network.

The reason for using RWR is that the similarity matrix obtained in the previous step only calculates the similarity of the two nodes in isolation. RWR can be used to consider global structure information in the network. If the distribution state of the two nodes is close, they can be considered to be in a similar position relative to other nodes in the network. According to the RWR principle, the greater the similarity between the two nodes, the higher the transition probability of them [22].

Taking the drug-disease similarity matrix $A_{i,j}$ as an example, we can get the drug-disease transition probability matrix $B$ according to the $A_{i,j}$, whose elements $B_{i,j}$ describe the transition probability from the drug node $i$ to the disease node $j$ [23], that is defined as follows:

$$B_{i,j} = \frac{A_{i,j}}{\sum_j A_{i,j}}. \tag{2}$$

Next, the final drug-disease diffusion state matrix can be obtained by iterative convergence as follows:

$$S_i^{t+1} = (1 - p_r)S_i^t B + p_r e_i. \tag{3}$$

Where $S_i^t$ is the result after $t$ iterations, and each element stores the probability of accessing a disease node from the drug node $i$ after iteration in the process of random walk, $p_r$ is the restart probability, and $e_i$ is an n-dimensional unit matrix.

After all the similarity matrices is transformed into diffusion state matrices, we splice the single diffusion state matrix of drug and protein networks, so that we can get two diffusion state matrices about drug and protein. The row of the drug diffusion matrix represents different drugs, and the column represents the four nodes of drugs, diseases, side effects and drugs, in which the element $D_(i,j)$ represents the transition probability between drug $i$ and node $j$. The row of the protein diffusion state matrix represents different proteins, and the column represents the three nodes of protein, disease and protein, in which the element $P_(i,j)$ represents the transfer probability of protein $i$ and node $j$.

### Denoising-autoencoder-based feature selector

The vector of diffusion state matrix obtained in the previous step is high-dimensional, noisy and incomplete. In order to obtain the essential features, we apply a DAE model which carry on the data operation on the basis of the autoencoder. The main idea of DAE is shown in Fig. 2. Taking the diffusion state matrix about drug as an input example, by adding noise to the input training data and making the self-encoder learn to remove this noise, the real input which has not been polluted by noise can be obtained [24]. Therefore, the encoder can obtain the most essential features from the original input to get more robust representation. This is why its generalization ability is better than that of the general encoder [25]. Autoencoders use automatic Encoders to obtain low-dimensional data through neural networks based on the input data. Similarly, the Decoders to recover the original input from low-dimensional data [26].

In the model, we reduce the dimension of drug features to 100 dimensions and protein features to 400 dimensions. We set the noise figure to 0.2, and use the softplus [27] and RMSProp fuction [28] to optimize the mean-square error (MSE) [29]. At last, the backpropagation (BP) algorithm is used to train our DAE [30].

### CNN-based interaction predictor

Convolutional Neural Networks is a classical and widely used structure since 1980s [31] and can greatly reduces the complexity of convolution neural network [32]. The parameters of each layer network are shared and the number of parameters to be trained

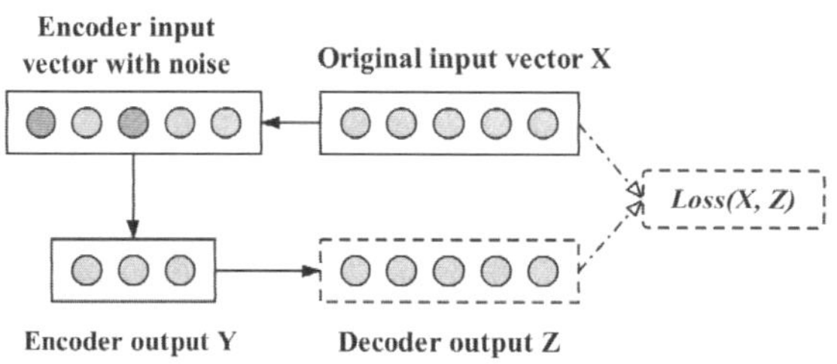


**Fig. 2** The schematic illustration of denoising autoencoder. The original input data is high-dimensional, noisy and incomplete, the DAE adds noise to it and makes the self-encoder learn to remove the noise, which makes the encoder learn more robust and low-dimensional representation in the input data. Then the decoder is used to recover the original input from low-dimensional data, the loss between the original input and the decoder output is optimized by the RMSProp algorithm

is reduced during training. Compared with the standard fully connected neural network, it has better performance in image classification, sentence classification and other classification tasks [33].

Inspired by the success of CNN in classification tasks [34], we use CNN as the supervised learning model and the structure of the prediction model is shown in Fig. 3. The prediction model contains the convolution, max-pooling, fully-connected and output layer. A convolution layer with a rectified linear unit (ReLU) activation fuction [34] is used as a feature extractor [35]. A max-pooling layer is employed to reduce the dimension of features, and the fully-connection layer and the output sigmoid layer are used to classify the tasks.

As the key component of a CNN, convolution layer can help the model to learn local and global structures from the input vector [35]. In our model, the convolutional layer consists of 4 kernels. Given the input vector $X$ of length $S$, and the length of weight vector is $4 * 1$. For each kernel, convolution operation is independent and thus we can obtain four particular feature with the length $(S - 4) + 1$, which was extracted from the input vector and named feature map $M$. The $M$ is obtained as follows:

$$M_i = \sum_{j=1}^{4} W_j X_{i+j}. \tag{4}$$

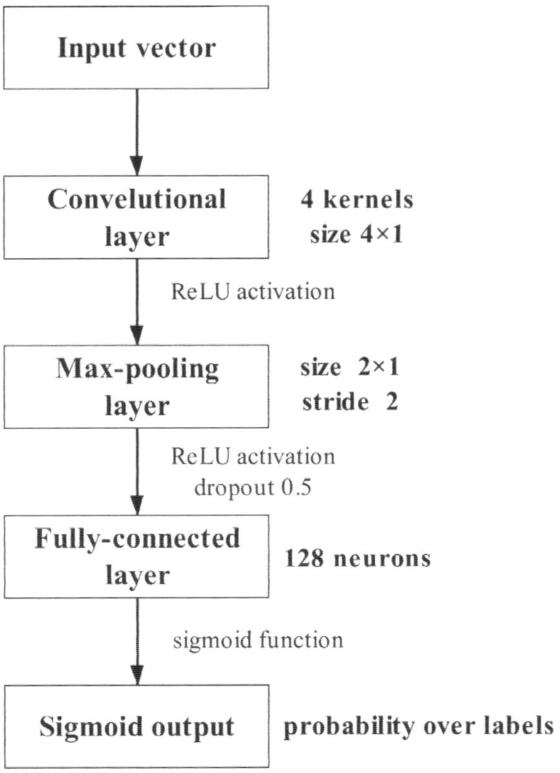


**Fig. 3** The structure of the convolutional neural network model

where ($i \in 0, 1, \ldots, S-4$), and $W$ is initialized by a truncated normal distribution and used as a weight vector. Then a ReLU function is used to the feature map obtained last step:

$$f(x) = max(0, x). \tag{5}$$

The ReLU function is selected for the excitation function, which can effectively simplify the calculation process and avoid the gradient explosion and disappearance [36].

Next layer is the max-pooling layer, which can extract the maximum value in the pooled region and the pooled region continues to move forward at a certain step size in an input sequence, thus reducing the dimension in each feature map [36]. In our model, the pooled size is $2 * 1$, and the step size is 2. Given an input vector $A_i$ ($i \in 0, 1, \ldots, S-4$), the length of output of this layer is $\frac{(S-2)+1}{2}$.

After the first two layers, we use a one-dimensional vector to connect the important features extracted from all the kernels and then pass them to the fully connected layer. The number of hidden units in this layer is 180 and the output of this layer is calculated as follows:

$$f = W * y. \tag{6}$$

where $W \in R^{(n*180)}$ is the weight matrix, $y$ is the output of pooling layer, and $f$ is the ReLU function.

The final output sigmoid layer is constructed for the binary classification. The Sigmoid function maps output values between 0 and 1 for classification, which is extracted by the following equation:

$$S(x) = \frac{1}{1 + e^{-x}}. \tag{7}$$

## Results

### Data preparation

To make the performance evaluationn, DTI-CNN was tested on the drug-target interactions prediction task.

We obtained the heterogeneous network from *Luo*'s paper, which include 12,015 nodes and 1,895,445 edges in total [2]. The isolated nodes are excluded. The heterogeneous network integrates four types of nodes (drugs, proteins, diseases and side-effects) and six type of edges (drug-protein interactions, drug-drug interactions, drug-disease associations, drug-side-effect associations, protein-disease associations and protein-protein interactions) [2]. Based on chemical structures of drugs and primary sequences of proteins, we also built up the multiple similarity networks [16].

The drug nodes, known DTIs and drug-drug interactions were extracted from the DrugBank database [37]. The protein nodes and protein–protein interactions were obtained from the HPRD database [38]. The disease nodes, drug–disease and protein–disease associations were extracted from the Comparative Toxicogenomics Database [39]. The side-effect nodes and drug–side-effect associations were collected from the SIDER database [40].

In our model, we first constructed seven similarity matrices after the Jaccard similarity algorithm. We obtained the drug-related similarity matrices including drug-drug similarity matrix, drug-disease similarity matrix, drug-side-effect similarity matrix and drug similarities matrix. The protein-related similarity matrices include protein-disease similarity matrix, protein-protein similarity matrix and protein similarities matrix. Secondly, we perform RWR algorithm for the two kinds of matrices respectively and splice the single diffusion state matrix of drug and protein networks. After this step, we get two diffusion state matrices corresponding to drug and protein respectively. The rows of the drug diffusion matrix represent different drugs, and the columns represent proteins, diseases, side effects and drugs nodes. The values in the matrix represent the associations between drugs and the four biological entities. The rows of the protein diffusion state matrix represent different proteins. The columns represent proteins, diseases and drugs nodes. The values in the matrix represent the associations between proteins and the three biological entities. Then, we reduce the dimension of drug diffusion state matrix and protein diffusion state matrix respectively by using DAE model. Finally, we obtain the drugs feature vector matrix of 100 dimensions and a total of 708 samples. Similarly, the proteins feature vector matrix is 400 dimensions and 1512 samples.

At last, we adopted the method of ten fold cross validation to divide the train set and test set, in which 90% of the positive and negative samples were used to train model and 10% of the positive and negative samples were used to test the model. According to the known drug-protein interactions matrix, we use the known drug-protein interaction pairs as positive samples. We randomly selected negative samples with the same number of positive samples. In total, we have 3846 samples. After splicing the corresponding protein vectors into drug vectors, we get drug-protein pair vectors of 500 dimensions.

#### Model parameters

For the RWR model, the restart probability is 0.5 and the number of iterations is 20.

The parameters used in the DAE model are as follows. For the drug features matrices, the original dimensions are 2832 and the DAE has one hidden layer with 100 units. For the protein features matrices, the original dimensions are 4536 and has one hidden layer with 400 units. For drug and protein features, there are 16 and 32 samples for each batch respectively. The number of epochs is 20. The noise scale value is 0.2 and we use Respro optimizer algorithm to train the model.

The parameters of the CNN model are as follows. We added a dropout layer before the fully-connected layer and the dropout percentage is 0.5. We run 35 epochs with 64 samples for each batch . We used Adam algorithm and set an initial learning rate as 0.001 to optimize the binary cross entropy loss [41].

#### Evaluation Metrics

The AUROC and AUPR scores were used to evaluate the model test and comparison. AUROC and AUPR scores are commonly used evaluation criteria for machine learning, which represent the area under the curve of ROC curve and PR curve respectively. The higher the score, the higher the prediction accuracy of the model and the better the performance of the model.

ROC curve refers to the curve with false positive probability (FPR) as horizontal axis and true positive rate (TPR) as longitudinal axis, in which $FPR = \frac{FP}{TN+FP}$ , $TPR = \frac{TP}{TP+FN}$. PR curve refers to the curve with Recall as horizontal axis and Precision as longitudinal axis, and $Recall = \frac{TP}{TP+FN}$, $Precision = \frac{TP}{TP+FP}$.

### Performance evaluation on predicting drug-target interactions

We compared the performance of DTI-CNN with three existing state-of-art methods (DTINet [2], CMF [17] and NRLMF [42]) on the task of predicting drug-target interactions.All models are trained and tested with a 10-fold cross validation. Our comparative results are shown in Fig. 4 and Table 1. Comparing with other methods, the result shows that DTI-CNN can perform best on both scores at the same time and DTINet is the second best method. The AUROC of DTI-CNN is 0.9416, which is 0.03 higher than DTINet. The AUPR of DTI-CNN is 0.02 higher than DTINet. In summary, DTI-CNN performs better on drug-target interactions prediction task than the other three state-of-the-art DTI prediction methods.

### Effects of DTI-CNN components

We choose two different approaches in feature selection module and interaction prediction module respectively to figure out the performance of each module of DTI-CNN. To test the effect of our feature selection model, we first use the singular value decomposition (SVD) method [43] to replace our DAE model and name this method as SVD-CNN. To test the effect of the CNN component, we substitute the CNN model with the support vector machine (SVM) model and name this method as DTI-SVM. All three methods are used on DTI prediction task to compare performance and the comparative results are shown in Fig. 5 and Table 2. Comparing with other two methods, the result shows that DTI-CNN can achieve higher AUROC and AUPR scores at the same time, indicating that the two modules of our model are appropriately designed.

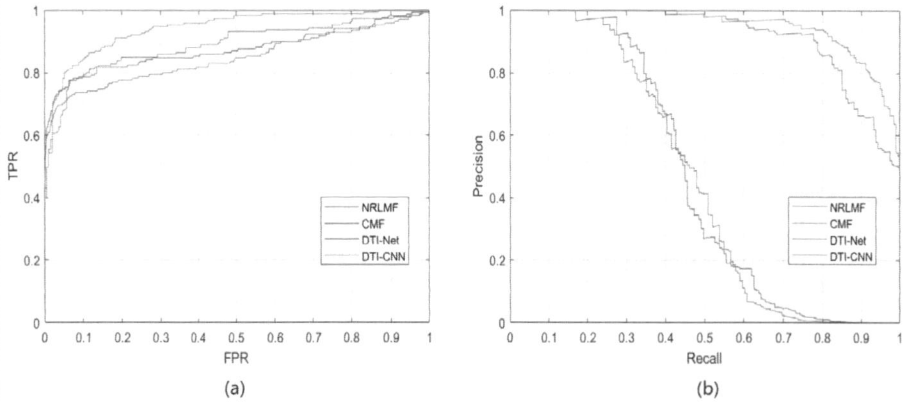


**Fig. 4** The ROC **(a)** and P-R **(b)** curves of DTI-CNN, DTINet, CMF and NRLMF on drug-target interactions prediction task. The AUROC of DTI-CNN is 0.9416 and the AUPR of DTI-CNN is 0.9499, which performs better on the drug-target interactions prediction task than the other three state-of-the-art DTI prediction methods

**Table 1** The AUROC, AUPR scores of DTI-CNN, DTINet, CMF and NRLMF on drug-target interactions prediction task

| Methods | AUROC | AUPR |
|---|---|---|
| DTINet | 0.9111 | 0.9290 |
| NRLMF | 0.8692 | 0.4411 |
| CMF | 0.9037 | 0.4173 |
| DTI-CNN | 0.9416 | 0.9499 |

### Case study in three drugs

We extracted the known DTIs from the DrugBank database. We choose the three drugs with the largest number of interactions in known DTIs, which are Quetiapine, Olanzapine and Meprobamate. In the train set, we exclude all the features and interactions between the three drugs and their related proteins to avoid logic circle. In the test set, we input the features of the three drugs and its related proteins. In the "Quetiapine set", 24 of 24 known interactions are identified. In the "Meprobamate set", 23 of 24 known interactions are discovered. And in the "Olanzapine set", 23 of 23 known interactions are recognized. These results indicate that the DTI-CNN method has a good performance on drug-target interactions prediction.

## Discussion

At present, there are three traditional kinds of DTI prediction approaches which contains molecular-based approach, ligand-based approach and network-based approach [10]. There are a variety of related models, and the demand for algorithm prediction accuracy is getting higher and higher. The feature learning approach based on deep learning is different from the traditional neural network [44]. Through layer-by-layer learning, we can learn the essential features of the data set without relying on a large scale of samples., so as to predict the unknown data more accurately [45].

The purpose of this research is to improve the prediction accuracy by using the CNN model based on depth learning on the basis of the developed method DTINet. The DTINet contains two modules which are feature extraction and classification. In the first

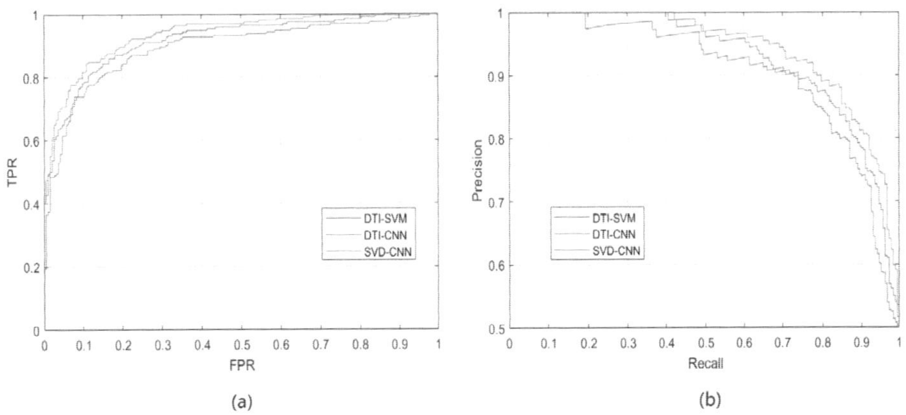


**Fig. 5** The ROC **(a)** and P-R **(b)** curves of DTI-CNN, SVD-CNN and DTI-SVM on drug-target interactions prediction task. The DTI-CNN achieves both higher AUROC and AUPR scores than the SVD-CNN and the DTI-SVM method, indicating that the DTI-CNN has been appropriately designed

**Table 2** The AUROC, AUPR scores of DTI-CNN, SVD-CNN and DTI-SVM on drug-target interactions prediction task

| Methods | AUROC | AUPR |
|---|---|---|
| DTI-SVM | 0.9068 | 0.9292 |
| SVD-CNN | 0.9306 | 0.9370 |
| DTI-CNN | 0.9416 | 0.9499 |

module, we replace the dimension reduction model based on SVD with DAE model to ensure that we can learn features that are more suitable for neural network learning. Then, we improve the classification prediction model of DTINet and use the CNN model as the new prediction model in the second module. Compared with the IMC approach used by DTINet, the CNN model can take into account the topological information and interrelation between the nodes in the network. In addition, we also choose three state-of-the-art DTI prediction methods as comparison. The experimental results show that the AUROC and AUPR scores of our model are both higher than DTINet, NRLMF and CMF.

In addition, we want to know which network contributes more to the DTI prediction. We sequentially remove a network from the original heterogeneous networks as new input data, and then use our DTI-CNN method to perform DTI prediction. The results are shown in Table 3. The result shows that the drug-drug interaction of drug networks and the protein similarities of protein networks contributed more to the DTI prediction. When the drug-drug interaction network was removed, the result achieves both the lowest AUROC and AUPR scores at the same time.

In the future, we will consider adding more relevant information to the heterogeneous network. For the CNN model, we can add the network structure appropriately to accommodate more complex input networks. In this work, although DTI-CNN is mainly designed to predict DTIs, it is an extendible method and can also be used to predict other related directions in the future, such as drug-drug, drug-side-effects and protein-disease.

## Conclusion

In this paper, we propose a learning-based method named DTI-CNN to predict the drug-target interactions. Firstly, the Jaccard similarity coefficient and RWR model are used to obtain the relevant features of drugs and targets from heterogeneous networks. Then, we use DAE model to reduce dimensions and identify the essential features. Thirdly, based on the features obtained from the last step, a CNN model is constructed to make a prediction of DTIs.To demonstrate the advantages of DTI-CNN, we compare it with three advanced methods. In addition, we also evaluate the effect of each DTI-CNN module. All

**Table 3** The AUROC, AUPR scores of sequentially strip out a network from the original heterogeneous network as new input data of DTI-CNN on drug-target interactions prediction task

| Networks | AUROC | AUPR |
|---|---|---|
| Without drug-drug | 0.9299 | 0.9370 |
| Without drug-disease | 0.9343 | 0.9416 |
| Without drug side-effect | 0.9344 | 0.9444 |
| Without drug similarities | 0.9345 | 0.9425 |
| Without protein-protein | 0.9418 | 0.9499 |
| Without protein-disease | 0.9364 | 0.9452 |
| Without protein similarities | 0.9327 | 0.9411 |

the experimental results show that the performance of DTI-CNN is better than that of the existing methods and the proposed method is appropriately designed. The case study also shows that DTI-CNN can be used to predict the drug-target interactions.

#### Abbreviations
DTIs: Drug-target interactions; IMC: Inductive matrix completion; CMF: Cooperation matrix factorization; CNN: Convolution neural networks; RWR: Random walk with restart; DAE: Denoising autoencoder; MSE: Mean-square error; BP: Backpropagation; ReLU: Rectified linear unit; FPR: False positive probability; TPR: True positive rate; SVD: Singular value decomposition; SVM: Support vector machine.

#### Acknowledgements
Not applicable.

#### About this supplement
This article has been published as part of BMC Bioinformatics Volume 21 Supplement 13, 2020: Selected articles from the 18th Asia Pacific Bioinformatics Conference (APBC 2020): bioinformatics. The full contents of the supplement are available online at https://bmcbioinformatics.biomedcentral.com/articles/supplements/volume-21-supplement-13 .

#### Authors' contributions
JP and XS designed the algorithm; JL implemented the algorithm; JP and JL wrote this manuscript. All authors read and approved the final manuscript.

#### Funding
Publication costs are funded by National Natural Science Foundation of China (No. 61702421, U1811262, 61772426), the international Postdoctoral Fellowship Program (no. 20180029), China Postdoctoral Science Foundation (No. 2017M610651), Top International University Visiting Program for Outstanding Young scholars of Northwestern Polytechnical University.

#### Availability of data and materials
The datasets generated during and analysed during the current study are available in the github repository, [https://github.com/MedicineBiology-AI/DTI-CNN.git].

#### Ethics approval and consent for publication
Not applicable.

#### Consent for publication
Not applicable.

#### Competing interests
The authors declare that they have no competing interests.

Published: 17 September 2020

#### References
1. Feng Y, Wang Q, Wang T. Drug target protein-protein interaction networks: A systematic perspective. BioMed Res Int. 2017;2017:1289259–1289259.
2. Luo Y, Zhao X, Zhou J, Yang J, Zhang Y, Kuang W, Peng J, Chen L, Zeng J. A network integration approach for drug-target interaction prediction and computational drug repositioning from heterogeneous information. Nature communications. 2017;8(1):573.
3. Huang Y, Zhu L, Tan H, Tian F, Zheng F, Mısır M, Wang X, Liu W. Predicting drug-target on heterogeneous network with co-rank. In: The 8th International Conference on Computer Engineering and Networks (CENet2018). Cham: Springer International Publishing; 2020. p. 571–81. ISBN: 978-3-030-14680-1.
4. Cheng F, Liu C, Jiang J, Lu W, Li W, Liu G, Zhou W, Huang J, Tang Y. Prediction of drug-target interactions and drug repositioning via network-based inference. PLoS Comput Biol. 2012;8(5):1002503.
5. Yamanishi Y, Kotera M, Kanehisa M, Goto S. Drug-target interaction prediction from chemical, genomic and pharmacological data in an integrated framework. Bioinformatics. 2010;26(12):246–54.
6. Whitebread S, Hamon J, Bojanic D, Urban L. Keynote review: in vitro safety pharmacology profiling: an essential tool for successful drug development. Drug Discov Today. 2005;10(21):1421–33.
7. Dominguez C, Boelens R, Bonvin A. Haddock: a protein- protein docking approach based on biochemical or biophysical information. J Am Chem Soc. 2003;125(7):1731–7.
8. Morris G, Huey R, Lindstrom W, Sanner M, Belew R, Goodsell D, Olson A. Autodock4 and autodocktools4: Automated docking with selective receptor flexibility. J Comput Chem. 2009;30(16):2785–91.
9. Keiser M, Roth B, Armbruster B, Ernsberger P, Irwin J, Shoichet B. Relating protein pharmacology by ligand chemistry. Nat Biotechnol. 2007;25(2):197.
10. Wan F, Hong L, Xiao A, Jiang T, Zeng J. NeoDTI: neural integration of neighbor information from a heterogeneous network for discovering new drug–target interactions. Bioinformatics. 2018;35(1):104–11. https://academic.oup.com/bioinformatics/article-pdf/35/1/104/27182907/bty543.pdf.
11. González-Díaz H, Prado-Prado F, García-Mera X, Alonso N, Abeijón P, Caamano O, Yanez M, Munteanu C, Pazos A, Dea-Ayuela MA, et al. Mind-best: Web server for drugs and target discovery; design, synthesis, and assay of mao-b inhibitors and theoretical- experimental study of g3pdh protein from trichomonas gallinae. J Proteome Res. 2011;10(4):1698–718.

12. Lv Z, Ao C, Zou Q. Protein function prediction: From traditional classifier to deep learning. PROTEOMICS. 2019;19(14):1900119. https://onlinelibrary.wiley.com/doi/pdf/10.1002/pmic.201900119.
13. Wu Z, Li W, Liu G, Tang Y. Network-based methods for prediction of drug-target interactions. Frontiers Media SA. 9: 1134.
14. Lan W, Wang J, Li M, Wu F-X, Pan Y. Predicting drug-target interaction based on sequence and structure information. IFAC-PapersOnLine. 2015;48(28):12–6.
15. Xia Z, Wu L-Y, Zhou X, Wong S. Semi-supervised drug-protein interaction prediction from heterogeneous biological spaces. In: BMC Systems Biology. BioMed Central; 2010. p. 6.
16. Wang W, Yang S, Zhang X, Li J. Drug repositioning by integrating target information through a heterogeneous network model. Bioinformatics. 2014;30(20):2923–30.
17. Zheng X, Ding H, Mamitsuka H, Zhu S. Collaborative matrix factorization with multiple similarities for predicting drug-target interactions. In: Proceedings of the 19th ACM SIGKDD International Conference on Knowledge Discovery and Data Mining. ACM; 2013. p. 1025–33.
18. Mei J-P, Kwoh C-K, Yang P, Li X-L, Zheng J. Drug–target interaction prediction by learning from local information and neighbors. Bioinformatics. 2012;29(2):238–45.
19. Niwattanakul S, Singthongchai J, Naenudorn E, Wanapu S. Using of jaccard coefficient for keywords similarity. In: Proceedings of the International Multiconference of Engineers and Computer Scientists; 2013. p. 380–4.
20. Jaccard P. Étude comparative de la distribution florale dans une portion des alpes et des jura. Bulletin del la Société Vaudoise des Sciences Naturelles. 1901;37:547–79.
21. Tong H, Faloutsos C, Pan J-Y. Random walk with restart: fast solutions and applications. Knowl Inf Syst. 2008;14(3): 327–46.
22. Lee S, Lee J, Lim J, Suh I. Robust stereo matching using adaptive random walk with restart algorithm. Image Vis Comput. 2015;37:1–11. https://doi.org/10.1016/j.imavis.2015.01.003.
23. Liu H, Song Y, Guan J, Luo L, Zhuang Z. Inferring new indications for approved drugs via random walk on drug-disease heterogenous networks. BMC Bioinformatics. 2016;17(17):539.
24. Peng J, Wang X, Shang X. Combining gene ontology with deep neural networks to enhance the clustering of single cell rna-seq data. BMC Bioinformatics. 2019;20(8):284. Springer.
25. Vincent P, Larochelle H, Bengio Y, Manzagol P-A. Extracting and composing robust features with denoising autoencoders. In: Proceedings of the 25th International Conference on Machine Learning, ICML '08. New York, NY, USA: ACM; 2008. p. 1096–103. https://doi.org/10.1145/1390156.1390294.
26. Peng J, Guan J, Shang X. Predicting parkinson's disease genes based on node2vec and autoencoder. Front Genet. 2019;10:226. https://doi.org/10.3389/fgene.2019.00226.
27. Ramachandran P, Zoph B, Le Q, Quoc V. Searching for activation functions. arXiv e-prints. 2017. https://ui.adsabs.harvard.edu/abs/2017arXiv171005941R. Provided by the SAO/NASA Astrophysics Data System.
28. Mukkamala M, Hein M. Variants of RMSProp and Adagrad with logarithmic regret bounds. In: Doina P, Yee Whye T, editors. Proceedings of the 34th International Conference on Machine Learning. vol. 70. Sydney: PMLR; 2017. p. 2545–53.
29. Allen D. Mean square error of prediction as a criterion for selecting variables. Technometrics. 1971;13(3):469–75.
30. LeCun Y, Boser B, Denker J, Henderson D, Howard R, Hubbard W, Jackel L. Backpropagation applied to handwritten zip code recognition. Neural Comput. 1989;1(4):541–51.
31. Ming Liang, Xiaolin Hu. Recurrent convolutional neural network for object recognition. 20153367–75. https://doi.org/10.1109/CVPR.2015.7298958.
32. Spanhol F, Oliveira L, Petitjean C, Heutte L. Breast cancer histopathological image classification using convolutional neural networks. In: 2016 International Joint Conference on Neural Networks (IJCNN). IEEE; 2016. p. 2560–7.
33. Ren F, Yang C, Qiu Q, Zeng N, Cai C, Hou C, Zou Q. Exploiting discriminative regions of brain slices based on 2d cnns for alzheimer's disease classification; 2019. p. 1–1. https://doi.org/10.1109/ACCESS.2019.2920241.
34. Clevert D-A, Unterthiner T, Hochreiter S. Fast and accurate deep network learning by exponential linear units (elus). arXiv e-prints. 2015arXiv:151107289. https://ui.adsabs.harvard.edu/abs/2015arXiv151107289C.
35. Luo P, Ding Y, Lei X, Wu F. deepdriver: predicting cancer driver genes by convolutional neural networks. Front Genet. 2019;10:13.
36. Peng J, Hui W, Li Q, Chen B, Hao J, Jiang Q, Shang X, Wei Z. A learning-based framework for miRNA-disease association identification using neural networks. Bioinformatics. 2019;35(21):4364–71. https://academic.oup.com/bioinformatics/article-pdf/35/21/4364/30330838/btz254.pdf.
37. Knox C, Law V, Jewison T, Liu P, Ly S, Frolkis A, Pon A, Banco K, Mak C, Neveu V, et al. Drugbank 3.0: a comprehensive resource for omics research on drugs. Nucleic Acids Res. 2010;39((suppl_1)):1035–41.
38. Keshava Prasad T, Goel R, Kandasamy K, Keerthikumar S, Kumar S, Mathivanan S, Telikicherla D, Raju R, Shafreen B, Venugopal A, et al. Human protein reference database 2009 update. Nucleic Acids Res. 2008;37((suppl_1)):767–72.
39. Davis A, Murphy C, Johnson R, Lay J, Lennon-Hopkins K, Saraceni-Richards C, Sciaky D, King B, Rosenstein M, Wiegers T, et al. The comparative toxicogenomics database: update 2013. Nucleic Acids Res. 2012;41(D1):1104–14.
40. Kuhn M, Campillos M, Letunic I, Jensen L, Bork P. A side effect resource to capture phenotypic effects of drugs. Mol Syst Biol. 2010;6(1):343.
41. Liu L, Qi H. Learning effective binary descriptors via cross entropy. IEEE. 20171251–8.
42. Liu Y, Min W, Chunyan M, Peilin Z, Xiao-Li L, M P. Neighborhood regularized logistic matrix factorization for drug-target interaction prediction; 2016. p 1004760.
43. GOLUB G, REINSCH C. Handbook series linear algebra. singular value decomposition and least squares solutions. Numer Math. 1970;14:403–20.
44. Arun K, Govindan V. A Hybrid Deep Learning Architecture for Latent Topic-based Image Retrieval. Data Sci Eng. 2018;3:166–95. https://doi.org/10.1007/s41019-018-0063-7.
45. Wen M, Zhang Z, Niu S, Sha H, Yang R, Yun Y-H, Lu H. Deep-learning-based drug-target interaction prediction. J Proteome Res. 201716. https://doi.org/10.1021/acs.jproteome.6b00618.

## Publisher's Note

Lin et al. BMC Bioinformatics 2020, 21(Suppl 13):389
https://doi.org/10.1186/s12859-020-03678-0

BMC Bioinformatics

RESEARCH Open Access

# Multi-omics profiling reveals microRNA-mediated insulin signaling networks

Yang-Chi-Dung Lin[1,2†], Hsi-Yuan Huang[1,2†], Sirjana Shrestha[3,4], Chih-Hung Chou[3,4], Yen-Hua Chen[5], Chi-Ru Chen[4], Hsiao-Chin Hong[1,2], Jing Li[1,2], Yi-An Chang[3], Men-Yee Chiew[4], Ya-Rong Huang[3], Siang-Jyun Tu[3], Ting-Hsuan Sun[4], Shun-Long Weng[6], Ching-Ping Tseng[4*] and Hsien-Da Huang[1,2,4*]

*From* The 18th Asia Pacific Bioinformatics Conference
Seoul, Korea. 18-20 August 2020

* Correspondence: cpts@cc.nctu.edu.tw; huanghsienda@cuhk.edu.cn
†Yang-Chi-Dung Lin and Hsi-Yuan Huang contributed equally to this work.
[4]Department of Biological Science and Technology, National Chiao Tung University, Hsinchu 300, Taiwan
[1]School of Life and Health Sciences, The Chinese University of Hong Kong, Longgang District, Shenzhen 518172, Guangdong Province, China
Full list of author information is available at the end of the article

## Abstract

**Background:** MicroRNAs (miRNAs) play a key role in mediating the action of insulin on cell growth and the development of diabetes. However, few studies have been conducted to provide a comprehensive overview of the miRNA-mediated signaling network in response to glucose in pancreatic beta cells. In our study, we established a computational framework integrating multi-omics profiles analyses, including RNA sequencing (RNA-seq) and small RNA sequencing (sRNA-seq) data analysis, inverse expression pattern analysis, public data integration, and miRNA targets prediction to illustrate the miRNA-mediated regulatory network at different glucose concentrations in INS-1 pancreatic beta cells (INS-1), which display important characteristics of the pancreatic beta cells.

**Results:** We applied our computational framework to the expression profiles of miRNA/mRNA of INS-1, at different glucose concentrations. A total of 1437 differentially expressed genes (DEGs) and 153 differentially expressed miRNAs (DEmiRs) were identified from multi-omics profiles. In particular, 121 DEmiRs putatively regulated a total of 237 DEGs involved in glucose metabolism, fatty acid oxidation, ion channels, exocytosis, homeostasis, and insulin gene regulation. Moreover, Argonaute 2 immunoprecipitation sequencing, qRT-PCR, and luciferase assay identified Crem, Fn1, and Stc1 are direct targets of miR-146b and elucidated that miR-146b acted as a potential regulator and promising target to understand the insulin signaling network.

**Conclusions:** In this study, the integration of experimentally verified data with system biology framework extracts the miRNA network for exploring potential insulin-associated miRNA and their target genes. The findings offer a potentially significant effect on the understanding of miRNA-mediated insulin signaling network in the development and progression of pancreatic diabetes.

**Keywords:** miRNA, Pancreatic beta-cell, RNA-seq, Multi-omics

## Background

MicroRNAs (miRNAs) are short, noncoding RNAs of approximately 22 nucleotides in length that function as important regulators at the post-transcriptional level either by degrading mRNA molecules or suppressing protein translation [1]. miRNAs play critical roles in regulating numerous physiological processes, including the cell cycle [2], cell growth, development, differentiation [3], apoptosis [4], and pathological processes, such as those associated with various cancers [5]. Recent advances in multi-omics profiling technologies have fundamentally changed the process of large-scale screening in miRNA studies. Different techniques, such as RNA sequencing (RNA-seq), small RNA sequencing (sRNA-seq), Argonaute immunoprecipitation sequencing (Ago IP-seq), Argonaute cross-linking and immunoprecipitation sequencing (CLIP-seq) and chromatin immunoprecipitation sequencing (ChIP-seq) can facilitate discovery of novel miRNAs, miRNA–target interactions (MTIs), and miRNA regulation by transcription factors (TF). In addition, miRTarBase 7.0 [6] houses 4076 mature miRNAs and about 422,517 MTIs that are supported by considerable strong experimental evidence.

Diabetes is a metabolic disorder occurring in patients with high blood glucose level or hyperglycemia, which brings serious complications, such as heart and kidney diseases. Pancreatic beta - cells are responsible for the secretion of insulin by which the blood glucose level is controlled [7]. High glucose infusion in rats can lead to an approximately 40–60% increment in mass and proliferation of beta-cells [8]. Several intracellular signaling pathways, such as the phosphatidylinositol 3-kinase (PI3K)/Akt pathway [9], can induce beta-cell proliferation and is correlated with blood sugar. Recently many studies have identified several miRNAs as pathological factors, which contribute to the development of diabetes [10–12]. For example, miR-375 is an abundantly found miRNA in beta-cells [13] and increased expression of this miRNA is observed in patients with type 2 diabetes (T2D) [14]. Moreover, overexpression of miR-375 down-regulates 3phosphoinositide-dependent protein kinase-1 [15] and decreases glucose-stimulated insulin secretion whereas its up-regulation is correlated with beta-cell mass loss [16]. Substantial accumulating evidence indicates that miRNAs can perform regulatory roles of cell proliferation and diabetes through glucose-stimulated response in pancreatic beta-cells [17, 18]. The secretion of insulin is negatively regulated by miR-375 and the repression of miR-375 through the cAMP-PKA pathway may enhance the insulin response in pancreatic beta-cells [19]. Due to repression in miR-7, mTOR-signaling gets activated, which in turn promotes beta-cell proliferation in both human and mouse [20]. However, miR-7 functions in supporting the survival of beta-cell at the embryonic developmental stage. The delivery of miR-7 morpholinos to early mouse embryos results in a decrease of the total beta-cells number and their dysfunction [21]. miR-124a-2, another important regulator of embryonic pancreatic development, targets Foxa2, the activator of pancreatic duodenal homeobox 1 involved in the differentiation and survival of pancreatic beta cells [22]. Defected miR-124a-2 may be associated with beta-cell dysfunction [23]. Even though numerous studies have provided evidence that glucose- stimulated insulin secretion (GSIS) can be mediated by miRNAs, the comprehensive molecular mechanisms of miRNA-mediated gene regulation in glucose-stimulated pancreatic beta cells remain poorly understood. MTIs analysis has become a widely adopted approach to unravel target genes and pathways in biological systems.

Recently several web servers and online bioinformatics tools have been developed for the study of MTIs and miRNA regulatory networks. For example, dChip-GemiNi [24] is a web server that can be used to identify miRNA, TF, and their target gene networks with gene and miRNA expression profiles. Another web server, mirConnX [25] provides mRNA and miRNA gene regulatory networks through the computational prediction of MTIs and TF-miRNA. Similarly, MAGIA [26] provides platforms to construct TF-miRNA regulatory networks. All of these tools offer an integrated gene and miRNA expression profiles with target prediction. For confirming the uncertainty of miRNA target prediction tools, high-throughput experimental techniques for MTIs with Ago IP-seq and TF-miRNA association studies with ChIP-seq are widely applied in the study of miRNA-gene regulatory networks [27, 28].

To facilitate the study of novel MTIs in glucose-stimulated beta cells and to discover new insights into diabetes, we used an integrative multi-omics and systems biology approach, including RNA sequencing (RNA-seq) and small RNA sequencing (sRNA-seq) at different glucose concentrations to reconstruct the miRNA-mediated gene regulatory network in INS-1 pancreatic beta cells (INS-1). INS-1 cells display important characteristics of the pancreatic beta cells, including insulin expression, stable pancreatic beta-cell phenotype over 116 passages, and responsiveness to glucose within the physiological range [29]. Due to the above reasons, we selected INS-1 as the model cell line to investigate the MTIs in this study.

In this study, we combined the multi-omics profiles, public MTI database, TF-miRNA database, and computational prediction tools to enhance the identification of MTIs. Based on our newly constructed miRNA-gene regulatory network, we further performed the large-scale MTIs validation based on Ago2 IP-seq methodology [30] to rule out the possibility of false-positive MTIs. In addition, we further validated that miR-146b which inhibited by high glucose concentration can promote the regulation of insulin release-related genes. In brief, this study reveals a novel MTIs in INS-1 under increasing glucose conditions and clarifies the molecular mechanisms associated with insulin signaling network in the pancreatic beta-cell.

## Results

Figure 1 presents an overview of the proposed computational framework to identify the novel MTIs in INS-1 under different glucose concentrations by high-throughput expression profiles. The method includes the following two steps: (1) identification of differentially expressed genes (DEGs) and differentially expressed miRNAs (DEmiRs) from RNA-seq and small RNA-seq data, respectively; (2) mapping of DEGs and DEmiRs to the curated novel MTIs and reconstruction of the regulatory network and functional annotation of the target genes.

### Effects of glucose concentration on INS-1 growth

INS-1 doubling time was between 50 and 100 h depending on the glucose concentration of the culture medium [31]. The speed of cell division was affected by glucose levels, and the average subculture time was 1 week. Insulin secretion and complete cell morphology were retained even after 80 generations of the successive subculture; thus, the cells were considered stable for the study. Subculture generations 4 to 10 were used

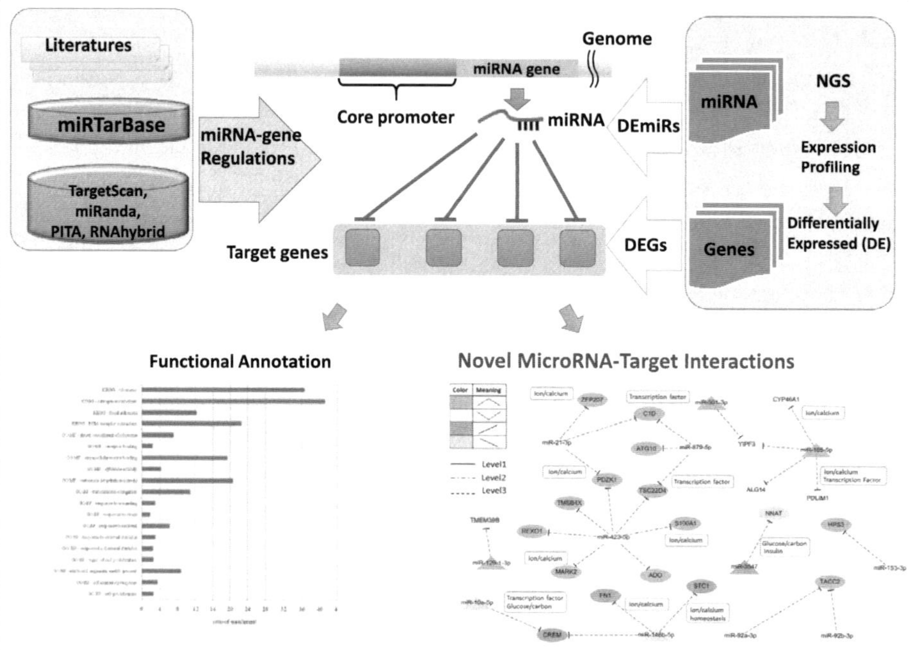


**Fig. 1** Schematic overview of the computational approaches to reconstruct the miRNA-gene regulatory network

in the study. Cell morphology and growth pattern were observed under a microscope (Figure S1). A high glucose level of 30 mM was beneficial to INS-1 adhesion, but a massive suspension of INS-1 was observed on day 4 under a low glucose level of 2 mM (Figure S1). To obtain a detailed understanding of the effect of glucose on INS-1 growth, we measured the cell growth curve and doubling time. The INS-1 doubling times were 102.76, 51.63, and 45.69 h at 2 mM, 11.1 mM, and 30 mM glucose concentration, respectively (Figure S2). The cell doubling time under moderate or high glucose level was found to be consistent with the values in the literature [31].

### Effect of glucose concentration on miRNAs and genes expression in INS-1

sRNA-seq and RNA-Seq are highly sensitive and accurate tools for measuring miRNAs and mRNAs expression across the transcriptome under different environmental conditions. In our study, there were three RNA samples under three increasing glucose conditions (2 mM, low glucose; 11.1 mM, moderate glucose; and 30 mM, high glucose) for sRNA and mRNA sequencing to observe the DEGs and DEmiRs. The read quality of the sRNA-seq and RNA-seq are shown in Fig. 2 and Figure S3. The read length distribution line chart indicates that all samples reached a peak miRNA length in Fig. 2a. More than 87% reads were mapped to the reference genome in Fig. 2b. Figure 2c displays more than 40% of reads were obtained from miRNAs. On the other hand, the read quality of RNA-seq present in Figure S3A and more than 91% reads were mapped to the reference genome. Figure S3B present the DEGs analyzed from RNA-seq data using Volcano plot. The detailed information about differentially expressed miRNAs and mRNAs are provided in Additional file 1 and Additional file 2. Furthermore, miRNAs were identified as significantly and differentially expressed relative to those at moderate glucose concentration if

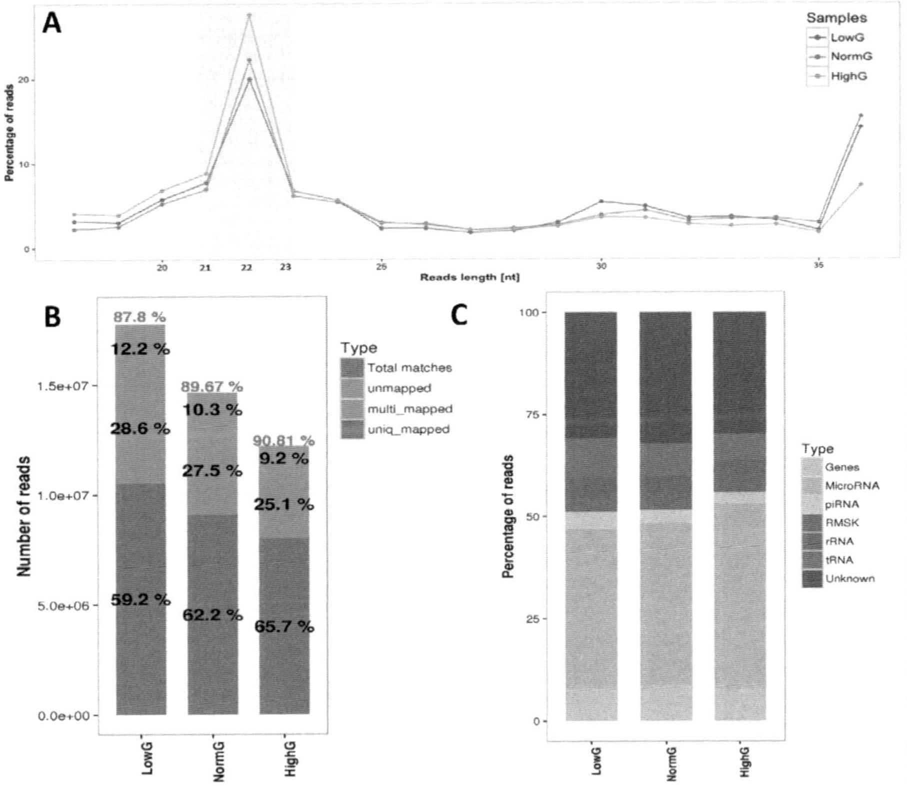


**Fig. 2** Read quality and distribution of small RNA-seq data. **a** The read length distribution shows that all samples reached a peak miRNA length **b** Mapping ratio of reads to the reference genome **c** Propor-tion of reads (> 40%) detected from miRNAs

the fold change exceeded or was equal to 1.4 (upregulated miRNAs) or was less than or equal to 0.71 (downregulated miRNAs). mRNAs were identified as significantly and differentially expressed relative to moderate glucose concentration if the fold change ≥ 1.5 (upregulated mRNAs) or ≤ 0.67 (downregulated mRNAs).

To analyze the experimental results of the miRNA expression and their target genes, we defined eight types of expression profiles according to their responses to different glucose concentrations. Figure 3 demonstrates different profile categories for miRNA (Fig. 3a) and mRNA (Fig. 3b) expression. The gene expression profiles of low and high glucose conditions were correlated with moderate glucose concentration. The eight types of expression profiles were named as follows: **I** (increased continuously); D (decreased continuously); M (decreased under both low and high glucose conditions); V (increased under both low and high glucose conditions); LD (decreased under low glucose and no significant expression change under high glucose conditions); LI (increased under low glucose and no significant expression change under high glucose conditions); HI (increased under high glucose conditions and no significant expression change under low glucose); HD (decreased under high glucose conditions and no significant expression change under low glucose).

Under our assumption that most miRNAs act as a negative regulator of their target mRNAs, each type of miRNA and its target should present an inverse expression profile. For example, the inverse expression of the I type was the D type and that of the HD type was the HI type. The number of genes and miRNAs in each type are shown in Fig. 3a

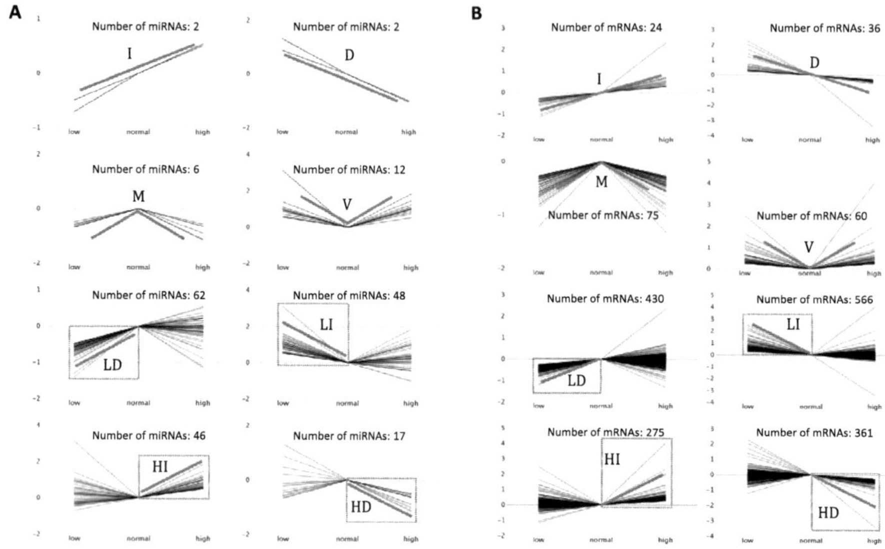


**Fig. 3** Eight types of (**a**) miRNA and (**b**) mRNA gene expression profiles in response to different glucose concentrations. The x-axis shows three different glucose concentrations (low, normal, and high). The y-axis shows the log2 fold change compared with normal glucose concentration. The red line represents the trends of the expression. I (increased continuously); D (decreased continuously); M (decreased under both low and high glucose conditions); V (increased under both low and high glucose conditions); LD (decreased under low glucose and no significant expression change under high glucose conditions); LI (increased under low glucose and no significant expression change under high glucose conditions); HI (increased under high glucose conditions and no significant expression change under low glucose); HD (decreased under high glucose conditions and no significant expression change under low glucose)

and b, respectively. To obtain high-confidence MTI pairs from the high-throughput data, we defined MTIs by using experimental or predicted evidence and miRNA and its target gene with the inverse expression level. Eight types of MTI pairs with high confidence were observed and further analyzed to construct the network and function.

In this study, RNA and small RNA sequencing techniques were applied to obtain DEGs and DEmiRs for each glucose concentration in which INS-1 cells were incubated. To validate the expression profiling of DEGs and DEmiRs, we selected nine genes (Glp1r, Calr, Pde4b, Cited2, Trpc4, Crem, Ins1, Rasgrp2, and Camk) and two miRNAs (miR-375 and miR-16) on the basis of the reports published in the literature [32]. Then, quantitative reverse-transcription PCR (qRT-PCR) was conducted to verify the expression profile of these genes in RNA samples used in next-generation sequencing and another three independent RNA samples under three culture conditions at different glucose concentrations. The results of qRT-PCR validation (Figure S4A) show that the experimental design is highly relevant, and the trend of differential gene expression induced by glucose is consistent with that of the RNA-seq profiles FPKM (fragments per kilobase of transcript per million mapped reads) value (Figure S4B).

### Novel MTIs analysis of miRNA-mediated gene regulatory in different glucose concentration stimulated

On the basis of the previous report [33], we selected genes with the following keywords: glucose/carbon/insulin (indicated as G), ion/calcium (indicated as O), exocytosis/

homeostasis (indicated as E), and CREB/cAMP/transcription factor (indicated as C) [34–39]; and then, we conducted network analysis to find the novel miRNA-mediated gene regulatory network under glucose variation in Figs. 4, 5, S5 and S6. In these figures, the solid line (Level 1) represents the interactions of miRNA targets with strong experimental evidence (supported by luciferase reporter assay or Western blot); dotted/dashed line (Level 2) represents the interactions of miRNA targets with substantial weaker experimental evidence (supported by CLIP-seq experiments); dashed line (Level 3) represents the interactions of the predicted miRNA targets.

The downregulated miRNA (DN-miRNA)-mediated gene regulatory network in high glucose involved 14 DN-miRNAs and 80 upregulated genes, which were associated with the selected gene function (G, O, E, and C) (Fig. 4). The results of our study indicated that miR-185-5p and miR-10a-5p were the key miRNAs that regulated more than 10 genes under high glucose conditions. The upregulated miRNA (UP-miRNA)-mediated gene regulatory network in high glucose involved 29 UP-miRNAs and 51 downregulated genes, which were associated with the selected gene function (G, O, E, and C) (Fig. 5). miR-423-5p, miR-3577, miR-21-3p, and miR-320-3p were the key regulators that suppressed more than 10 genes. The differentially expressed miRNA (DE-miRNA)-mediated gene regulatory network at low glucose concentration are shown in Figures S5 and S6. We used gene functional annotation and analyzed the network under different conditions because of the large number of genes involved in different types of networks.

We summarized the top 20 gene ontology (GO) results of different glucose-stimulated types of the miRNA-mediated gene regulatory network. We found that DN-miRNA-mediated genes under high glucose conditions (Figure S7) were associated with the organelle membrane ($p$-value: 4.01E− 4), response to starvation ($p$-value: 0.001) and

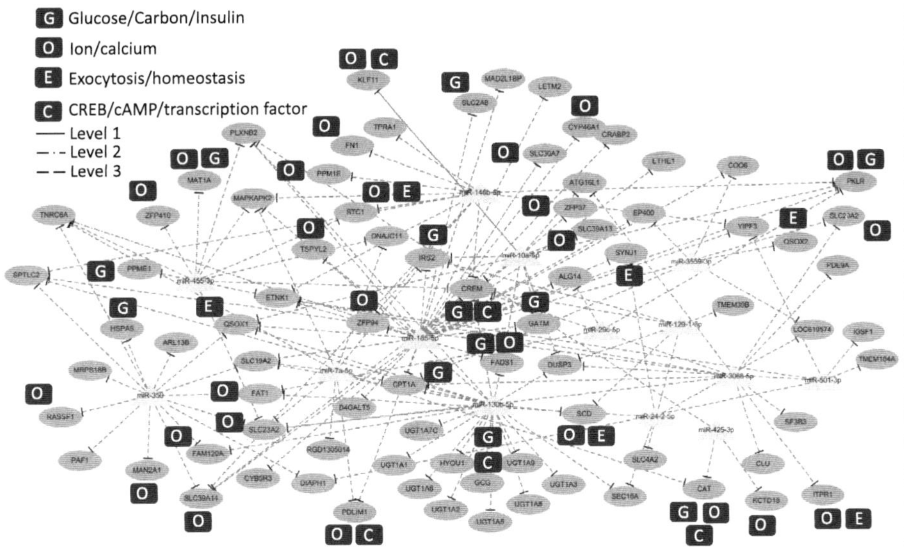


**Fig. 4** DN-miRNA-mediated gene regulatory network in high glucose concentration. Level 1 represents the interactions of miRNA targets with strong experimental evidence (supported by luciferase reporter assay or Western blot). Level 2 represents the interactions of miRNA targets with substantial weaker experimental evidence (supported by CLIP-seq experiments). Level 3 represents the interactions of the predicted miRNA targets. The oval denotes mRNA, the triangle represents miRNA

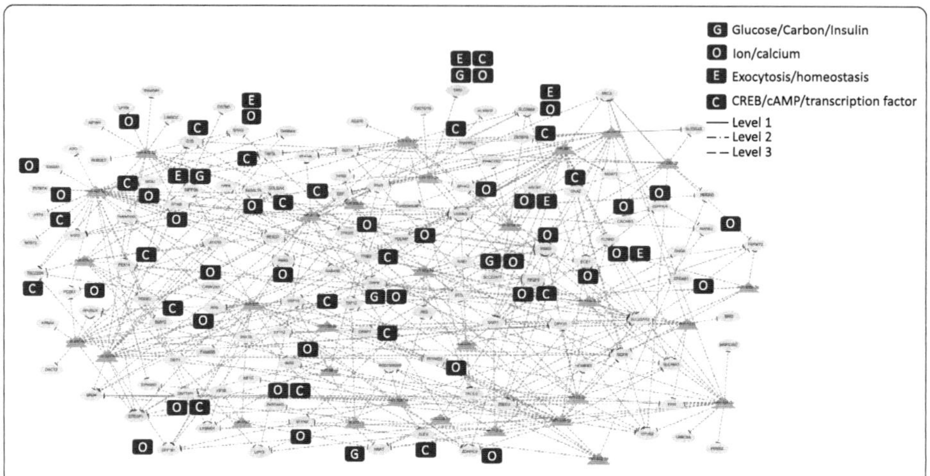


**Fig. 5** UP-miRNA-mediated gene regulatory network at high glucose concentration. Level 1 represents the interactions of miRNA targets with strong experimental evidence (supported by luciferase reporter assay or Western blot). Level 2 represents the interactions of miRNA targets with substantial weaker experimental evidence (supported by CLIP-seq experiments). Level 3 represents the interactions of the predicted miRNA targets. The oval denotes mRNA, the triangle represents miRNA

nutrient levels (p-value: 0.002) under biological process categories. UP-miRNA-mediated genes under high glucose conditions (Figure S8) were majorly associated with the transcription regulator activity ($p$-value: 8.30E− 4) under the categories of biological process and molecular function. Cell cycle-related genes ($p$-value: 1.12E− 10) were downregulated at low glucose levels (DN-miRNA-mediated genes in low glucose concentration, Figure S9); this result is in accordance with the slow growth rate of INS-1 cells when cultured under low glucose condition. The genes were induced by glucose starvation (UP-miRNA-mediated genes in low glucose concentration Figure S10) via the endoplasmic reticulum-unfolded protein response, which protects cells from cell death ($p$-value: 2.90 E $\times 10^5$).

In addition, we established TF-miRNA regulatory networks in pancreatic beta cells. Thirty-four TF-miRNA interactions, including 10 TFs and 28 DE-miRNAs, were identified using TransmiR and gene/miRNA expression profiling Fig. 6 and these results indicated EGR1 as a key transcription factor regulating 19 DE-miRNAs.

## High-throughput identification of miRNA-target interactions by Ago2 Immunoprecipitation (Ago2 IP)

In silico prediction of miRNA targets is an extremely useful approach to identify potential mRNA targets from sequencing data, however, the large number of MTIs still need further experimental validation. The identification of miRNA targets is experimentally laborious and time-consuming. Recently, high-throughput sequencing methods are combined with Ago2 IP-seq to identify miRNAs associated with Ago2. In order to reduce the number of possibilities of MTIs, we further validated the MTIs using Ago2 IP-seq under high or low glucose conditions. Combining the Ago2 IP-seq results, we successfully reduced the number of MTIs in INS-1 pancreatic beta-cells under high glucose concentration. The original number of MTIs from Fig. 4 (151 MTIs), Fig. 5 (331 MTIs), Figure S4 (668 MTIs), and Figure S5 (1013 MTIs), has been reduced to 11

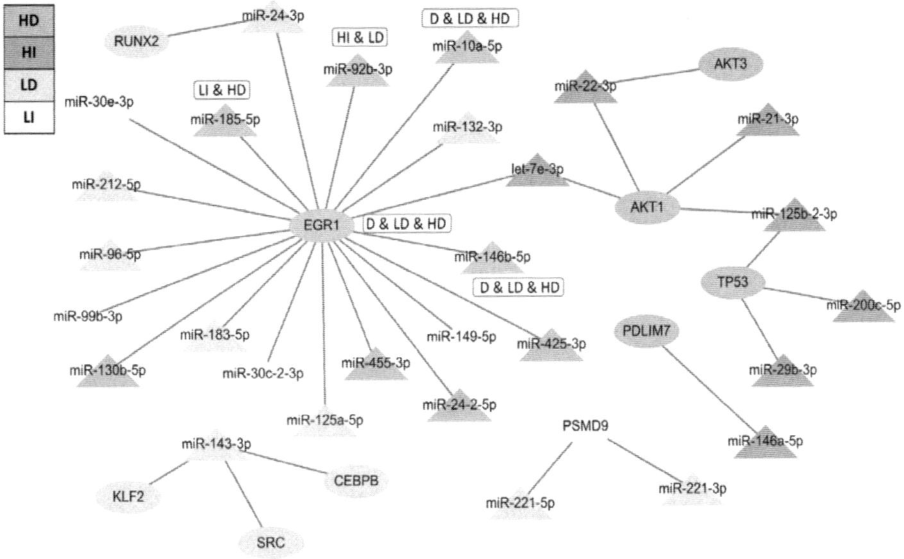

**Fig. 6** TF-miRNA regulatory networks in pancreatic beta cells. (HD- decreased under high glucose condition, HI-increased under high glucose condition, LD-decreased under low glucose condition, LI-increased under low glucose condition). The oval denotes mRNA, the triangle represents miRNA

(Additional file 3: Table S1), 85 (Additional file 3: Table S2), 197 (Additional file 3: Table S3), and 314 (Additional file 3: Table S4), respectively. In Fig. 7, we summarized the gene analysis results of different glucose-stimulated types of miRNA-mediated gene regulatory network. The dashed (Level 3) represents the interactions of the predicted miRNA targets under glucose stimulation and the solid red line (Level 4) indicated the MTIs which validated by Ago2 IP-seq experiment.

### Regulatory role of miR-146b on insulin-related genes

On the basis of the novel reconstructed miRNA-mediated gene regulatory network (Fig. 7), we further selected and validated two MTIs of Crem and Fn1 to miR-146b by miR-146b mimic transfection, independent Ago2 IP, and qRT-PCR experiments. Notably, the two genes are related to the GSIS mechanism, but no evidence supports that miR-146b can participate in the simultaneous regulation of these genes. The potential target sites of Crem and Fn1 of miR-146b are shown in Fig. 8a. To investigate the regulation of miR-146b target genes, we overexpressed miR-146b and measured the expression levels of Crem and Fn1 using qRT-PCR. The qRT-PCR results confirmed that the mRNA expression of the targets genes was downregulated by miR-146b mimics in comparison with the scramble control (Fig. 8b). Furthermore, as mentioned above, RNA fractions were concentrated using the Ago2 IP method and subjected to miRNA target validation. Either miR-146b mimic or the miR-146b scrambled control (src) was delivered to INS-1 cells with low miR-146b expression level under high glucose condition. Then, qRT-PCR was used to compare the expression levels of Crem and Fn1 mRNAs. The results show that the expression levels of Crem and Fn1 were enriched in the Ago2 IPRNA fractions, whereas low expression of Crem and Fn1 was observed in the total RNA fractions, probably due to the RISC cleavage (Fig. 8c). Moreover, to

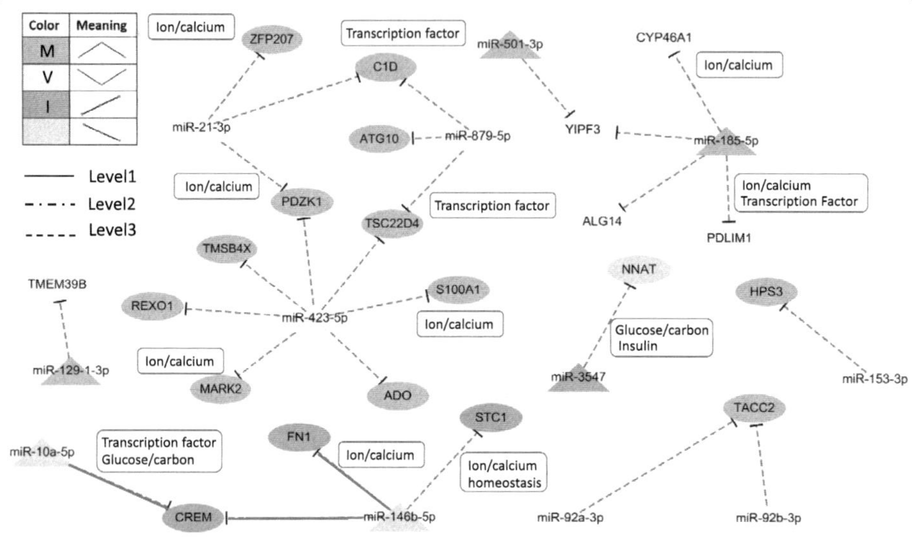


**Fig. 7** DE-miRNA-mediated gene regulatory network under glucose stimulation. The dashed (Level 3) represents the interactions of the predicted miRNA targets under glucose stimulation and the solid red line (Level 4) indicated the MTIs which validated by Ago2 IP-seq experiment. The oval denotes mRNA, the triangle represents miRNA

determine whether miR-146b can repress Crem and Fn1 by targeting its binding site at 3'UTR, we inserted the PCR product containing full length of 3'UTR of two target genes into pmiRGLO luciferase reporter vector. Considering that the translation of target genes are affected by miRNA, we performed dual luciferase reporter assay with INS1 cells co-transfected with either empty luciferase vector or the construct containing 3'UTR of five target genes individually and miR-146b mimics or scramble control. The luciferase assay data indicated that the luciferase reporter activities of Crem and Fn1 were markedly repressed by miR-146b mimics in comparison to scramble control (Fig. 8d). The repression of luciferase reporter assay might be due to binding of miR-

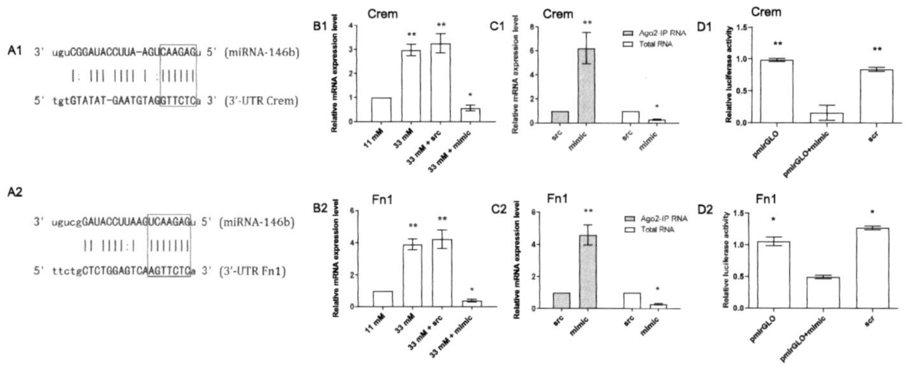


**Fig. 8** Validation of the two target genes Crem and Fn1 of miR-146b that are common in prediction and RNA-seq analysis. **a** Predicted target sites of miR-146b in 3'-UTR of rat Crem and Fn1. **b** mRNA expression analysis of two target genes of miR-146b by qRT-PCR **c** Ago2-IP analysis was carried out to measure relative enrichment of Crem and Fn1 genes of miR-146b as measured by qRT-PCR. src-scramble. **d** Luciferase activity in INS1 cells cotransfected with either empty luciferase vector pmiR-GLO as positive control or the construct containing 3'UTR of two target genes Crem and Fn1 and miR-146b mimics (mimic) or scramble control (scr)

146b to the 3'UTR of these target genes. These results suggest that miR-146b might target Crem and Fn1 via their 3'UTR.

## Discussion

Hypoglycemia is a metabolic disorder condition characterized by abnormally low blood glucose levels of usually less than 70 mg/dL. Hypoglycemia is referred to as an insulin reaction or insulin shock. By contrast, hyperglycemia is a medical condition characterized by high blood glucose. When diabetes occurs the blood sugar level fluctuates periodically [40, 41]. Several glucose-dependent genes in pancreatic beta cells have been reported to date. For example, Txnip is a key regulatory factor for apoptosis and diabetic beta-cell dysfunction and prevents T1D and T2D in a rat model. Previous studies showed that Txnip is strongly induced by glucose. In this study, RNA-seq analysis showed that Txnip gene expression increased by 17 times with elevated glucose level from 2 mM to 30 mM. The regulatory mechanism of Txnip shows that transcription regulation initiated by glucose is highly correlated with miRNA [42].

Accumulating data suggest the important role of miRNA in diabetes [43]. We found that the glucose level-associated miRNAs identified using our workflow matched the previously reported diabetes-associated miRNA, such as miR-34, miR-141, and miR-200c [44]. The expression levels of miR-200 family miRNAs, which include miR-141 and miR-200c, were reportedly upregulated in *db/db* mice and induce the upregulation of apoptosis genes. In addition to its function in diabetes, miR-200c facilitates the regulation of the expression of class III beta-tubulin (TUBB3) under the hypoglycemic condition in ovarian cancer [45]. Our present study found that the miR-200 family miRNAs in the INS-1 cell line were downregulated with the increase in environmental glucose concentration, suggesting that the miR-200c family may also play a role in hypoglycemia.

Several miRNAs identified in T2D patients may be useful biomarkers for the disease progression of T2D and treatment response to antidiabetic medications. miR-185-5p is one of the most significantly decreased miRNAs. The effect of decreased miR-185 in diabetes patients remains unclear [46]. In our study, we found that low miR-185-5p expression level at both low and high glucose concentrations. According to the miRNA–mRNA–target interaction analysis, we found four predicted target genes that were regulated by miR-185-5p. Three of the genes (Cyp46a1, Pdlim1, and Alg14) control the synthesis of cholesterol and long-chain fatty acids. The dominant factor in the pathogenesis of T2D is insulin resistance which in turn increases fatty acid and cholesterol of plasma membranes [47].

Moreover, Cyp46a1 encodes a member of the cytochrome P450 superfamily of enzymes. Cytochrome P450 proteins have monooxygenase activity and catalyze a large number of reactions which are involved in drug metabolism as well as the synthesis of cholesterol, steroids, and other lipids [48]. p75 neurotrophin receptor (p75NTR/CD271) mediates glioma invasion that requires regulated interaction with PDLIM1 [49]. Similar to other tumors, gliomas are believed to primarily metabolize glucose for energy production. However, the dependence of gliomas on glycolysis has recently gained attention because fatty acid oxidation enzymes are present and active within glioma tissues. Targeting of this metabolic pathway can reduce energy production and cellular proliferation in glioma cells [50]. The endoplasmic reticulum is the predominant site of the elongation of long-chain fatty acids. A study in cultured rat hepatocytes showed that the suppression of N-linked glycosylation enhances the activation of the sterol regulatory element-binding protein-1 following increase in the downstream expression of DNL enzymes, including fatty acid synthase [51].

It should be noticed that the V-type miRNA (Fig. 3b) was expressed regardless of glucose level and may be involved in maintaining constant blood sugar levels. The expression of inverted V-type miRNA was inhibited regardless of glucose level. These results suggest that the target gene is inhibited at the normal glucose level and provide a good basis for future research on miRNAs related to glucose metabolism disorders. By contrast, according to the known T2D-related miRNA target or pathway genes (such as p38, PTEN, PI3K/AKt, Bax, Foxa22, and Glut4) [52, 53], we found that none of these genes were screened probably because the short-term glucose-stimulated culture cannot rapidly induce changes in gene expression.

In this study, we also used the Ago2 IP-seq experiment to observe and reduce the number of possibilities of MTIs of the rat INS-1 pancreatic beta-cells under high glucose concentration. It is noteworthy that in this Ago2 IP-seq experiment, we observed several miRNAs which have been confirmed to be associated with diabetes or metabolic abnormalities, such as miR-30b, miR-143, and miR-27b. miR-30b is one of miRNA which correlates with human obesity and T2D [54]. In addition, miR-143 is correlated with metabolic stress in the liver, resulting in insulin resistance on organismal glucose metabolism [55]. Furthermore, glucocorticoids transcriptionally regulate the miR-27b expression and promote body fat accumulation [56]. In brief, through Ago2 IP-seq, we can observe the changes of MTIs under different experimental conditions accurately and extensively. The identified novel MTIs in this study may provide valuable insights into blood glucose homeostasis.

The two candidate direct target genes (Crem and Fn1) of miR-146b under glucose stimulation are presented in Fig. 7. Notably, the two genes are related to the insulin release mechanism [57], Insulin receptor signaling cascade [58], and the correlation of glycemia and glycosylated hemoglobin among patients with T2D [59], respectively, but no evidence supports that miR-146b can participate in the simultaneous regulation of these genes. Two target genes of miR-146b were still not annotated in miRTarBase [60]; thus, these target genes were potential candidates for experimental validation. According to computational prediction, overexpressed miR-146b mimic, and Ago2IP with qRT-PCR results (Figs. 7 and 8a-c), Crem and Fn1 were highly associated with miR-146b. Further validation by luciferase reporter assay proved that the direct integration between miR-146b and Crem or Fn1 genes (Fig. 8D).

In addition, Yipf3 is a predicted target gene of miR-501-3p and miR-185-5p (Fig. 7). Yipf3 is involved in the maintenance of the Golgi structure and plays a role in hematopoiesis. Due to hyperglycemia, dysregulated hematopoietic cells move to those organs which are affected by diabetes and cause different effects like inflammation, cellular dysfunction, and accelerated apoptosis [61].

## Conclusions

In summary, we combined our computational framework with wet-laboratory experiments to uncover the effect of environmental glucose level on the dynamic of miRNA–mRNA regulatory network in the INS-1 cells. The expression profiles of miRNA and mRNA under various experimental conditions were measured by high-throughput sequencing and the differential expression of the selected genes were validated using qRT-PCR. Our data enabled us to reconstruct the miRNA-mediated regulatory network and disclosed previously unidentified miRNA-mediated regulation in adapting to different glucose levels. Notably, GO analysis demonstrated that genes with similar biological functions were regulated by the same miRNAs and shared similar expression profiles; these findings suggest that miRNA-mediated gene regulation can be used as a biomarker for examination or treatment. Moreover, we found that more than 30% of reported miRNAs associated with diabetes show differential expression with the change in glucose level. In addition, using Ago2 IP experiment and luciferase reporter assay, we showed that miR-146b can inhibit the mRNA levels of Crem and Fn1 genes under high glucose condition and these results provide the first evidence of the involvement of miR-146b in the glucose-stimulated insulin secretion mechanism. These findings offer a potentially significant effect on the understanding of miRNA-mediated regulatory networks in the development and progression of pancreatic diabetes. The integration of experimentally verified data with this computational framework can effectively and efficiently extract the miRNA network and is thus suitable for exploring potential pancreatic disease-associated miRNA and their target genes.

## Methods

### Cell culture, RNA isolation, and library preparation for RNA and small RNA sequencing

INS-1 cells (AddexBio, Taoyuan, Taiwan) were routinely cultured in RPMI-1640 (Thermo Fisher, Waltham, MA, USA) supplemented with 10% FBS, 1 mM pyruvate, 10 mM HEPES, and 50 mM 2-mercaptoethanol at 37 °C in a humidified atmosphere at 5% $CO_2$. The cells were detached using Trypsin–EDTA and were passaged once per week. For RNA isolation, INS-1 cells were plated at a density of $10^5$ cells per T-75 flask and incubated in RPMI-1640 containing 2 mM (low glucose), 11.1 mM (moderate glucose), and 30 mM glucose (high glucose). As osmolality control, mannose was added to the culture medium to avoid osmotic effects resulting from low glucose levels [62]. On the following day, the cells were collected, and RNA extraction was carried out. mRNA from the INS-1 cells was extracted using RNAZol kit (SigmaAldrich, Merck KGaA), and miRNA was extracted using miRNeasy Mini Kit (Qiagen, Hilden, Germany). The extracted RNA was dissolved in RNase-free water and stored at − 80 °C. The concentration of RNA was measured using a Nanodrop spectrophotometer (Thermo Fisher, Waltham, MA, USA). The integrity of total RNA was assessed using Agilent 2200 TapeStation-RNA R6K assay (Agilent, Santa Clara, CA). RNA-seq and small RNA-seq cDNA libraries were constructed following the standard Illumina (Illumina, San Diego, CA) protocol. All resulting cDNA library fragments were validated using Agilent 2200 TapeStation-D1000 assay (Agilent). The cDNA libraries were measured using quantitative reverse-transcription PCR (qRT-PCR) (Roche, LightCycler® 480 system, Basel, Switzerland) and a Qubit fluorometer (Invitrogen, Carlsbad, California, USA), pooled, and sequenced on an Illumina NextSeq 500 platform in single-ended mode with 75 bps. Depth of coverage of approximately 20 million or 5 million reads was obtained in RNA-seq or small RNA-seq, respectively. The expression levels of selected target genes in RNA sequencing samples and other independent samples were validated via qRT-PCR method using a general SYBR Green PCR Kit (Roche).

### RNA sequencing data analysis

Quality control of RNA-seq reads was conducted by using the FASTX-Toolkit version 0.0.13.2. Nucleotides with high-quality base calling (Phred quality score ≥ 20, which represents 99% accuracy of a base call) were included, and RNA reads exceeding 35 nucleotides were retained. Then, the reads were mapped to the UCSC *Rattus norvegicus* genome rn6 by using TopHat2 [63] version 2.1.0 with the option --b2-very-sensitive and –G for transcriptome annotations from the UCSC Genome Browser [64]. The transcripts were assembled, and the estimated abundances were normalized by Cufflinks version 2.2.1 [65]. Gene expression profiles were estimated using Cuffdiff version 2.2.1. Genes with FPKM values of less than 10 in both compared samples were filtered. Genes were identified as upregulated or downregulated if the fold change exceeded 1.2 or was less than 0.83, respectively.

### Small RNA sequencing data analysis

The 3′-adapters of the small RNA-seq reads were trimmed using Cutadapt v1.5 [66]. Quality control and data preprocessing were done using the FASTX-Toolkit. Only nucleotides with Phred quality score ≥ 20 and reads longer than 18 were included. To assess the quality of the small RNA distribution and miRNA in the sequencing data, ncPRO-seq [67] package (version 1.6.1) was used to confirm the read distribution in the reference genome. Only reads that were mapped with a maximum of one mismatch and 20 locations in the genome were used (Bowtie [68] v1.1.2 parameter: -v1 -a -m20 --best --strata --nomaqround -f -y). Five databases, namely, the UCSC reference genome (rn5), miRBase v21, UCSC refGene, RFam v11.0, and UCSC RepeatMasker (rn5) were employed for annotation. To quantify the miRNA profiles, miRDeep2 [69] package (version 2.0.0.5) and same bowtie parameter as ncPRO-seq were used. To standardize miRNA expression across different samples, we normalized the read count in each sample into reads per million (RPM). To categorize extremely low-expressed miRNA, we filtered miRNAs with RPM lower than 10 in both compared samples. miRNAs were identified as upregulated or downregulated if the fold change exceeded 1.4 or was less than 0.71, respectively.

### A computational approach for reconstruction of miRNA-mediated gene regulatory network

An overview of the proposed computational approach to reconstruct the miRNA-gene regulatory network is shown in Fig. 1. Information on TF-miRNA regulation was derived from the TransmiR [70] database and review of the literature. DEGs and miRNAs were obtained from expression profiles. Experimental MTIs were obtained from miRTarBase [71]. The predicted MTIs were obtained using four prediction tools, namely, miRanda [72], PITA [73], TargetScan [74], and RNAhybrid [75]. Only MTIs with minimum free energy ≤10 and miRanda score ≥ 140 and predicted by at least three tools were selected. For each MTI, the miRNA and its target gene with inverse expression level (upregulated genes with downregulated miRNAs or downregulated genes with upregulated miRNAs) were selected to further reconstruct the miRNA-mediated regulatory network. Each type of miRNA and its target interaction was based on inverse expression level, whereas each type of TF and its potential association with miRNA were defined on the basis of regulatory type by TransmiR or by the literature. Cytoscape (version 3.4.0) software [76] was used to visualize the association of TF and miRNA or MTI. To discuss the function of the miRNA regulatory network, we further analyzed the miRNA target gene functional annotation by using DAVID [77].

#### Argonaute 2 immunoprecipitation sequencing (Ago2 IP-seq)

The detailed information of cell crosslinking, cell lysis and Ago2 IP are described in the Supplementary Information. Purified total RNA from INS-1 cell lysate was used as a template for subsequent RNA-seq and small RNA-seq experiments. The mRNAs with FPKM values of less than 8 and the miRNAs with RPM lower than 100 were filtered. The expression of target genes of miRNA was analyzed using qRT-PCR.

#### Experimental validation of miR-146b targets

To investigate the regulation of target genes by miR-146b, we overexpressed the miR-146b, and the levels of expression of two target genes Crem and Fn1 were measured using qRT-PCR. Prior to qRT-PCR, INS-1 cells were transfected with the miR-146b mimics and scrambled control. After 48 h of transfection, total RNA was isolated from the INS-1 cell line, and the mRNA expression levels of four target genes were measured. In addition, the RNA fractions were concentrated using the Ago2 IP method (Supplementary Information) and used to validate the target mRNA of miR-146b by using qRT-PCR.

#### Plasmid construction and dual luciferase reporter assay

The 3'UTR of Crem and Fn1 were amplified using genomic DNA and cloned downstream of the Renilla luciferase open reading frame in the pmiRGLO vector (Promega) using PmeI and XbaI restriction sites. The primers for 3'UTR amplification are Crem-F: 5'CTAGCTAGCTAGAGAATAGCCTGACACGGCTA3'; Crem-R: 5'TGCTCTAGAGCAAACTCATTCCAGATAATAA3';Fn1-F:5'CTAGCTAGCTAGCAGCCCAAGC CAACAAGTGT3'; Fn1-R: TGCTCTAGAGCAAGACAATAATACTGAGCAGT3'. The day before performing the luciferase reporter assay, 12 well plate with $1 \times 10^4$ INS1 cells were co-transfected with either empty luciferase vector or the construct containing 3'UTR of two target genes Crem and Fn1 and miR-146b mimics or scramble control. pmiRGLO vector with no insert was used as control. The plates were removed from the incubator next day and 200 μl of Dual-Glo reagent was added to each well and mixed and allowed to wait for 10 min for cell lysis to occur, then firefly luminescence was measured using Lumat 9507 LB (Berthold Technologies). One hundred microliter of Dual-Glo Stop & Glo was added to the cell lysate and mixed well and then renilla luminescence was measured. Normalized luciferase activity (firefly luciferase activity/ renilla luciferase activity) for each construct was calculated.

## Supplementary information

**Supplementary information** accompanies this paper at https://doi.org/10.1186/s12859-020-03678-0.

**Additional file 1.** RNA-seq data.

**Additional file 2.** sRNA-seq data.

**Additional file 3.** Ago-IP seq data.

**Additional file 4.** Supplementary Information.

### Abbreviations
miRNAs: MicroRNAs; RNA-seq: RNA sequencing; sRNA-seq: Small RNA sequencing; DEmiRs: Differentially expressed miRNAs; Ago IP-seq: Argonaute immunoprecipitation sequencing; CLIP-seq: Cross-link immunoprecipitation sequencing; ChIP-seq: Chromatin immunoprecipitation sequencing; MTIs: MiRNA–target interactions; PI3K: Phosphatidylinositol 3-kinase; DEGs: Differentially expressed genes; src: Scrambled control

### Acknowledgements
We thank all individuals who participated in this study.

### Availability and implementation
The dataset and results are available at Supplementary information.

### About this supplement
This article has been published as part of *BMC Bioinformatics Volume 21 Supplement 13, 2020: Selected articles from the 18th Asia Pacific Bioinformatics Conference (APBC 2020): bioinformatics.* The full contents of the supplement are available online at https://bmcbioinformatics.biomedcentral.com/articles/supplements/volume-21-supplement-13 .

### Authors' contributions
H.D.H. and C.P.T. supervised the study; Y.C.D. L. and H.Y.H. designed the study, wrote and prepared the manuscript; Y.C.D. L. and C.R.C. carried cell culture and performed cell functional assays; Y.C.D.Y.L. and C.R.C. performed RNA isolation and qPCR experiment; H.C.H., Y.A.C., and M.Y.C., carried RNA isolation, RNA-seq, and miRNA-seq experiment; C.H.C., Y.R.H., S.J.T., and H.Y.H. performed bioinformatics analysis of RNA-seq and sRNA-seq data; S.S., Y.H.C., T.H.S. S.L.W., J.L. and all the above authors revised and approved the final manuscript.

### Funding
This work and publication costs are funded by the Warshel Institute for Computational Biology funding from Shenzhen City and Longgang District under award number C10120180043 and D10120180089. Labor allowance for doctoral students was supported by the Ganghong Young Scholar Development Fund of Shenzhen Ganghong Group under award number F10120190096. Experimental materials based upon supported by the Ministry of Science Technology, Taiwan, under award number MOST 106–2627-M-009-002, MOST 106–2319-B-400-001, and MOST 106–2633-B-009-001.

### Availability of data and materials
The additional files and supplementary information mentioned in the article are all availability of data and materials within this study.

### Ethics approval and consent to participate
Not applicable.

### Consent for publication
Not applicable.

### Competing interests
The authors declare that they have no competing interests.

### Author details
[1]School of Life and Health Sciences, The Chinese University of Hong Kong, Longgang District, Shenzhen 518172, Guangdong Province, China. [2]Warshel Institute for Computational Biology, The Chinese University of Hong Kong, Longgang District, Shenzhen 518172, Guangdong Province, China. [3]Institute of Bioinformatics and Systems Biology, National Chiao Tung University, Hsinchu 300, Taiwan. [4]Department of Biological Science and Technology, National Chiao Tung University, Hsinchu 300, Taiwan. [5]Department of Microbiology and Immunology, Weill Cornell Medicine, Cornell University, New York, NY 10021, USA. [6]Department of Obstetrics and Gynecology, Hsinchu Mackay Memorial Hospital, Hsinchu 300, Taiwan.

Published: 17 September 2020

### References
1. Bartel DP. MicroRNAs: genomics, biogenesis, mechanism, and function. Cell. 2004;116(2):281–97.
2. Carleton M, Cleary MA, Linsley PS. MicroRNAs and cell cycle regulation. Cell Cycle. 2007;6(17):2127–32.
3. Harfe BD. MicroRNAs in vertebrate development. Curr Opin Genet Dev. 2005;15(4):410–5.
4. Lynam-Lennon N, Maher SG, Reynolds JV. The roles of microRNA in cancer and apoptosis. Biol Rev Camb Philos Soc. 2009;84(1):55–71.
5. Soifer HS, Rossi JJ, Saetrom P. MicroRNAs in disease and potential therapeutic applications. Mol Ther. 2007;15(12):2070–9.
6. Chou CH, Shrestha S, Yang CD, Chang NW, Lin YL, Liao KW, Huang WC, Sun TH, Tu SJ, Lee WH, et al. miRTarBase update 2018: a resource for experimentally validated microRNA-target interactions. Nucleic Acids Res. 2017;4;46(D1):D296-302.

7. Schuit FC, Huypens P, Heimberg H, Pipeleers DG. Glucose sensing in pancreatic beta-cells: a model for the study of other glucose-regulated cells in gut, pancreas, and hypothalamus. Diabetes. 2001;50(1):1–11.
8. Paris M, Bernard-Kargar C, Berthault MF, Bouwens L, Ktorza A. Specific and combined effects of insulin and glucose on functional pancreatic beta-cell mass in vivo in adult rats. Endocrinology. 2003;144(6):2717–27.
9. Heit JJ, Karnik SK, Kim SK. Intrinsic regulators of pancreatic beta-cell proliferation. Annu Rev Cell Dev Biol. 2006;22:311–38.
10. Wu L, Dai X, Zhan J, Zhang Y, Zhang H, Zhang H, Zeng S, Xi W. Profiling peripheral microRNAs in obesity and type 2 diabetes mellitus. APMIS. 2015;123(7):580–5.
11. Bai C, Li X, Gao Y, Wang K, Fan Y, Zhang S, Ma Y, Guan W. Role of microRNA-21 in the formation of insulin-producing cells from pancreatic progenitor cells. Biochim Biophys Acta. 2016;1859(2):280–93.
12. Floris I, Descamps B, Vardeu A, Mitic T, Posadino AM, Shantikumar S, Sala-Newby G, Capobianco G, Mangialardi G, Howard L, et al. Gestational diabetes mellitus impairs fetal endothelial cell functions through a mechanism involving microRNA-101 and histone methyltransferase enhancer of zester homolog-2. Arterioscler Thromb Vasc Biol. 2015;35(3): 664–74.
13. Poy MN, Eliasson L, Krutzfeldt J, Kuwajima S, Ma X, Macdonald PE, Pfeffer S, Tuschl T, Rajewsky N, Rorsman P, et al. A pancreatic islet-specific microRNA regulates insulin secretion. Nature. 2004;432(7014):226–30.
14. Avnit-Sagi T, Kantorovich L, Kredo-Russo S, Hornstein E, Walker MD. The promoter of the pri-miR-375 gene directs expression selectively to the endocrine pancreas. PLoS One. 2009;4(4):e5033.
15. El Ouaamari A, Baroukh N, Martens GA, Lebrun P, Pipeleers D, van Obberghen E. miR-375 targets 3'-phosphoinositide-dependent protein kinase-1 and regulates glucose-induced biological responses in pancreatic beta-cells. Diabetes. 2008; 57(10):2708–17.
16. Hashimoto N, Kido Y, Uchida T, Asahara S, Shigeyama Y, Matsuda T, Takeda A, Tsuchihashi D, Nishizawa A, Ogawa W, et al. Ablation of PDK1 in pancreatic beta cells induces diabetes as a result of loss of beta cell mass. Nat Genet. 2006; 38(5):589–93.
17. Walker MD. Role of MicroRNA in pancreatic beta-cells: where more is less. Diabetes. 2008;57(10):2567–8.
18. LaPierre MP, Stoffel M. MicroRNAs as stress regulators in pancreatic beta cells and diabetes. Mol Metab. 2017;6(9):1010–23.
19. Keller DM, Clark EA, Goodman RH. Regulation of microRNA-375 by cAMP in pancreatic beta-cells. Mol Endocrinol. 2012; 26(6):989–99.
20. Wang Y, Liu J, Liu C, Naji A, Stoffers DA. MicroRNA-7 regulates the mTOR pathway and proliferation in adult pancreatic beta-cells. Diabetes. 2013;62(3):887–95.
21. Nieto M, Hevia P, Garcia E, Klein D, Alvarez-Cubela S, Bravo-Egana V, Rosero S, Damaris Molano R, Vargas N, Ricordi C, et al. Antisense miR-7 impairs insulin expression in developing pancreas and in cultured pancreatic buds. Cell Transplant. 2012;21(8):1761–74.
22. Baroukh N, Ravier MA, Loder MK, Hill EV, Bounacer A, Scharfmann R, Rutter GA, Van Obberghen E. MicroRNA-124a regulates Foxa2 expression and intracellular signaling in pancreatic beta-cell lines. J Biol Chem. 2007;282(27):19575–88.
23. Wang P, Chen L, Zhang J, Chen H, Fan J, Wang K, Luo J, Chen Z, Meng Z, Liu L. Methylation-mediated silencing of the miR-124 genes facilitates pancreatic cancer progression and metastasis by targeting Rac1. Oncogene. 2014;33(4):514–24.
24. Yan Z, Shah PK, Amin SB, Samur MK, Huang N, Wang X, Misra V, Ji H, Gabuzda D, Li C. Integrative analysis of gene and miRNA expression profiles with transcription factor-miRNA feed-forward loops identifies regulators in human cancers. Nucleic Acids Res. 2012;40(17):e135.
25. Huang GT, Athanassiou C, Benos PV. mirConnX: condition-specific mRNA-microRNA network integrator. Nucleic Acids Res. 2011;39(Web Server issue):W416–23.
26. Bisognin A, Sales G, Coppe A, Bortoluzzi S, Romualdi C. MAGIA(2): from miRNA and genes expression data integrative analysis to microRNA-transcription factor mixed regulatory circuits (2012 update). Nucleic Acids Res. 2012;40(Web Server issue):W13–21.
27. Hsu SD, Huang HY, Chou CH, Sun YM, Hsu MT, Tsou AP. Integrated analyses to reconstruct microRNA-mediated regulatory networks in mouse liver using high-throughput profiling. BMC Genomics. 2015;16(Suppl 2):S12.
28. Gosline SJ, Gurtan AM, JnBaptiste CK, Bosson A, Milani P, Dalin S, Matthews BJ, Yap YS, Sharp PA, Fraenkel E. Elucidating MicroRNA regulatory networks using transcriptional, post-transcriptional, and histone modification measurements. Cell Rep. 2016;14(2):310–9.
29. Skelin M, Rupnik M, Cencic A. Pancreatic beta cell lines and their applications in diabetes mellitus research. ALTEX. 2010; 27(2):105–13.
30. Kanematsu S, Tanimoto K, Suzuki Y, Sugano S. Screening for possible miRNA-mRNA associations in a colon cancer cell line. Gene. 2014;533(2):520–31.
31. Asfari M, Janjic D, Meda P, Li G, Halban PA, Wollheim CB. Establishment of 2-mercaptoethanol-dependent differentiated insulin-secreting cell lines. Endocrinology. 1992;130(1):167–78.
32. Andrali SS, Sampley ML, Vanderford NL, Ozcan S. Glucose regulation of insulin gene expression in pancreatic beta-cells. Biochem J. 2008;415(1):1–10.
33. Esguerra JL, Mollet IG, Salunkhe VA, Wendt A, Eliasson L. Regulation of pancreatic Beta cell stimulus-secretion coupling by microRNAs. Genes (Basel). 2014;5(4):1018–31.
34. Chakraborty C, Doss CG, Bandyopadhyay S, Agoramoorthy G. Influence of miRNA in insulin signaling pathway and insulin resistance: micro-molecules with a major role in type-2 diabetes. Wiley Interdiscip Rev RNA. 2014;5(5):697–712.
35. Regazzi R, Rodriguez-Trejo A, Jacovetti C. Insulin secretion in health and disease: nutrients dictate the pace. Proc Nutr Soc. 2016;75(1):19–29.
36. Salunkhe VA, Esguerra JL, Ofori JK, Mollet IG, Braun M, Stoffel M, Wendt A, Eliasson L. Modulation of microRNA-375 expression alters voltage-gated Na(+) channel properties and exocytosis in insulin-secreting cells. Acta Physiol (Oxf). 2015;213(4):882–92.
37. Sebastiani G, Po A, Miele E, Ventriglia G, Ceccarelli E, Bugliani M, Marselli L, Marchetti P, Gulino A, Ferretti E, et al. MicroRNA-124a is hyperexpressed in type 2 diabetic human pancreatic islets and negatively regulates insulin secretion. Acta Diabetol. 2015;52(3):523–30.
38. Shaer A, Azarpira N, Karimi MH, Soleimani M, Dehghan S. Differentiation of human-induced pluripotent stem cells into insulin-producing clusters by MicroRNA-7. Exp Clin Transplant. 2016;14(5):555-63.

39. Xu H, Abuhatzira L, Carmona GN, Vadrevu S, Satin LS, Notkins AL. The Ia-2beta intronic miRNA, miR-153, is a negative regulator of insulin and dopamine secretion through its effect on the Cacna1c gene in mice. Diabetologia. 2015;58(10):2298–306.
40. Taubes G. Diabetes. Paradoxical effects of tightly controlled blood sugar. Science. 2008;322(5900):365–7.
41. Winter T. Increased blood sugar in type 2 diabetes mellitus: what now? Praxis (Bern 1994). 2002;91(12):498–502.
42. XG HK, Grayson TB, Shalev A. Cytokines Regulate β-Cell Thioredoxin-interacting Protein (TXNIP) via Distinct Mechanisms and Pathways. J Biol Chem. 2016;291(16):8428–39.
43. Zhu H, Leung SW. Identification of microRNA biomarkers in type 2 diabetes: a meta-analysis of controlled profiling studies. Diabetologia. 2015;58(5):900–11.
44. Simpson K, Wonnacott A, Fraser DJ, Bowen T. MicroRNAs in diabetic nephropathy: from biomarkers to therapy. Curr Diab Rep. 2016;16(3):35.
45. Belgardt BF, Ahmed K, Spranger M, Latreille M, Denzler R, Kondratiuk N, von Meyenn F, Villena FN, Herrmanns K, Bosco D, et al. The microRNA-200 family regulates pancreatic beta cell survival in type 2 diabetes. Nat Med. 2015;21(6):619–27.
46. Delic D, Eisele C, Schmid R, Luippold G, Mayoux E, Grempler R. Characterization of Micro-RNA changes during the 652 progression of type 2 diabetes in Zucker Diabetic fatty rats. Int J Mol Sci. 2016;17(5):665.
47. Bakan E, Yildirim A, Kurtul N, Polat MF, Dursun H, Cayir K. Effects of type 2 diabetes mellitus on plasma fatty acid composition and cholesterol content of erythrocyte and leukocyte membranes. Acta Diabetol. 2006;43(4):109–13.
48. Russell DW, Halford RW, Ramirez DM, Shah R, Kotti T. Cholesterol 24-hydroxylase: an enzyme of cholesterol turnover in the brain. Annu Rev Biochem. 2009;78:1017–40.
49. Ahn BY, Saldanha-Gama RF, Rahn JJ, Hao X, Zhang J, Dang NH, Alshehri M, Robbins SM, Senger DL. Glioma invasion mediated by the p75 neurotrophin receptor (p75(NTR)/CD271) requires regulated interaction with PDLIM1. Oncogene. 2016;35(11):1411–22.
50. Lin H, Patel S, Affleck VS, Wilson I, Turnbull DM, Joshi AR, Maxwell R, Stoll EA. Fatty acid oxidation is required for the respiration and proliferation of malignant glioma cells. Neuro Oncol. 2017;19(1):43-54.
51. Sanders FW, Griffin JL. De novo lipogenesis in the liver in health and disease: more than just a shunting yard for glucose. Biol Rev Camb Philos Soc. 2016;91(2):452–68.
52. Gong W, Xiao D, Ming G, Yin J, Zhou H, Liu Z. Type 2 diabetes mellitus-related genetic polymorphisms in microRNAs and microRNA target sites. J Diabetes. 2014;6(4):279–89.
53. Guay C, Roggli E, Nesca V, Jacovetti C, Regazzi R. Diabetes mellitus, a microRNA-related disease? Transl Res. 2011;157(4): 253–64.
54. Hu F, Wang M, Xiao T, Yin B, He L, Meng W, Dong M. Liu F: miR-30 promotes thermogenesis and the development of beige fat by targeting RIP140. Diabetes. 2015;64(6):2056–68.
55. Jordan SD, Kruger M, Willmes DM, Redemann N, Wunderlich FT, Bronneke HS, Merkwirth C, Kashkar H, Olkkonen VM, Bottger T, et al. Obesity-induced overexpression of miRNA-143 inhibits insulin-stimulated AKT activation and impairs glucose metabolism. Nat Cell Biol. 2011;13(4):434–46.
56. Kong X, Yu J, Bi J, Qi H, Di W, Wu L, Wang L, Zha J, Lv S, Zhang F, et al. Glucocorticoids transcriptionally regulate miR-27b expression promoting body fat accumulation via suppressing the browning of white adipose tissue. Diabetes. 2015; 64(2):393–404.
57. Favre D, Niederhauser G, Fahmi D, Plaisance V, Brajkovic S, Beeler N, Allagnat F, Haefliger JA, Regazzi R, Waeber G, et al. Role for inducible cAMP early repressor in promoting pancreatic beta cell dysfunction evoked by oxidative stress in human and rat islets. Diabetologia. 2011;54(9):2337–46.
58. Benyoucef S, Surinya KH, Hadaschik D, Siddle K. Characterization of insulin/IGF hybrid receptors: contributions of the insulin receptor L2 and Fn1 domains and the alternatively spliced exon 11 sequence to ligand binding and receptor activation. Biochem J. 2007;403(3):603–13.
59. Lopez JJ, Jardin I, Chamorro CC, Duran ML, Tarancon Rubio MJ, Reyes Panadero M, Jimenez F, Montero R, Gonzalez MJ, Martinez M, et al. Involvement of stanniocalcins in the deregulation of glycaemia in obese mice and type 2 diabetic patients. J Cell Mol Med. 2018;22(1):684–94.
60. Chou CH, Chang NW, Shrestha S, Hsu SD, Lin YL, Lee WH, Yang CD, Hong HC, Wei TY, Tu SJ, et al. miRTarBase 2016: updates to the experimentally validated miRNA-target interactions database. Nucleic Acids Res. 2016;4;44(D1):D239-47.
61. Kojima H, Kim J, Chan L. Emerging roles of hematopoietic cells in the pathobiology of diabetic complications. Trends Endocrinol Metab. 2014;25(4):178–87.
62. Wu L, Derynck R. Essential role of TGF-beta signaling in glucose-induced cell hypertrophy. Dev Cell. 2009;17(1):35–48.
63. Kim D, Pertea G, Trapnell C, Pimentel H, Kelley R, Salzberg SL. TopHat2: accurate alignment of transcriptomes in the presence of insertions, deletions and gene fusions. Genome Biol. 2013;14(4):R36.
64. Karolchik D, Baertsch R, Diekhans M, Furey TS, Hinrichs A, Lu Y, Roskin KM, Schwartz M, Sugnet CW, Thomas DJ. The UCSC genome browser database. Nucleic Acids Res. 2003;31(1):51–4.
65. Trapnell C, Hendrickson DG, Sauvageau M, Goff L, Rinn JL, Pachter L. Differential analysis of gene regulation at transcript resolution with RNA-seq. Nat Biotechnol. 2013;31(1):46–53.
66. Martin M. Cutadapt removes adapter sequences from high-throughput sequencing reads. EMBnet J. 2011;17(1):10–2.
67. Chen CJ, Servant N, Toedling J, Sarazin A, Marchais A, Duvernois-Berthet E, Cognat V, Colot V, Voinnet O. Heard E *et al*: ncPRO-seq: a tool for annotation and profiling of ncRNAs in sRNA-seq data. Bioinformatics. 2012;28(23):3147–9.
68. Langmead B, Trapnell C, Pop M, Salzberg SL. Ultrafast and memory-efficient alignment of short DNA sequences to the human genome. Genome Biol. 2009;10(3):R25.
69. Friedlander MR, Mackowiak SD, Li N, Chen W. Rajewsky N: miRDeep2 accurately identifies known and hundreds of novel microRNA genes in seven animal clades. Nucleic Acids Res. 2012;40(1):37–52.
70. Wang J, Lu M, Qiu C, Cui Q. TransmiR: a transcription factor-microRNA regulation database. Nucleic Acids Res. 2010; 38(Database issue):D119–22.
71. Chou CH, Chang NW, Shrestha S, Hsu SD, Lin YL, Lee WH, Yang CD, Hong HC, Wei TY. Tu SJ *et al*: miRTarBase 2016: updates to the experimentally validated miRNA-target interactions database. Nucleic Acids Res. 2016;44(D1):D239–47.
72. Enright AJ, John B, Gaul U, Tuschl T, Sander C, Marks DS. MicroRNA targets in drosophila. Genome Biol. 2003;5(1):R1.
73. Kertesz M, Iovino N, Unnerstall U, Gaul U, Segal E. The role of site accessibility in microRNA target recognition. Nat Genet. 2007;39(10):1278–84.

74. Lewis BP, Burge CB, Bartel DP. Conserved seed pairing, often flanked by adenosines, indicates that thousands of human genes are microRNA targets. Cell. 2005;120(1):15–20.
75. Kruger J, Rehmsmeier M. RNAhybrid: microRNA target prediction easy, fast and flexible. Nucleic Acids Res. 2006;34(Web Server issue):W451–4.
76. Shannon P, Markiel A, Ozier O, Baliga NS, Wang JT, Ramage D, Amin N, Schwikowski B, Ideker T. Cytoscape: a software environment for integrated models of biomolecular interaction networks. Genome Res. 2003;13(11):2498–504.
77. Huang da W, Sherman BT, Lempicki RA. Systematic and integrative analysis of large gene lists using DAVID bioinformatics resources. Nat Protoc. 2009;4(1):44–57.

## Publisher's Note

Springer Nature remains neutral with regard to jurisdictional claims in published maps and institutional affiliations.

Chen *et al. BMC Bioinformatics* 2020, **21**(Suppl 13):392
https://doi.org/10.1186/s12859-020-03679-z

BMC Bioinformatics

RESEARCH Open Access

# EnClaSC: a novel ensemble approach for accurate and robust cell-type classification of single-cell transcriptomes

Xiaoyang Chen, Shengquan Chen and Rui Jiang*

*From* The 18th Asia Pacific Bioinformatics Conference
Seoul, Korea. 18-20 August 2020

* Correspondence: ruijiang@tsinghua.edu.cn
MOE Key Laboratory of Bioinformatics, Bioinformatics Division and Center for Synthetic and Systems Biology, Beijing National Research Center for Information Science and Technology, Department of Automation, Tsinghua University, Beijing 100084, China

## Abstract

**Background:** In recent years, the rapid development of single-cell RNA-sequencing (scRNA-seq) techniques enables the quantitative characterization of cell types at a single-cell resolution. With the explosive growth of the number of cells profiled in individual scRNA-seq experiments, there is a demand for novel computational methods for classifying newly-generated scRNA-seq data onto annotated labels. Although several methods have recently been proposed for the cell-type classification of single-cell transcriptomic data, such limitations as inadequate accuracy, inferior robustness, and low stability greatly limit their wide applications.

**Results:** We propose a novel **en**semble approach, named EnClaSC, for accurate and robust cell-type **cla**ssification of **s**ingle-**c**ell transcriptomic data. Through comprehensive validation experiments, we demonstrate that EnClaSC can not only be applied to the self-projection within a specific dataset and the cell-type classification across different datasets, but also scale up well to various data dimensionality and different data sparsity. We further illustrate the ability of EnClaSC to effectively make cross-species classification, which may shed light on the studies in correlation of different species. EnClaSC is freely available at https://github.com/xy-chen16/EnClaSC.

**Conclusions:** EnClaSC enables highly accurate and robust cell-type classification of single-cell transcriptomic data via an ensemble learning method. We expect to see wide applications of our method to not only transcriptome studies, but also the classification of more general data.

**Keywords:** Single-cell, scRNA-seq, Cell types, Classification, Feature selection, Few-sample classes, Neural networks

## Background

Recent advances in single-cell RNA-sequencing (scRNA-seq) techniques make it possible to reveal previously unknown heterogeneity and functional diversity at a microscopic resolution [1–3]. The exponential growth of the number of cells profiled in individual scRNA-seq experiments has shed light on the studies aiming to identify new cell types [4, 5], reveal regulatory mechanisms [6, 7], assess tissue composition [1, 8, 9], investigate cell development and lineage processes [10–12], and many others. Currently, most of the analysis methods for scRNA-seq data are commenced with unsupervised clustering, which highly relies on the investigator's background knowledge about the signature molecules, and is not efficient and accurate enough for the cell-type assignment of clusters [13]. Therefore, there is a demand for novel computational methods for classifying newly-generated scRNA-seq data onto annotated labels.

A variety of methods have recently been proposed for the cell-type classification of single-cell transcriptomes. For example, an unsupervised approach, named scmap [14], projects cells to the identified cell-types based on the similarity between query and reference cells. Conventional supervised learning-based methods, such as Random Forest and Support Vector Machine, though can be used for the cell-type classification, have been proven that their performance is not comparable to scmap. SuperCT, a supervised neural network framework, characterizes cell types of single-cell transcriptomic profiles with transformed binary features [13]. Nevertheless, there are still several limitations to be addressed in the proposed methods for the cell-type classification of single-cell transcriptomes. First, even the state-of-the-art methods have achieved encouraging performance, the classification performance and stability can still be further improved for various datasets and tasks as shown in the Results Section. Second, to select informative features of the query and reference datasets, there is a demand for a tailored feature selection approach for the cell-type classification. Third, both the query and reference datasets may contain cell types that have only a small number of samples. A superior classification method should be able to effectively characterize the cell types with only a small number of samples. Last but equally important, a method that can be generally applied to data of various dimensions or dropout rates is desirable for the broader application scenarios.

Motivated by the above understanding, we propose in this paper EnClaSC, a novel **en**semble approach for accurate and robust cell-type **cla**ssification of **s**ingle-**c**ell transcriptomes. Through comprehensive experiments, we illustrate that our method is superior to existing methods in not only the self-projection within a specific dataset, but also the cell-type classification across datasets. With a few-sample learning strategy, EnClaSC can effectively characterize the cell types that have only a small number of samples. We further show the robustness of our method for various data dimensionality and different data sparsity. Through joint analysis of classification results with scRNA-seq datasets of different species, we demonstrate the ability of our method to make the cross-species cell-type classification.

## Methods

### Design of EnClaSC

As illustrated in Fig. 1, EnClaSC consists of four modules. First, a feature selection module finds informative genes which benefit the cell-type classification from the

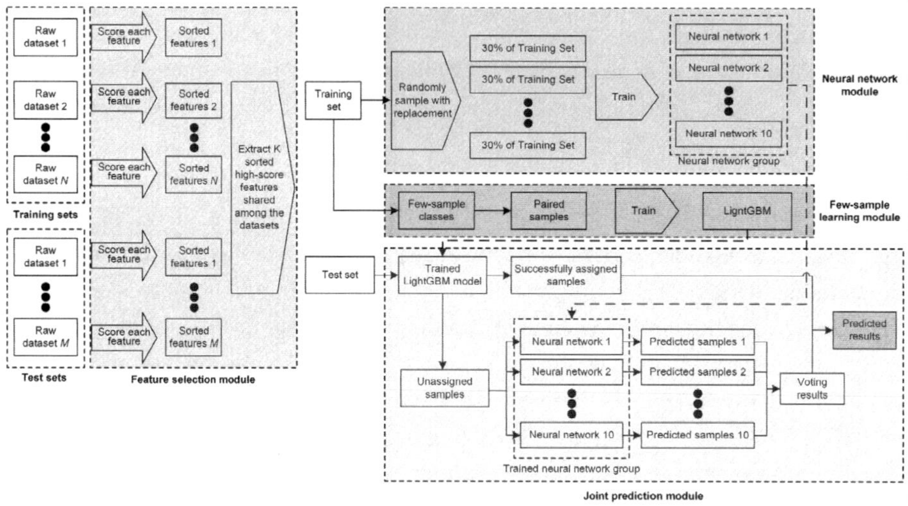


**Fig. 1** The graphical illustration of EnClaSC. EnClaSC consists of four modules, including feature selection, few-sample learning, neural network and joint prediction modules

common genes of the query and reference sets. Second, a few-sample learning module is adopted to sufficiently learn the characteristics of classes with few samples, and thus improve the performance in identifying the few-sample classes. Third, a neural network module uses artificial neural networks with an ensemble learning strategy to stabilize the performance of the cell-type classification. Finally, a joint prediction module integrates outputs of the few-sample learning and neural network modules to predict the class that a query cell belongs to. In general, EnClaSC draws on the idea of ensemble learning in the feature selection, few-sample learning, neural network and joint prediction modules, respectively, and thus constitutes a novel ensemble approach for cell-type classification of single-cell transcriptomes.

### Feature selection

Let $E$ be the value of gene expression in scRNA-seq, $n$ the number of the cells in the dataset. The dropout rate of the $j$ $th$ feature is abbreviated as $D_j$. For the $i$ $th$ cell (sample), the expression of the $j$ $th$ feature (gene) of is abbreviated as $E_{ij}$. For the $j$ $th$ feature, the level of gene expression $F(j)$ is calculated by the logarithm-plus-one of its arithmetic mean value as

$$F(j) = \log\left(\frac{\sum_{i=1}^{n} E_{ij}}{n} + 1\right)$$

For the estimation of gene expression, we first use the dropout rate to fit a linear model using the least square method

$$\tilde{F}_1(j) = aD_j + b$$

We then estimate $F(j)$ by the mean of the logarithm-plus-one of each feature, namely,

$$\tilde{F}_2(j) = \frac{\sum_{i=1}^{n} log\left(E_{ij} + 1\right)}{n}$$

The residuals of these two approaches between the true gene expression level are recorded as $\Delta F_1(j)$ and $\Delta F_2(j)$, respectively. Briefly, $\Delta F_1(j) = F(j) - \tilde{F}_1(j)$, and $\Delta F_2(j) = F(j) - \tilde{F}_2(j)$. $\Delta F_1(j)$ reflects the rate of gene expression, that is, whether the gene can be expressed or not. $\Delta F_2(j)$ is the information entropy of each gene, which reflects the residual degree of each gene expression. We can use the integrated score $G(j) = \alpha\Delta F_1(j) + (1 - \alpha)\Delta F_2(j)$ to consider the above two factors synthetically. Here, $\alpha$ is the control coefficient, which is responsible for regulating the influence of the two calculated factors. We set $\alpha$ to 0.5 in all the experiments of this work.

For a specific scRNA-seq dataset, we compute $G(j)$ as the scores of features, and then sort the features according to the scores in descending order. We extract $K$ sorted high-score features shared among the training and test sets. This feature selection approach can effectively maximize the common high-score features of different scRNA-seq datasets, and thus contribute to the cell-type classification across datasets.

### Few-sample learning strategy

We define few-sample classes as the cell types whose number of samples does not exceed 0.5% of the total number of training samples. In order to fully extract features of the few-sample classes, we perform data augmentation and pre-train a few-sample classification model using samples of the few-sample classes in the training set. In more detailed, let $N$ be the number of samples in the training set, n the number of samples of the few-sample classes in the training set, $M$ the number of samples in the test set, and $D$ the number of features after feature selection. We pair the samples of these few-sample classes one by one to form a few-sample training set with $n \times n$ samples and $2 \times D$ features. If the two cells of a paired sample belong to the same cell type, we mark the label of the paired sample as 1; otherwise, we mark it as 0. We then pair the samples of the few-sample classes in the original training set with each sample in the original test set to form a few-sample test set with $n \times M$ samples and $2 \times D$ features.

We adopt LightGBM, a gradient boosting framework that uses tree-based learning algorithms, to perform the few-sample training [15]. We use the default setting of parameters except for the parameters listed in Table 1. To obtain the probability scores that an original test sample belongs to the few-sample classes, we weight the prediction results of the few-sample test set by the *Pearson* correlation coefficients between the expression values of the two cells in corresponding few-sample test samples drawing on the idea of ensemble learning, namely,

$$score(x, i) = \frac{\sum_{j=1}^{N_x} r_{ij} c_{ij}}{\sum_{j=1}^{N_x} r_{ij}}$$

**Table 1** Parameters of the LightGBM model

| Parameters | Setting |
|---|---|
| boosting_type | 'gbdt' |
| objective | 'regression' |
| metric | {'l2', 'auc'} |
| learning_rate | 0.05 |
| verbose | 1 |

where $x$ is one of the classes in the original training set, $N_x$ the number of samples of class $x$ in the original training set, $j$ the $j$ $th$ sample of class $x$ in the original training set, $i$ the $i$ $th$ sample in the original test set, and $r_{ij}$ the *Pearson* correlation coefficients between $i$ and $j$, $c_{ij}$ the prediction result of LightGBM for the paired sample of $i$ and $j$.

If the predicted maximum score of an original test sample is greater than $\gamma$, then we assign the sample as the class that has the maximum score, otherwise, the sample is predicted as "unassigned" and should be further classified using the subsequent artificial neural networks.

### Artificial neural networks for the cell-type classification

The neural network module uses artificial neural networks with ensemble learning strategy to classify the test samples which are predicted as "unassigned" by the few-sample learning module. We first design an artificial neural network framework with parameters as shown in Table 2, and implement it using Keras with Tensorflow as the backend. In order to improve the stability of the classification performance, drawing on the Bootstrap strategy, we generate 10 new training sets by selecting 30% of the original training set at random with replacement, and then use them to train 10 neural networks, respectively. For each trained neural network, if the predicted maximum score of a test sample exceeds $\gamma$, the sample is assigned as the class that has the maximum score, otherwise, it is classified as "unassigned". We set $\gamma$ to 0.7 in all the experiments of this work. For each test sample, we classify it as "unassigned" unless there are more than half of the 10 neural networks predict the test sample as a specific same class, and we thus classify the sample based on the voting result of the 10 neural networks.

### Assessment of performance

We adopt the widely used *kappa* value and *assigned rate* value to assess the classification performance of the models. Briefly, we calculate the *kappa* values of the classification results of assigned samples and the real classes of corresponding samples using the following formula

$$kappa = \frac{p_o - p_e}{1 - p_e}$$

where $p_0$ is the sum of the number of samples correctly classified divided by the number of assigned samples in the test set, that is, the overall classification accuracy of assigned samples in the test set. Assuming that the number of real samples which are assigned by the model in each class of test set is $a_1$, $a_2 ... a_c$ respectively, the number of

**Table 2** The architecture of the neural network

| Layers | Setting |
|---|---|
| Dense_1 | {(feature numbers,128), 'relu'} |
| Dropout_1 | 0.25 |
| Dense_2 | {(128,64), 'relu'} |
| Dropout_2 | 0.5 |
| Dense_3 | {(64,32), 'relu'} |
| Dense_4 | {(32, number of classes in training sets), 'softmax'} |

predicted samples in each class of the test set is $b_1$, $b_2$...$b_C$ respectively, and the number of assigned samples in the test set is $n$. We have

$$p_e = \frac{a_1 \times b_1 + a_2 \times b_2 + ... + a_C \times b_C}{n \times n}$$

For the assessment of the ability of a model to recognize the cells in the test set, we assume that the number of samples in the test set is $N$ and use *assigned rate* to represent the proportion of the number of samples that can be assigned to a cell type by the classifier, namely,

$$assigned\ rate = \frac{b_1 + b_2 + ... + b_C}{N}$$

Because the classes of test sets and training sets may not be completely consistent, the *assigned rate* should be maintained at a relatively high level, but not the higher the better, and thus the classification performance is primarily related to *kappa* values.

### Data collection

We downloaded three humans pancreatic scRNA-seq dataset provided by Baron, M. et al., Muraro, MJ et al., and Xin, Y. et al. (hereinafter are abbreviated as Baron Dataset, Muraro Dataset, and Xin Dataset, respectively) from NCBI Gene Expression Omnibus via accession GSE84133, GSE85241, and GSE81608, respectively [16–18]. The human pancreatic scRNA-seq dataset provided by Segerstolpe, Å. et al. (hereinafter is abbreviated as Segerstolpe Dataset) was downloaded from EMBL-EBI ArrayExpress via accession E-MTAB-5061 [5]. For the subsequent usability of the dataset, we removed the cells of the "unclear" class in Muraro Dataset, the cells of the "not applicable", "unclassified endocrine" and "unclassified" classes in Segerstolpe Dataset and the cells of the "alpha.contaminated", "beta.contaminated", "delta.contaminated", and "gamma.contaminated" classes in Xin Dataset. After preprocessing, Baron Dataset contains 8569 samples with 20, 125 features, Muraro Dataset contains 2122 samples with 19, 127 features, Xin Dataset contains 1492 samples with 39, 851 features, and Segerstolpe Dataset contains 2166 samples with 25, 525 features.

We also downloaded two mouse retina scRNA-seq datasets provided by Macosko, EZ et al. and Shekhar, K. et al., (hereinafter are abbreviated as Macosko Dataset and Shekhar Dataset, respectively) from NCBI Gene Expression Omnibus via accession GSE63473 and GSE81904 [2, 19]. Macosko Dataset measures the expression of 23, 288 genes in 44, 808 cells, while Shekhar Dataset measures the expression of 13, 166 genes in 27, 499 cells.

Two mouse brain cell scRNA-seq datasets provided by Romanov, RA et al. and Zeisel, A. et al. (hereinafter are abbreviated as Romanov Dataset and Zeisel Dataset, respectively) and one human brain cell scRNA-seq dataset collected by Darmanis, S. et al. (hereinafter is abbreviated as Darmanis Dataset) were downloaded from the NCBI Gene Expression Omnibus via accession GSE74672, GSE60361 and GSE67835, respectively [9, 20, 21]. To unify the label information, we replaced the cell type label 'oligos' in the Remanov Dataset with 'oligodendrocytes', and the cell type labels 'ca1pyramidal', 's1pyramidal' and 'interneurons' in the Zeisel Dataset with 'neurons'. After preprocessing, Romanov Dataset contains 2881 samples with 24, 341 features, Zeisel Dataset contains

3005 samples with 19, 972 features, and Darmanis Dataset contains 466 samples with 22, 088 features.

## Results

### EnClaSC achieves high performance for self-projection within a dataset

In order to illustrate that our method can effectively perform self-projection within a scRNA-seq dataset, we conducted a series of self-projection experiments using 6 datasets, including Baron Dataset, Muraro Dataset, Xin Dataset, Segerstolpe Dataset, Macosko Dataset and Shekhar Dataset. We compared the performance of our method with scmap, a similarity-based method, and SuperCT, an ANN-based method (we self-implemented the method according to the paper because SuperCT is not open source) [13, 14]. Using the same training (reference) and test (query) sets with EnClaSC, we run the two baseline methods with parameters or structures proposed by the respective authors.

Both EnClaSC and scmap provide a feature selection method, while SuperCT uses all features. When running EnClaSC and scmap, we selected 100 features, which is considered to be able to better demonstrate the advantages of scmap in their paper. We performed 5-fold cross-validation within each of the six datasets. As shown in Fig. 2, EnClaSC model has better self-projection performance than SuperCT and scmap. At the same time, in terms of self-projection stability, EnClaSC is significantly superior to SuperCT. The variance of the *kappa* value of EnClaSC is 71.09% lower than SuperCT, and the variance of the *assigned rate* of EnClaSC is 68.03% lower than SuperCT.

### EnClaSC outperforms other methods in the cell-type classification across datasets

We use six scRNA-seq datasets to demonstrate the superior cell-type classification performance of EnClaSC. Using Baron Dataset, Muraro Dataset, Xin Dataset, and Segerstolpe Dataset these four human pancreatic datasets, we selected each one of them as the test set, while the integration of remaining three datasets serves as the training set, respectively. In addition, we used two mouse retina datasets, namely, Macosko Dataset and Shekhar Dataset, to form two symmetric training-test sets. We used 100 selected features in EnClaSC and scmap, while all features in SuperCT, and repeated each experiment five times.

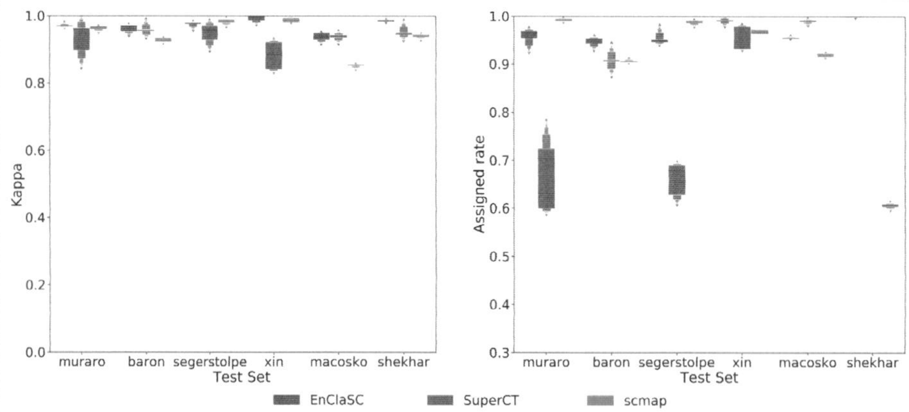


**Fig. 2** Performance comparison of the self-projection within a specific dataset

As shown in Fig. 3, the *kappa* value of EnClaSC is higher than that of scmap given comparable *assigned rates*, except for the group where Macosko Dataset serve as the test set and scmap achieves much lower *assigned rates*. Compared with SuperCT, which also uses a neural network architecture, *assigned rates* of EnClaSC has little difference except for Baron Dataset as the test set, while most *kappa* values of EnClaSC outperform SuperCT. When Baron Dataset serves as the test set, the *assigned rates* of SuperCT is much lower than that of scmap and EnClaSC. In addition, it can be seen from the figure that the classification performance of EnClaSC is much more stable than that of SuperCT. Therefore, even though EnClaSC only uses 100 features, the classification performance of EnClaSC is still far better than that of SuperCT. In general, EnClaSC has a more outstanding performance in cell-type classification compared with other two base-line methods.

## Contribution of each module

### *Feature selection module*

We validated the contribution of the feature selection module using the four human pancreatic datasets including Baron Dataset, Muraro Dataset, Xin Dataset and Segerstolpe Dataset, and two mouse retina datasets including Macosko Dataset and Shekhar Dataset. We performed six experiments as the above section. We first took each one of the four human pancreas datasets as a test set and integrated the remaining three datasets as a training set. Then we took each one of the two mouse retina datasets as a test set, and the remaining one was served as a training set. We replaced the feature selection method of our feature selection module with PCA, which is commonly used for scRNA-seq dimensionality reduction, feature selection methods of Seurat v3.0 and that of scmap, and compared them with our original method [22].

The results are shown in Fig. 4. Compared to PCA, our feature selection method is more stable. The *kappa* and the *assigned rate* of EnClaSC is much better than the EnClaSC framework using the feature selection method of Seurat v3.0 (the average of *kappa* values of EnClaSC is 69.65% higher than Seurat v3.0, and the average of *assigned rates* is 3.95% higher than Seurat v3.0). EnClaSC has slightly higher *kappa* values with comparable *assigned rates* compared with the EnClaSC framework using the feature

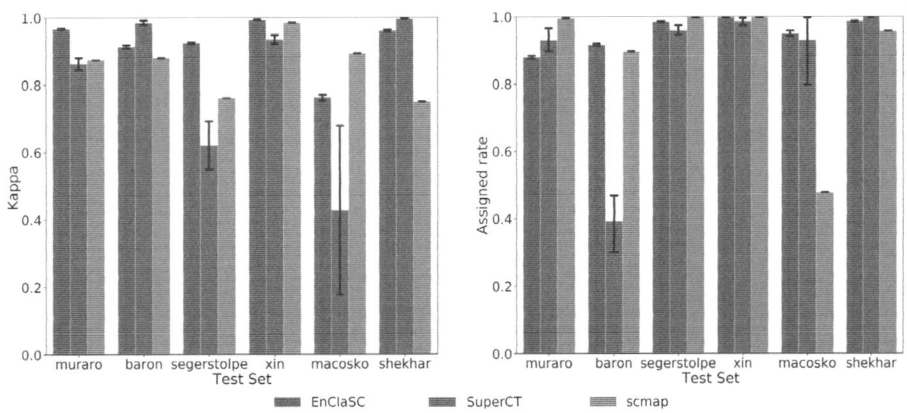


**Fig. 3** Performance comparison of the cell-type classification across different datasets

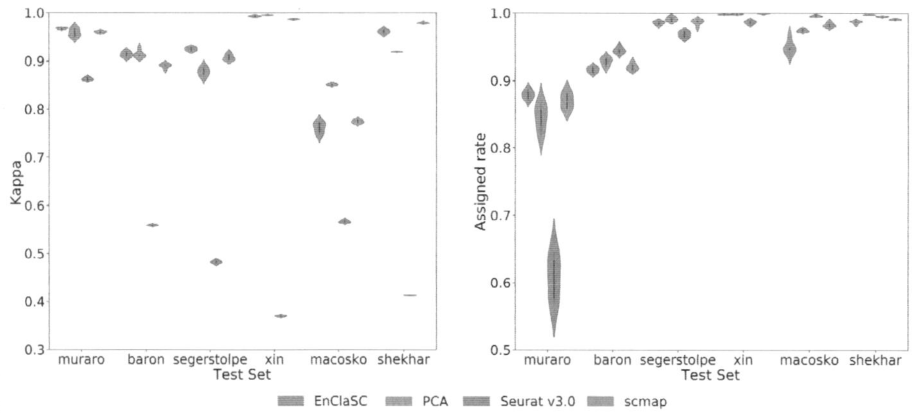


**Fig. 4** Performance of EnClaSC using different feature selection strategy

selection method of scmap. In summary, the feature selection module of EnClaSC is superior to other widely used feature selection approaches in the EnClaSC framework, and thus serves as the effective feature selection approach for the cell-type classification.

### *Neural network module*

We validated the contribution of the neural network module using the four human pancreatic datasets and the two mouse retina datasets, and we again performed six experiments as the "*Feature selection module*" section. We replaced the neural networks in EnClaSC with a single neural network that does not use the ensemble learning strategy. We compared the performance of the above approach with that of our complete model, and show the results in Fig. 5. With 64.14% reduced variance of *kappa* and 71.97% reduced variance of *assigned rate* on average, the main contribution of ensemble learning strategy with several neural networks is significantly improving the stability of classification performance. Besides, on most datasets, the ensemble learning strategy slightly improves the classification performance.

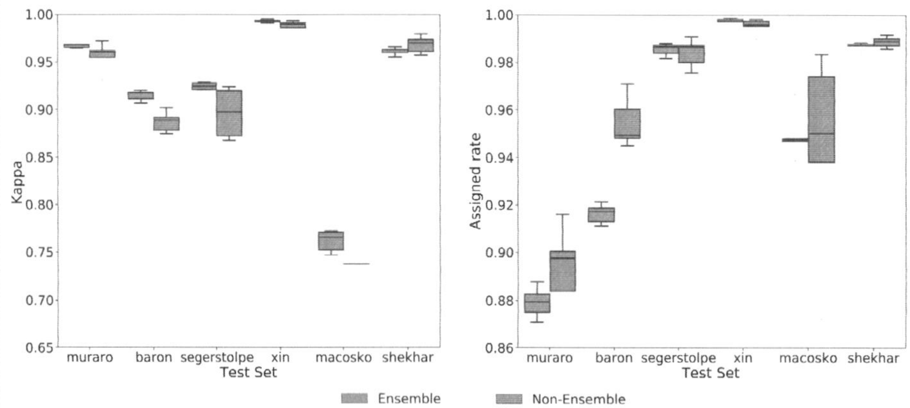


**Fig. 5** Performance of EnClaSC with or without ensemble learning in the neural network module

#### *Few-sample learning module*

To demonstrate the contribution of the few-sample learning module, we compare the performance of EnClaSC with or without the few-sample learning module. We first used Shekhar Dataset as the training set, Macosko Dataset as the test set to evaluate the contribution of the few-sample learning module. This training-test set has a very distinct characteristic compared to other sets we have used, because there are massive samples of few-sample classes of the training set in the test set, which makes the classification task much more challenging. According to the definition in the Methods Section, we consider "rods" and "cones" as the few-sample classes in the training set. These two cell types account for 0.34 and 0.18% respectively in the training set, and account for 65.61 and 4.17% respectively in the test set.

As shown in Fig. 6a and b, EnClaSC with the few-sample learning module can identify 66.34% samples in the test set with the classification accuracy rate of 95.92%. However, the classification accuracy rate of EnClaSC without the few-sample learning module is only 87.33%. As shown in Fig. 6a and b, the number of incorrectly classified samples of the "rods" class is less than without the few-sample classification model, and the samples of the "cones" class in the test set are all classified incorrectly.

We also evaluated the contribution of the few-sample learning module using training and test sets whose cell types have similar sample distributions. We integrated Baron Dataset, Muraro Dataset, and Xin Dataset as the training set, and used Segerstolpe Dataset as the test set, and Segerstolpe Dataset, Muraro Dataset and Xin Dataset as training sets, Baron Dataset as the test set to form two training-test sets for the experiment.

As shown in Fig. 6c-f, the training of few-sample classes in the neural network is insufficient with limited samples, resulting in the worse classification performance, while the few-sample learning module can effectively characterize the few-sample classes, and thus improve the classification performance.

In addition, we found that in the Segerstolpe Dataset, which contains "MHC class II" cell type but not 'macrophage' cell type. In the case that Baron Dataset or Segerstolpe Dataset serves as the test set, the few-sample learning module can associate the 'macrophage' class with the "MHC class II" class as shown in Fig. 6c-f. The data showed that "MHC class II" is a protein that is mainly secreted by the macrophage cell, indicating that the few-sample learning module possesses the potential ability to recognize new cell types and assign them to relative cell types. In summary, the few-sample learning module endows EnClaSC the ability to effectively resolve few-sample classes of the training set, while providing the potential ability to correlate new cell types with known relative cell types.

### EnClaSC scales up well with various data dimensionality

To illustrate the robustness of our method for various data dimensionality, we selected different numbers of features among the four human pancreatic datasets including Muraro Dataset, Baron Dataset, Segerstolpe Dataset, and Xin Dataset. We took each one of the four human pancreas datasets as a test set and integrated the remaining three datasets as a training set, and repeated each experiment five times. In order to illustrate the superior data-dimension adaptability of our method, we selected 100, 300, 500, 700 and 900 as the feature numbers of our method and scmap. It is worth noting that instead of designing a feature selection module,

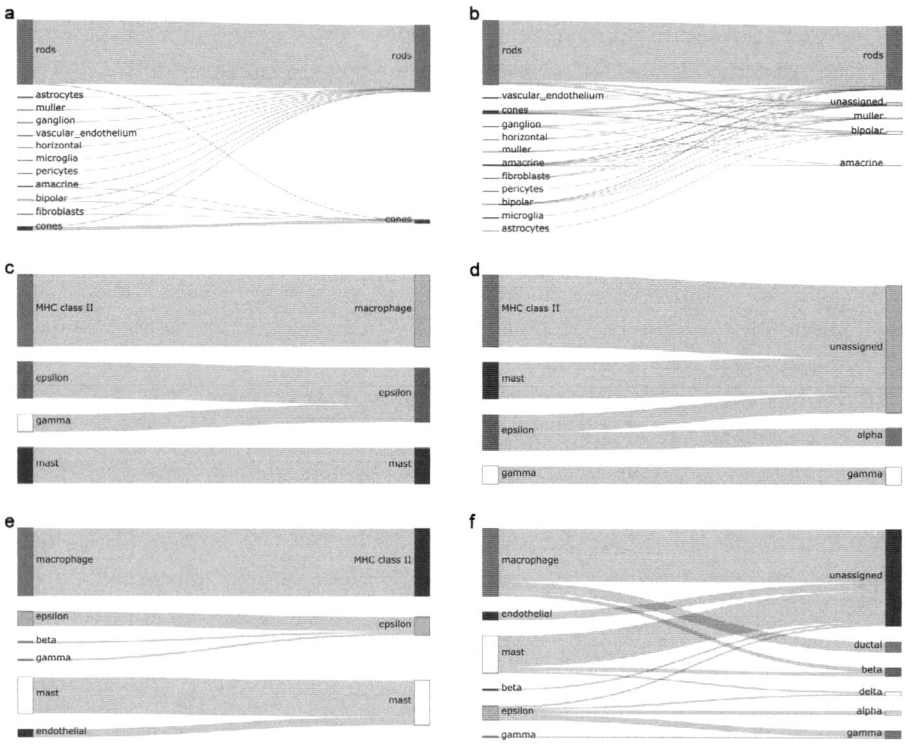


**Fig. 6** Performance of EnClaSC with or without the few-sample learning module. **a** The results of EnClaSC with few-sample learning module on Macosko Dataset. **b** The results of EnClaSC without few-sample learning module on Macosko Dataset. **c** The results of EnClaSC with few-sample learning module on Segerstolpe Dataset. **d** The results of EnClaSC without few-sample learning module on Segerstolpe Dataset. **e** The results of EnClaSC with few-sample learning module on Baron Dataset. **f** The results of EnClaSC without few-sample learning module on Baron Dataset

SuperCT directly binarizes all features, and thus uses all features in these experiments.

As shown in Fig. 7, the performance of EnClaSC slightly improves with the increase of data dimensionality. When the data dimensionality is 300, the *kappa* value of scmap rises to the peak and remains basically unchanged, while the *assigned rates* of scmap drops dramatically with the increase of data dimensionality. At the same time, we can see that when the data dimensionality reaches 500 or becomes higher, the *kappa* values and the *assigned rate* of EnClaSC are consistently higher than that of scmap and SuperCT. In summary, EnClaSC scales up well with various data dimensionality, and the higher the data dimensionality, the better the performance of EnClaSC.

### EnClaSC scales up well with different data sparsity

Considering that scRNA-seq data suffer from the dropout phenomenon, we further demonstrated that our method can scale up well with different data sparsity using the four human pancreatic datasets to perform four experiments. We randomly set 0, 10, 20, 30, 40, and 50% of the non-zero elements in the raw expression matrix, and repeated each experiment five times. As illustrated in Fig. 8, EnClaSC is much more

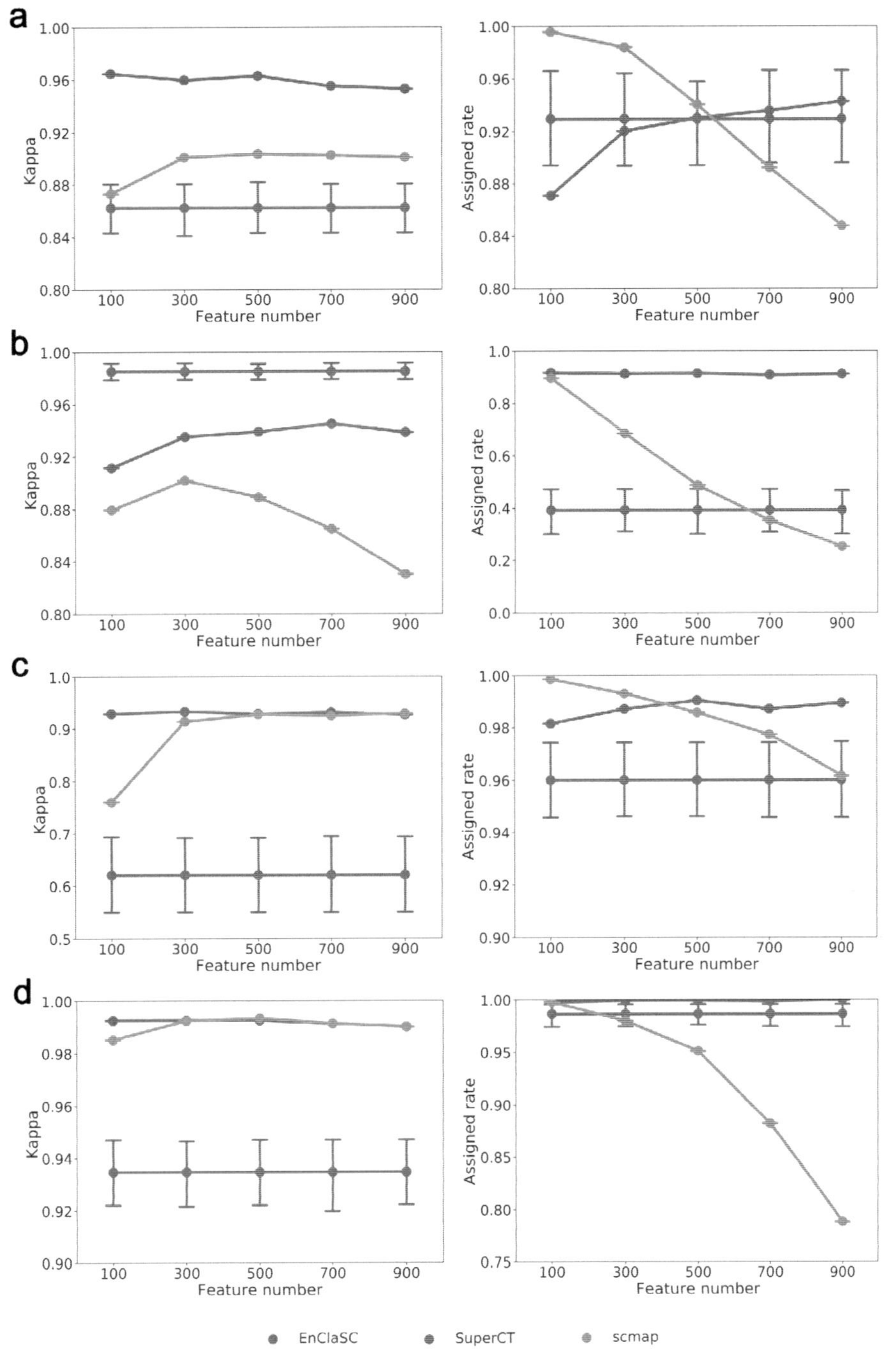


**Fig. 7** Performance comparison on datasets of various dimensionality. **a-d** The performance on Muraro Dataset, Baron Dataset, Segerstolpe Dataset and Xin Dataset, respectively

robust to the dropout rate compared with other two methods. The *kappa* values hardly change with the increasing of the dropout rate, even though the *assigned rate* decrease slightly. Nevertheless, the performance of scmap becomes worse because the randomly zeroing of the feature matrix could cause significant changes in cell-to-cell similarity. The performance of SuperCT fluctuates irregularly with the increasing dropout rate. In

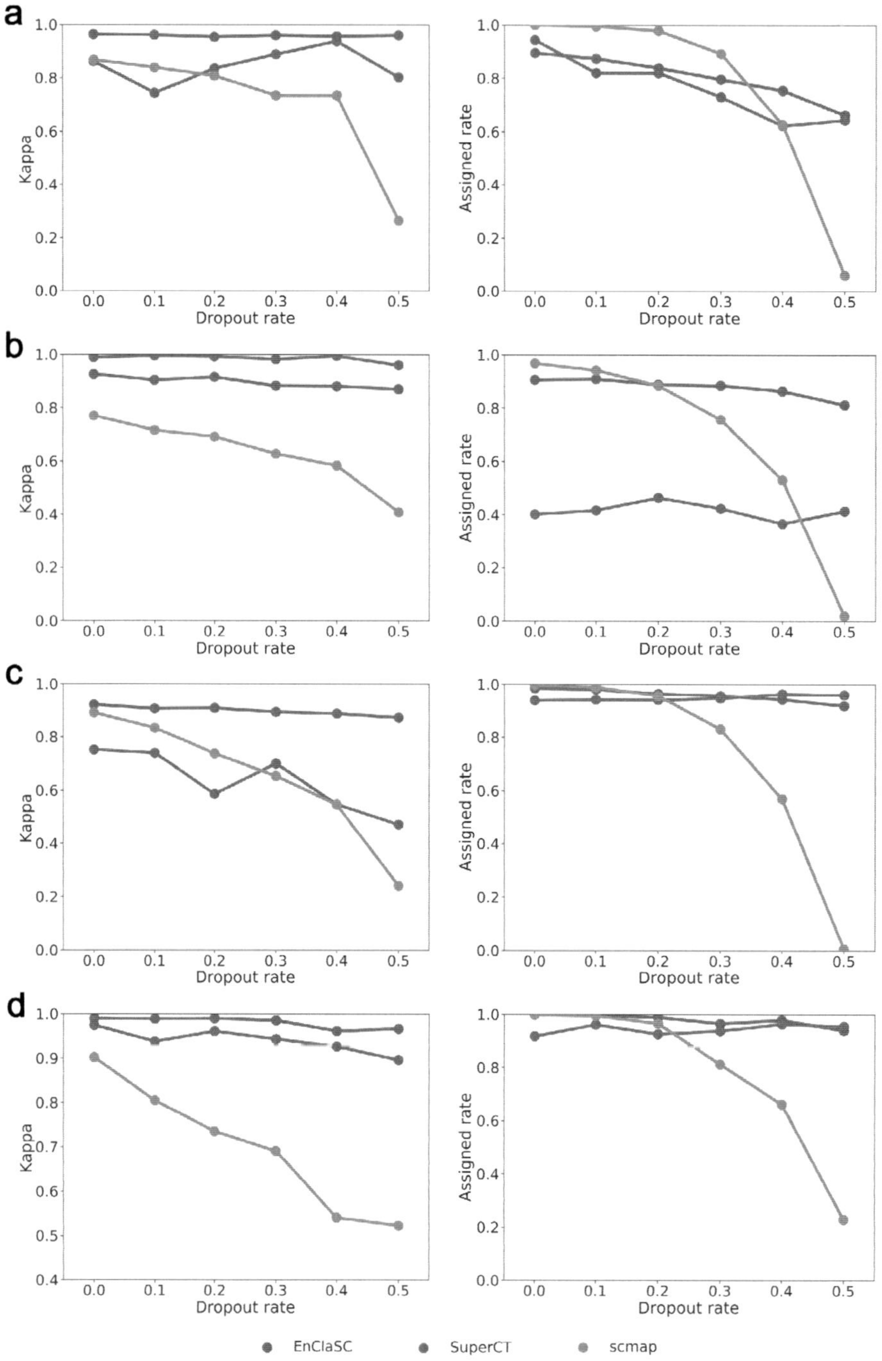


**Fig. 8** Performance comparison on datasets of different sparsity. **a-d** The performance on Muraro Dataset, Baron Dataset, Segerstolpe Dataset and Xin Dataset, respectively

some cases, the *kappa* value may decrease, which indicating that the increasing dropout rate could cause further deterioration of the stability of SuperCT. The results demonstrate that EnClaSC can account for the sparsity of single-cell gene expression data, and thus benefits the cell-type classification tasks with the exponential growth of the Drop-seq based scRNA-seq data.

### EnClaSC enables cross-species classification

To illustrate that EnClaSC can effectively make **cross-species** classification, we used datasets of mouse brain cells provided by Romanov, R. A. et al. (Romanov Dataset for short) and Zeisel, A. et al. (Zeisel Dataset for short) as the training set, and a dataset of human brain cells provided by Darmanis, S. et al. (Darmanis Dataset for short) as the test set for the cross-species classification. Because there are only 466 samples on Darmanis Dataset, we did not use the Darmanis Dataset as the training set. As shown in Fig. 9, our method not only achieves the highest *assigned rate* (37.83% higher than scmap, and 2.54% higher than SuperCT), but also has the highest *kappa* (79.13% higher than scmap and 80.06% higher than SuperCT) in this cross-species classification task. Similarly, using our feature selection module in EnClaSC also provides the highest *kappa* value (211.60% higher than PCA, 11.89% higher than Seurat v3.0, and 4.94% higher than scmap), and *assigned rates* (0.51% higher than PCA, 2.84% higher than Seurat v3.0, and 2.46% higher than scmap). The results demonstrate that EnClaSC has superior classification performance both in the classification of homologous cells of the same species and in the classification of homologous cells of different species.

## Discussion

scRNA-Seq techniques have advanced rapidly in recent years and enable the quantitative characterization of cell types at a single-cell resolution. EnClaSC is proposed to classify cell types using supervised learning. With the contribution of the tailored feature selection, neural network and few-sample learning modules, our method is superior to other baseline methods, such as scmap and SuperCT, with regard to not only accuracy and robustness, but also the performance of classifying few-sample classes.

Our method can certainly be improved in some aspects. First, a more well-crafted module, such as modules considering dropout evens, can be introduced to better characterize the scRNA-seq data, and thus further improves the performance of our method. Second, our method can be extended to incorporate other types of functional genomics data such as chromatin accessibility [23, 24]. Finally, drawing on the idea of VPAC, we can integrate the feature selection module with other modules to endow the method with the ability to balance the feature selection and prediction steps, and thus extract features that are more conducive to the cell type classification [25].

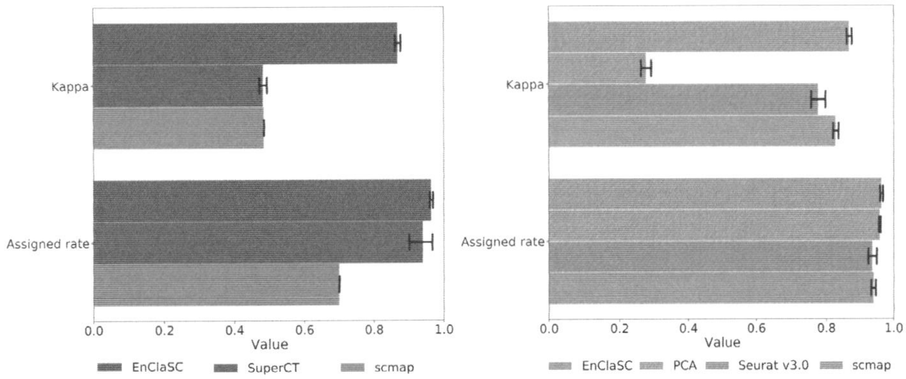


**Fig. 9** Performance comparison of cross-species classification

## Conclusions

We have proposed a supervised learning method, named EnClaSC, for accurate and robust cell-type classification of single-cell transcriptomes. Each of the well-crafted feature selection, neural network and few-sample learning modules draws on the idea of ensemble learning, which makes EnClaSC superior to existing methods in the self-projection within a specific scRNA-seq dataset, the cell-type classification across different scRNA-seq datasets, various data dimensionality, and different data sparsity. We have further demonstrated the ability of EnClaSC to effectively make cross-species classification, which may shed light on the studies in the correlation of different species. Eventually, we expect that such a supervised learning approach will be widely applicable for the cell-type classification with the explosive growth of scRNA-seq data.

### Abbreviations

scRNA-seq: single-cell RNA-sequencing; ANN: Artificial neural network; PCA: Principal component analysis

### Acknowledgements

Rui Jiang is a RONG professor at the Institute for Data Science, Tsinghua University.

### About this supplement

This article has been published as part of *BMC Bioinformatics Volume 21 Supplement 13, 2020: Selected articles from the 18th Asia Pacific Bioinformatics Conference (APBC 2020): bioinformatics*. The full contents of the supplement are available online at https://bmcbioinformatics.biomedcentral.com/articles/supplements/volume-21-supplement-13 .

### Authors' contributions

RJ designed the research. XC and SC designed and implemented the models. XC collected data and analyzed the results. XC, SC and RJ wrote the manuscript. All authors read and confirmed the manuscript.

### Funding

Publication costs are funded by the National Key Research and Development Program of China (No. 2018YFC0910404), the National Natural Science Foundation of China (Nos. 61873141, 61721003, 61573207), Shanghai Municipal Science and Technology Major Project (No. 2017SHZDZX01), and the Tsinghua-Fuzhou Institute for Data Technology.

### Availability of data and materials

The datasets supporting the conclusions of this article are publicly available from the NCBI Gene Expression Omnibus (https://www.ncbi.nlm.nih.gov/geo/), the NCBI Sequence Read Archive (http://www.ncbi.nlm.nih.gov/Traces/sra/), and the EMBL-EBI ArrayExpress (https://www.ebi.ac.uk/arrayexpress/).

### Ethics approval and consent to participate

Not applicable.

### Consent for publication

Not applicable.

### Competing interests

The authors declare that they have no competing interests.

Published: 17 September 2020

### References

1. Jaitin DA, Kenigsberg E, Keren-Shaul H, Elefant N, Paul F, Zaretsky I, Mildner A, Cohen N, Jung S, Tanay A, et al. Massively parallel single-cell RNA-Seq for marker-free decomposition of tissues into cell types. Science. 2014;343(6172):776–9.
2. Macosko EZ, Basu A, Satija R, Nemesh J, Shekhar K, Goldman M, Tirosh I, Bialas AR, Kamitaki N, Martersteck EM, et al. Highly parallel genome-wide expression profiling of individual cells using Nanoliter droplets. Cell. 2015;161(5):1202–14.
3. Picelli S, Bjorklund AK, Faridani OR, Sagasser S, Winberg G, Sandberg R. Smart-seq2 for sensitive full-length transcriptome profiling in single cells. Nat Methods. 2013;10(11):1096–8.
4. Grun D, Lyubimova A, Kester L, Wiebrands K, Basak O, Sasaki N, Clevers H, van Oudenaarden A. Single-cell messenger RNA sequencing reveals rare intestinal cell types. Nature. 2015;525(7568):251.
5. Segerstolpe A, Palasantza A, Eliasson P, Andersson EM, Andreasson AC, Sun XY, Picelli S, Sabirsh A, Clausen M, Bjursell MK, et al. Single-cell Transcriptome profiling of human pancreatic islets in health and type 2 diabetes. Cell Metab. 2016; 24(4):593–607.
6. Jaitin DA, Weiner A, Yofe I, Lara-Astiaso D, Keren-Shaul H, David E, Salame TM, Tanay A, van Oudenaarden A, Amit I. Dissecting Immune Circuits by Linking CRISPR-Pooled Screens with Single-Cell RNA-Seq. Cell. 2016;167(7):1883.
7. Xue ZG, Huang K, Cai CC, Cai LB, Jiang CY, Feng Y, Liu ZS, Zeng Q, Cheng LM, Sun YE, et al. Genetic programs in human and mouse early embryos revealed by single-cell RNA sequencing. Nature. 2013;500(7464):593.
8. Scialdone A, Natarajan KN, Saraiva LR, Proserpio V, Teichmann SA, Stegle O, Marioni JC, Buettner F. Computational assignment Of cell-cycle stage from single-cell transcriptome data. Methods. 2015;85:54–61.

9. Zeisel A, Munoz-Manchado AB, Codeluppi S, Lonnerberg P, La Manno G, Jureus A, Marques S, Munguba H, He L, Betsholtz C, et al. Cell types in the mouse cortex and hippocampus revealed by single-cell RNA-seq. Science. 2015; 347(6226):1138–42.
10. Marco E, Karp RL, Guo GJ, Robson P, Hart AH, Trippa L, Yuan GC. Bifurcation analysis of single-cell gene expression data reveals epigenetic landscape. Proc Natl Acad Sci U S A. 2014;111(52):E5643–50.
11. Satija R, Farrell JA, Gennert D, Schier AF, Regev A. Spatial reconstruction of single-cell gene expression data. Nat Biotechnol. 2015;33(5):495–U206.
12. Treutlein B, Brownfield DG, Wu AR, Neff NF, Mantalas GL, Espinoza FH, Desai TJ, Krasnow MA, Quake SR. Reconstructing lineage hierarchies of the distal lung epithelium using single-cell RNA-seq. Nature. 2014;509(7500):371.
13. Xie P, Gao M, Wang C, Zhang J, Noel P, Yang C, Von Hoff D, Han H, Zhang MQ, Lin W. SuperCT: a supervised-learning framework for enhanced characterization of single-cell transcriptomic profiles. Nucleic Acids Res. 2019;47(8):e48.
14. Kiselev VY, Yiu A. Hemberg M: scmap: projection of single-cell RNA-seq data across data sets. Nat Methods. 2018;15(5): 359–62.
15. Ke G, Meng Q, Finley T, Wang T, Chen W, Ma W, Ye Q, Liu T-Y. LightGBM: A Highly Efficient Gradient Boosting Decision Tree. In: Guyon I, Luxburg UV, Bengio S, Wallach H, Fergus R, Vishwanathan S, Garnett R, editors. *Advances in Neural Information Processing Systems 30*, vol. 30; 2017.
16. Baron M, Veres A, Wolock SL, Faust AL, Gaujoux R, Vetere A, Ryu JH, Wagner BK, Shen-Orr SS, Klein AM, et al. A Single-Cell Transcriptomic Map of the Human and Mouse Pancreas Reveals Inter- and Intra-cell Population Structure. Cell Syst. 2016;3(4):346.
17. Muraro MJ, Dharmadhikari G, Gruen D, Groen N, Dielen T, Jansen E, van Gurp L, Engelse MA, Carlotti F, de Koning EJP, et al. A Single-Cell Transcriptome Atlas of the Human Pancreas. Cell Syst. 2016;3(4):385.
18. Xin Y, Kim J, Okamoto H, Ni M, Wei Y, Adler C, Murphy AJ, Yancopoulos GD, Lin C, Gromada J. RNA sequencing of single human islet cells reveals type 2 diabetes genes. Cell Metab. 2016;24(4):608–15.
19. Shekhar K, Lapan SW, Whitney IE, Tran NM, Macosko EZ, Kowalczyk M, Adiconis X, Levin JZ, Nemesh J, Goldman M, et al. Comprehensive Classification of Retinal Bipolar Neurons by Single-Cell Transcriptomics. Cell. 2016;166(5):1308.
20. Darmanis S, Sloan SA, Zhang Y, Enge M, Caneda C, Shuer LM, Gephart MGH, Barres BA, Quake SR. A survey of human brain transcriptome diversity at the single cell level. Proc Natl Acad Sci U S A. 2015;112(23):7285–90.
21. Romanov RA, Zeisel A, Bakker J, Girach F, Hellysaz A, Tomer R, Alpar A, Mulder J, Clotman F, Keimpema E, et al. Molecular interrogation of hypothalamic organization reveals distinct dopamine neuronal subtypes. Nat Neurosci. 2017; 20(2):176–88.
22. Stuart T, Butler A, Hoffman P, Hafemeister C, Papalexi E, Mauck WM, Stoeckius M, Smibert P, Satija R. Comprehensive integration of single cell data. bioRxiv. 2018;177(7):1888.
23. Chen S, Wang Y, Jiang R. OPENANNO: annotating genomic regions with chromatin accessibility. bioRxiv. 2019.
24. Duren ZN, Chen X, Zamanighomi M, Zeng WW, Satpathy AT, Chang HY, Wang Y, Wong WH. Integrative analysis of single-cell genomics data by coupled nonnegative matrix factorizations. Proc Natl Acad Sci U S A. 2018;115(30):7723–8.
25. Chen SQ, Hua K, Cui HF, Jiang R. VPAC: Variational projection for accurate clustering of single-cell transcriptomic data. Bmc Bioinformatics. 2019;20:0.

## Publisher's Note

Lü et al. BMC Bioinformatics 2020, 21(Suppl 13):391
https://doi.org/10.1186/s12859-020-03680-6

BMC Bioinformatics

METHODOLOGY **Open Access**

# Spark-based parallel calculation of 3D fourier shell correlation for macromolecule structure local resolution estimation

Yongchun Lü[1,2*], Xiangrui Zeng[3], Xinhui Tian[1], Xiao Shi[1], Hui Wang[1], Xiaohui Zheng[1,2], Xiaodong Liu[1], Xiaofang Zhao[1,2], Xin Gao[4] and Min Xu[3*]

*From* The 18th Asia Pacific Bioinformatics Conference
Seoul, Korea. 18–20 August 2020

*Correspondence:
lvyongchun@ncic.ac.cn;
mxu1@cs.cmu.edu
[1]Institute of Computing Technology of the Chinese Academy of Sciences, Beijing, China
[3]Computational Biology Department, School of Computer Science, Carnegie Mellon University, Pittsburgh, USA
Full list of author information is available at the end of the article

## Abstract

**Background:** Resolution estimation is the main evaluation criteria for the reconstruction of macromolecular 3D structure in the field of cryoelectron microscopy (cryo-EM). At present, there are many methods to evaluate the 3D resolution for reconstructed macromolecular structures from Single Particle Analysis (SPA) in cryo-EM and subtomogram averaging (SA) in electron cryotomography (cryo-ET). As global methods, they measure the resolution of the structure as a whole, but they are inaccurate in detecting subtle local changes of reconstruction. In order to detect the subtle changes of reconstruction of SPA and SA, a few local resolution methods are proposed. The mainstream local resolution evaluation methods are based on local Fourier shell correlation (FSC), which is computationally intensive. However, the existing resolution evaluation methods are based on multi-threading implementation on a single computer with very poor scalability.

**Results:** This paper proposes a new fine-grained 3D array partition method by key-value format in Spark. Our method first converts 3D images to key-value data (K-V). Then the K-V data is used for 3D array partitioning and data exchange in parallel. So Spark-based distributed parallel computing framework can solve the above scalability problem. In this distributed computing framework, all 3D local FSC tasks are simultaneously calculated across multiple nodes in a computer cluster. Through the calculation of experimental data, 3D local resolution evaluation algorithm based on Spark fine-grained 3D array partition has a magnitude change in computing speed compared with the mainstream FSC algorithm under the condition that the accuracy remains unchanged, and has better fault tolerance and scalability.

(Continued on next page)

(Continued from previous page)

**Conclusions:** In this paper, we proposed a K-V format based fine-grained 3D array partition method in Spark to parallel calculating 3D FSC for getting a 3D local resolution density map. 3D local resolution density map evaluates the three-dimensional density maps reconstructed from single particle analysis and subtomogram averaging. Our proposed method can significantly increase the speed of the 3D local resolution evaluation, which is important for the efficient detection of subtle variations among reconstructed macromolecular structures.

**Keywords:** 3D local Fourier shell correlation, 3D local resolution map, Key-value data, Spark, 3D array partition

## Background

Macromolecular complexes are nano-machines involved in a wide range of biological cellular processes. The biological functions of macromolecular complexes are determined by their 3D structures. Therefore, it is important to know the structures of these macromolecule complexes to fully understand these cellular processes [1].

In recent years, electron cryomicroscopy (cryo-EM) has played an increasingly important role in 3D structure recovery of macromolecular complexes. 3D structures of biological macromolecules and complexes can be further studied to reveal their specific function, which may lead to major breakthroughs in biological cellular mechanisms, pharmacy and disease treatment, etc.

The single particle analysis (SPA) techniques in cryoelectron microscopy (cryo-EM) and subtomogram averaging (SA) techniques in electron cryotomography (cryo-ET) are widely applied to obtain 3D density maps of macromolecular complexes [2, 3]. The quality of reconstructed 3D density maps needs to be evaluated normally in terms of the structural resolution. The resolution can be directly utilized to interpret the extent of the structural details in the density map. A number of different methods have been proposed on resolution evaluation, among which popular ones are Fourier shell correlation (FSC) [4–6], spectral signal-to-noise ratio (SSNR) [7], R measure [8], likelihood-ratio hypothesis testing (ResMap) [9], and monogenic signal transformation (MonoRes) [10].

In this paper we focus on FSC, the most widely used resolution evaluation method [6, 11, 12]. It is normally employed in the SPA and SA for resolution evaluation. The golden standard procedure is to divide the data into two independent halves and recover separately. Then the normalized cross correlation between the two results with different shell in Fourier space is calculated.

In such resolution estimation, all particles are considered structurally homogeneous under ideal conditions in the SPA technique and SA technique, save for different orientations and positions, so the FSC can measure resolution of density maps of reconstruction. However, in practice, particles of a macromolecular complex can often be structurally heterogeneous [13–15]. The traditional global FSC measure cannot be used to accurately detect the subtle structural variations in the reconstructions of SPA and SA. In order to evaluate local structural heterogeneity, a local resolution evaluation has been proposed [6, 9, 10]. The first 3D local resolution evaluation method is blocres [6], which employed windowed FSC to calculate local density maps. The method utilizes a density map of windows to calculate corresponding FSC value. The blocres method extends 2D FSC evaluation to 3D local FSC evaluation. Implementation of blocres utilizes Open Multi-Processing (OpenMP) to parallelize calculation on a single compute node.

The existing methods of FSC-based local resolution evaluation are single-node multi-threaded implementation. Part of the reason is that single-node multi-threading implementation can satisfy local resolution evaluation, but multi-threaded implementation often considers thread security and is not scalable. Although it is relatively accurate to obtain 3D local resolution map by calculating 3D local FSC, the computations are enormous, mainly due to a large number of coupled 3D small windows calculated. For volume density map of 300 × 300 × 300 voxel with 0.982Å resolution, computational time is 123.5 minutes for the most accurate algorithm, blocres, at criterion of FSC 0.5 [10]. Hence, it is necessary to develop a distributed parallel method of calculating 3D local FSC for 3D local resolution evaluation. Specifically, by applying distributed computing algorithms, one can quickly chalk up 3D local resolution map through parallel computing on a cluster, not just on one compute node. Nevertheless, the main challenge for distributed algorithms design and development is how to effectively divide a 3D array (volume density map) into 3D sub-arrays (sub-volume density map), distribute them on the computer cluster, and result of each pair 3D sub-array's calculation is not influenced by other threads.

In this paper, we present a Spark-based parallel method to calculate 3D local FSC for 3D local resolution estimation. Our method converses 3D images to K-V data. It then performs fine-grained 3D array partition by K-V format of Spark in memory. This method utilizes Spark [16] to implement parallel and distributed computation on a cluster. Our method can obtain the 3D local resolution map significantly faster than blocres method without losing accuracy. With a sufficiently large number of compute nodes, even in a few seconds, the 3D local resolution map can be computed. In order to divide 3D array into plenty of 3D sub-arrays for parallel computation, we propose a K-V format based basic fine-grained 3D array partition method in Spark. To further reduce memory space, data shuffling and transfer, we propose a K-V format based optimized fine-grained 3D array partition method in Spark. The 3D density map is essentially a 3D gray image represented as a 3D array. The 3D array partition method can be easily extended to other 3D array operations. A detailed comparison of different local resolution methods are performed to illustrate the advantages of our Spark-based 3D local resolution algorithm.

Our Spark-based 3D local resolution estimation algorithm can achieve completely automatic calculation of 3D local FSC values and resolution with suitable window size and specified FSC threshold, and is significantly faster than the blocres algorithm. In addition, our algorithm is flexible and can be run on a single compute node with specified number of cores.

## Related work

The 3D local resolution evaluation methods [6, 9, 10] can inspect the subtle changes during reconstruction of particles in SPA and SA, however, these methods can only be used on a single compute node (Table 1), and cannot quickly compute the 3D local resolution results. Some of these methods are based on the FSC method, while others are based on hypothesis testing.

**Table 1** Introduction of different 3D local resolution evaluation methods

| Methods | Characteristics | Is distributed? |
|---|---|---|
| blocres | based on FSC, window sampling | no |
| ResMap | based on 3D sinusodial wave, hypothesis testing | no |
| MonoRes | based on monogenic signal, hypothesis testing | no |

### FSC-based methods for 3D local resolution estimation

**blocres algorithm**: For the blocres [6] algorithm, the moving window is used to traverse the half-map respectively, and the area of two 3D windows is calculated using FSC to gain local resolution. Finally, the local resolution is filled into the new 3D density map corresponding to coordinate, and the new 3D density map is the same size of half-maps. Because FSC measures are calculated without any transformation on the original density maps, blocres is the most accurate algorithm in calculating local resolution. In the blocres method, many windows will be generated, and it is very time-consuming to calculate these windows using FSC and acquire local resolution. This poses an issue when coupled with the fact that the blocres method runs on a single compute node.

### Hypothesis testing based methods for 3D local resolution estimation

Besides local FSC, another kind of local resolution evaluation is based on likelihood-ratio hypothesis testing of density map approximated by basis function versus noise or signal transformation, such as ResMap [9] and MonoRes [10].

**ResMap algorithm**: For ResMap [9] algorithm, the following conditions need to be followed: if a 3D local sinusoid of wavelength $\lambda$ can be detected statistically higher than noise at this point, then there exists a $\lambda$-Å characteristic at this point in the volume. The likelihood ratio hypothesis testing of the local sinusoid curve and noise is employed to detect the feature when $P$ value is 0.5. The local resolution at this point is the minimum $\lambda$ that can detect the local sinusoid curve. A set of the second-order Hermite polynomials function expresses the local sinusoids of wavelength $\lambda$. However, the ResMap algorithm has some limits, such as the need for an initial step and running on single compute node.

**MonoRes algorithm**: MonoRes [10] is the intensity map of the signal, which is converted by the initial electron density map at that location, and also called monogenic magnitude map. The transformation of signal is Riesz transform (or Hibert transform of 1D signals). Riesz transform is a generalization of Hibert transform's multidimensional (N-D), defined only for 1D signals. The map generated by MonoRes algorithm is the same size as the original map, where each voxel participates in the local resolution estimation. For each voxel of the mask volume, MonoRes algorithm use hypothesis testing to detect whether the local monogenic amplitude is greater than the $1 - \alpha$ percentile of the monogenic amplitude in noise. MonoRes algorithm runs on a single compute node.

Since the hypothesis testing is used in ResMap and MonoRes algorithms, the density map approximated by transformation is faster in calculation, but still costs a few minutes for density maps of size $200^3$ voxels [10]. Generally, the density map approximated by transformation is constrained with various factors, and one of the key factors is resolution of density map. Due to the low signal-to-noise ratio (SNR), using transformation to approximate density map of reconstruction cannot improve the local resolution estimation under low resolution conditions. The method of local resolution evaluation based on FSC directly employs two half-maps to calculate, and doesn't need to approximate the density map using specific basis functions, so that the details of the reconstructed density map can be evaluated more accurately. Although the calculations of local resolution evaluation used in FSC are larger, these methods reflect the resolution of density map more objectively and are more suitable for the resolution evaluation of SA in cryo-ET.

#### Distributed computation frameworks

Due to the low SNR of particles in SPA and SA, FSC is still widely applied [17] for 3D local resolution estimation to assess the details during reconstruction of SPA and SA. In order to gain 3D local resolution map faster, a distributed version of 3D local resolution evaluation algorithm is implemented. Currently, there exist many mainstream distributed computation frameworks, such as Message Passing Interface (MPI), Hadoop MapReduce and Apache Spark.

**Message Passing Interface (MPI)**: MPI [18] is a standardized portable messaging standard, which can be used in various parallel computing architectures. MPI defines the grammar and semantics of the library routines' core, and is also a communication protocol of programming parallel computers, and styled for supporting point-by-point and collective communication. MPI is an application programming interface of messaging, together with the protocols and semantic specifications that its features must represent in any implementation. MPI has the features of high performance, scalability and portability. MPI remains the main model in high performance computing [19], but it is not easy to implement and has no fault-tolerant mechanism and redundancy mechanism.

**Hadoop MapReduce**: MapReduce [20] is a programming model. It is a related implementation of using parallel distributed algorithms to process and generate large data sets in a cluster. MapReduce programs consist of map processes that perform filtering and sorting, and reduce processes that perform summary operations. MapReduce combines the distributed serves, runs the different tasks in parallel, manages all communication and data transmission in a different system, and provides redundancy and fault tolerance [21, 22]. MapReduce processes K-V data. MapReduce framework handles data (e.g input and output data) from disk.

**Apache Spark**: Apache Spark [16] is a generic open source distributed cluster computing framework. Spark provides an interface to implement data parallelism and programming on a cluster. Spark is based on the resilient distributed datasets (RDD), which is a read-only multi-set data item distributed on a set of machines and maintained in a fault-tolerance manner. The Dataframe Application Programming Interface (API) is published as an abstraction over RDD, followed by the Dataset API. RDD technology remains the foundation of the Dataset API, and the operations of RDD rely on K-V data. Spark framework only deals with data in memory. Spark framework is better suited for fast data processing and iterative processing [23].

A 3D local resolution evaluation algorithm based on FSC needs a large number of sub-volumes to participate in calculation, which contains iterative processes [24, 25]. Hence, Spark framework is better suited for fast data processing and iterative processing. Because Spark is only a coarse-grained distributed framework, for a specific task such as local FSC computation, we need to design a K-V form based novel fine-grained 3D data partition and transformation algorithm for distribution parallel computing in Spark.

## Results and discussions

To inspect the performance of our Spark-based optimized 3D local resolution estimation algorithm, this method was applied to two experimental density maps, and compared with other methods in computing time and accuracy.

### Comparison of time cost between different methods of 3D local resolution evaluation

First, we compare time cost between our Spark-based optimized 3D local resolution estimation algorithm and blocres [6] algorithm. In order to objectively evaluate both algorithms, we followed the same parameter conditions with two experimental data sets respectively.

To gain 3D local resolution map faster, our Spark-based optimized 3D local resolution estimation method can extend compute node freely until all tasks can parallel work simultaneously. But blocres method only works under single compute node. For distinguishing time cost, we just employ four compute nodes and eight compute nodes for Spark parallel algorithm.

Under two 3D local resolution methods, we used windows size of $64^3$ and FSC of 0.143 criteria to traverse the half-maps (EMDB:EBD-3802), and calculated 3D local FSC and resolution respectively. From Fig. 1, blocres method consumed about 106 minutes for 3D local resolution estimation during two half map of the human TFIIH (EMDB: EMD-3802), however, our Spark-based optimized 3D local resolution estimation method only utilized 25 minutes and 13 minutes under the four and eight computing node respectively.

The above observations indicate that our Spark-based optimized 3D local resolution estimation method is significantly faster than the blocres method. Assuming more computing nodes, even virtual machine nodes, our Spark-based optimized 3D local resolution

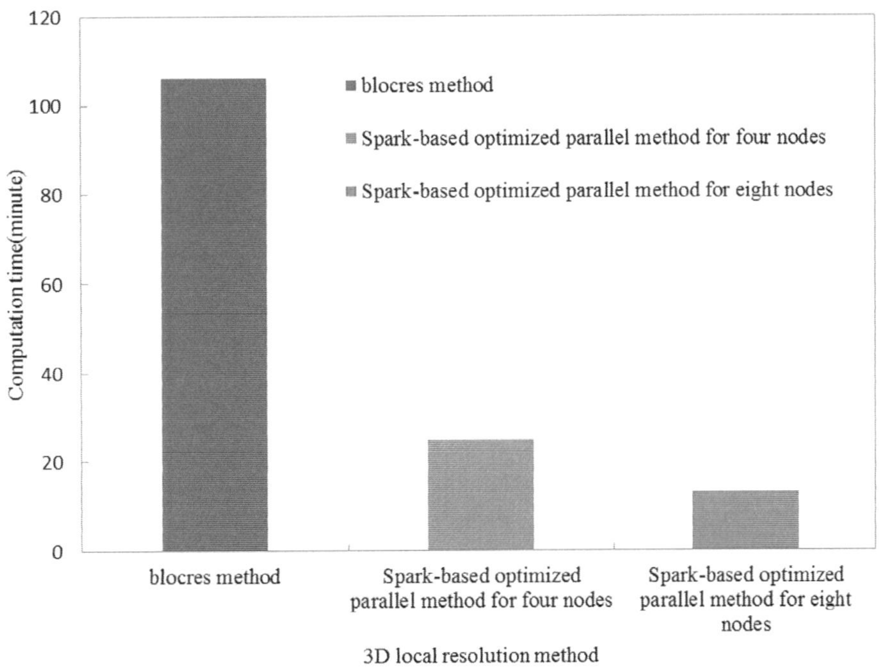


**Fig. 1** The time cost of Spark-based optimized 3D local resolution estimation method and blocres method

estimation method can perform all tasks simultaneously, whereas the blocres algorithm cannot be implemented.

## Comparison of accuracy between different method of 3D local resolution evaluation

In order to compare accuracy of 3D local resolution map resulting from blocres method and Spark-based optimized 3D local resolution estimation method, we tested two experimental data: one is half map from SPA, one is half map from SA.

First, through applying two half-maps of cryo-EM structure of $\beta$-galactosidase, the 3D local resolution maps gained by our Spark-based optimized 3D local resolution estimation method and blocres method are shown in Fig. 2. Distribution by histogram, 3D local resolution estimation of Spark-based optimized 3D local resolution estimation method range from 4.4 to 26.2Å, and 3D local resolution estimation of blocres method range from 4.4 to 26.2Å. Because a large number of resolution values were 0Å in the edge of the volume, the median of 3D resolution of our Spark-based optimized 3D local resolution estimation method and blocres method were 0 and 0Å respectively. The color range of slice of our Spark-based optimized 3D local resolution estimation method and blocres method were a little different, which were 4.4 to 26.2Å and 4.4 to 8.0Å respectively. The color range of slice and histogram in our Spark-based optimized 3D local resolution estimation method was almost the same. The original resolution of cryo-EM reconstruction of human TFIIH published in [26] was 4.4Å. All methods were under the FSC of 0.143 criterion.

Through the histogram of our Spark-based optimized 3D local resolution estimation method and blocres method, we can found that the resolution value distribution of Spark-based optimized 3D local resolution estimation method was similar to that of the blocres method. So combining the histogram and slice of two methods, the two 3D local resolution maps were almost similar, and from the 3D local resolution maps, the detail

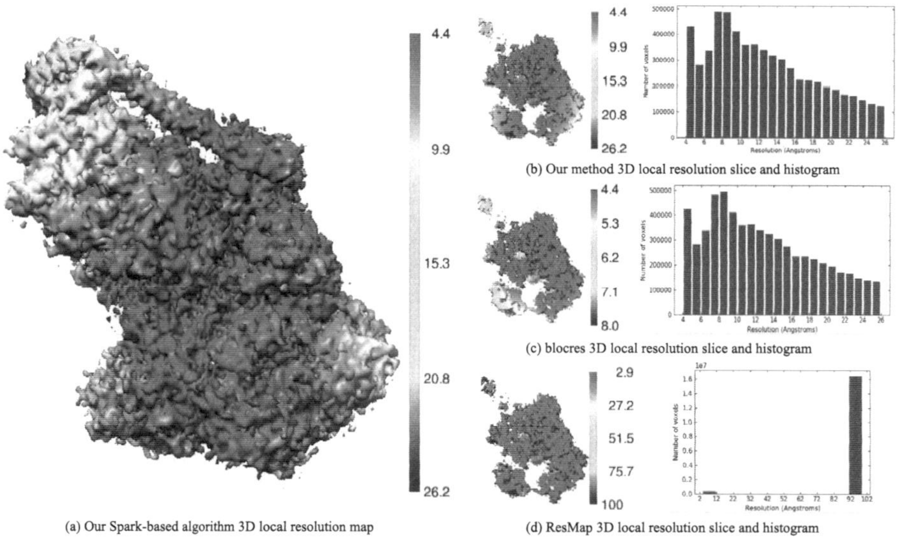


**Fig. 2** Spark-based 3D local resolution density map. Spark-based, blocres and ResMap slices, and histogram for experimental maps. **a** Spark-based 3D local resolution density map. **b** Spark-based method resolution slice and histogram for the cryo-EM structure of human TFIIH (EMDB: EMD-3802). **c** blocres method resolution slice and histogram for the cryo-EM structure of human TFIIH. **d** ResMap method resolution slice and histogram for the cryo-EM structure of human TFIIH

of cryo-EM reconstruction of human TFIIH can be reflected. We also utilized ResMap method to obtain 3D local resolution map, which were colored by Chimera with the final reconstruction (EMDB: EDB-3802) in Fig. 2d, 3D local resolution estimation of ResMap method range from 2.9 to 100Å, the median of ResMap was 100Å. Although the histogram of our Spark-based optimized 3D local resolution estimation method and ResMap method was different, the color range of slice of our Spark-based optimized 3D local resolution estimation method was similar to that of ResMap method.

Next, using two half-maps of SA of Sar1-Sec23-Sec24 (EMDB: EMD-0044), the 3D local resolution maps were obtained by our Spark-based optimized 3D local resolution estimation method and blocres method respectively (Fig. 3). Distribution by histogram, the 3D local resolution range estimated by our Spark-based optimized 3D local resolution estimation method and the blocres method was between 0 and 23.4Å with the median of 0.0Å. By using the tool Surface Color in Chimera, the color range of slice of Spark-based optimized 3D local resolution estimation method was 0 to 23.4Å, and the color range of slice of the blocres method was 0 to 7.0Å. Although the color range of two methods were a little different, the histograms of the two methods were almost the similar. The covered resolution of the SA of Sar1-Sec23-Sec24 was 4.9Å [27]. From the Fig. 3. The 3D local resolution map of our Spark-based optimized 3D local resolution estimation method achieved similar results compared with the 3D local resolution map of blocres method, and our Spark-based optimized 3D local resolution map had the same resolution on the boundary.

## Conclusion

In this paper, we focus on reducing the time cost of the local FSC calculation for macromolecule 3D structure local resolution estimation. The traditional 3D local resolution algorithm by calculating 3D FSC are computationally intensive and normally only execute under single compute node, therefore we consider distributed 3D local resolution estimation algorithm in order to significantly increase the speed of obtaining 3D local resolution map.

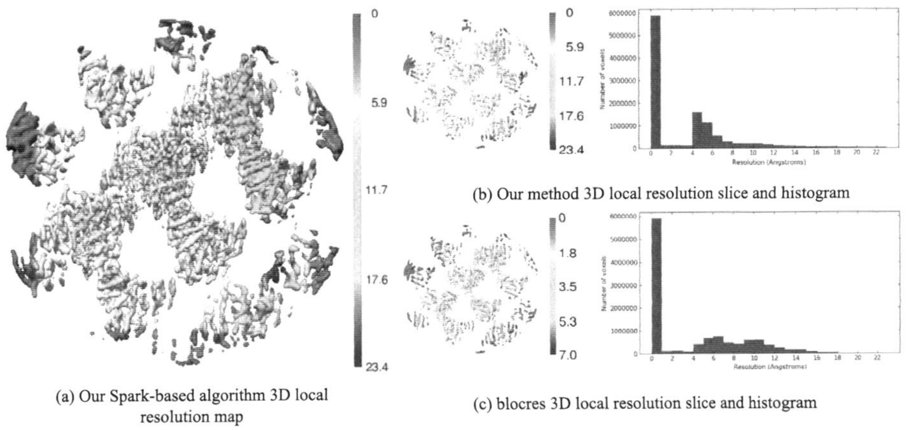


**Fig. 3** Spark-based 3D local resolution density map. Spark-based and blocres slices, and histogram for experimental maps. **a** Spark-based 3D local resolution density map. **b** Spark-based method resolution slice and histogram for SA of Sar1-Sec23-Sec24 (EMDB: EMD-0044). **c** blocres method resolution slice and histogram for SA of Sar1-Sec23-Sec24

By analyzing the mainstream distributed architecture, we choose Spark architecture for efficient K-V data processing to design a distributed 3D local resolution evaluation algorithm. Specifically, Spark architecture is more suitable for iterative algorithm because K-V data operations are implemented in memory, which does not need frequent IO exchange. Also, Spark structure has strong extensibility and fault tolerance.

In this paper, we propose two kinds K-V format based fine-grained 3D array partition algorithms in Spark for computing 3D FSC in parallel. Compared with the K-V format based basic fine-grained 3D array partition algorithm, the K-V format based optimized fine-grained 3D array partition algorithm is more memory-saving, faster in calculation and more concise in expression. The K-V format based optimized fine-grained 3D array partition algorithm in Spark as the core of the Spark-based optimized 3D local resolution estimation method. Spark-based optimized 3D local resolution estimation method can quickly get the 3D local resolution map and evaluate the details of single particle analysis reconstruction in cryo-EM and average reconstruction of subtomogram in cryo-ET.

Through the calculation of experimental data, Spark-based optimized 3D local resolution estimation method are faster than blocres method to gain 3D local resolution map, but still lack some optional functions, such as filtering and mask etc. We plan to develop more full-featured algorithms in the future. Integrating graphics processing unit (GPU) into our current framework may further accelerate computation. Spark-based optimized 3D local resolution estimation method can be applied in cloud services and cluster environments. When users are computing 3D local resolution in a cloud server or cluster with multiple idle computing nodes, Spark-based optimized 3D local resolution estimation method can make full use of these computing nodes to quickly calculate 3D local resolution. Spark-based optimized 3D local resolution estimation method can also support on-line fast calculation of 3D local resolution and avoid the user's online waiting time. In the future, we will optimize the scheduling algorithm, through analyzing the size of the three-dimensional volume submitted by the user, and reasonably allocate the calculation nodes, so as to quickly realize the calculation of the 3D local resolution.

## Methods

In this section, we first introduce the resolution, the 3D FSC, the 3D local FSC and 3D local resolution estimation method. Then, we describe two kinds of K-V format based fine-grained 3D array partition algorithms for 3D local resolution evaluation. We use Spark's efficient K-V data batch processing platform to realize the distributed parallel computing of 3D local FSC. This algorithm extends the multithreading of a single computing node to the distributed parallel computing in spark. Finally, we depict the algorithm implementation and experimental datasets.

### The resolution and 3D fourier shell correlation

Resolution in the electron density expresses a measure of electron density resolvability. In cryo-EM, resolution is a comparison of frequency space correlation of two density maps of the data divided into two halves. Resolution is typically measured by the spatial frequency correlation function in cryo-EM.

The method of 3D FSC, initially proposed by Harauz and van Heel [5] in 1986, is frequently applied as quantitative evaluation of the reconstruction volume resolution in SPA and SA. The 3D FSC method mainly obeys the following process: two density maps of 3D volumes are reconstructed with one half data set in SPA or SA respectively, two 3D volumes are transformed from real space into Fourier space and form a series of shells, and then the corresponding two series of shells in Fourier space are calculated and gained normalization cross correlation coefficient. Hence, the 3D FSC effectively evaluates the resolution of reconstruction density maps with the normalization cross correlation coefficient between two series of shells, which are generated by 3D volumes in Fourier space.

The 3D FSC [6] between the two density maps in Fourier space, $F_1$ and $F_2$, is represented by

$$FSC_{1,2}(s) = \frac{\sum_{s_i \in s} F_1(s_i) \cdot F_2(s_i)^\star}{\sqrt{\sum_{s_i \in s} |F_1(s_i)|^2 \cdot \sum_{s_i \in s} |F_2(s_i)|^2}} \tag{1}$$

where $s_i$ is corresponding a series of shells $s$ in Fourier space, $F(s_i)$ is complex structure factor at radial frequency $s_i$ in Fourier space, $\sum_{s_i \in s}$ is summation over all Fourier space voxels $s_i$ in shell s. $F(s_i)^\star$ denotes a conjugate complex of $F(s_i)$.

The 3D FSC method extends the pristine two-dimensional Fourier Ring Correlation (FRC) algorithm [4], and obtains density map of 3D FSC to evaluate the resolution of reconstructed density map more visually and scientifically.

### 3D local fourier shell correlation and 3D local resolution map estimation

To better evaluate details of resolution of reconstructed density map from SPA in cryo-EM and SA in cryo-ET, the 3D local resolution evaluation method based on 3D FSC has been proposed in [6], which is implemented by 3D local FSC.

For given two half-maps (3D density maps), two small cube windows are used to traverse their respective 3D density maps. Each time a small cube window accesses the 3D density maps, the 3D density map of the corresponding region of the two small cube windows are calculated to generate 3D FSC value, and then are converted to the corresponding resolution value by specified FSC threshold and voxel value. The resolution value is filled into the corresponding center position of the small cube window in the new 3D volume, and the corresponding other position of the small cube window in the new 3D volume does not fill in the value, and then operate in turn until the small cube window ends traversing the 3D density map. In the new resolution 3D, the corresponding center position of traversing small cube window will be filled in the value in turn, so that the new resolution 3D (the size and the 3D density map are the same) generate discrete resolution values, and then 3D cube interpolation methods are used to interpolate coordinate positions without values in turn. When the new resolution 3D volume is filled completely with resolution values, this resolution 3D volume is called 3D local resolution map.

On the 3D local resolution map, the size of the accessed small cube window and the step size of traversal are the pivotal factors affecting the 3D local resolution map. If the size of the small cube window is too small, the calculation of smaller FSC threshold (e.g. FSC = 0.143) cannot be performed for local resolution estimation. Similarly, if the size of the small cube window is too large, 3D local resolution map will not be veritable. The step size of traversal also influences the 3D local resolution map. The larger the step size of the traversal, the fewer the 3D local resolution values to be calculated, which need to be interpolated in the new 3D resolution volume map and cannot veritably reflect the resolution situation. The smaller the step size of the traversal, the more the calculated 3D local resolution values, and thus this is closer to the real 3D local resolution map, However, the amount of computation is too large. As a result, it is highly prudent to determine the appropriate size of the small cube window and the step size of the traversal.

However, 3D local resolution evaluation method not only calculates 3D FSC in small windows, but also considers global resolution calibrating the average value of all local resolutions for accurate estimating local resolution. Hence, 3D local resolution evaluation method is influenced by windows side length, step of the traversal, threshold of the FSC, global resolution and so on.

### Spark-based parallel calculation of 3D local FSC for estimation of 3D local resolution

To significantly reduce the time used for computing 3D local FSC, we propose a Spark-based distributed 3D local resolution evaluation method with a parallel and distributed computing architecture.

In our 3D local resolution evaluation algorithm, the K-V format based fine-grained 3D array partition can facilitate the distributed parallel computing of 3D FSC, so as to obtain the 3D local resolution map quickly. As discussed in Distributed computation frameworks Section, Apache Spark has more advantages than MPI and MapReduce, processes K-V data in memory, owns redundancy mechanism and fault tolerant mechanism, and is suitable for high frequency data exchange and large iterative processing.

Based on Apache Spark distributed framework, we design two methods of fine-grained 3D array partition algorithms by K-V format in Spark. Two methods can calculate 3D local FSC in parallel, so as to acquire 3D local resolution estimation map faster. This algorithm can be applied not only to SA and SPA, but also to other 3D image operations.

The work flow for evaluation of 3D local resolution on Spark framework is shown in Fig. 4. Our Spark-based parallel computing algorithm adopts the standalone cluster mode. The cluster manager is the Master. The Master is responsible for allocating resources, the Worker is responsible for monitoring the memory, CPU and other conditions of its own node and reporting to the Master. The Master allocates resources to Workers and monitors the resource status of the Workers at any time. Program Spark, package submitted to the Driver side, constitutes a Driver. When Spark program runs, the Driver requires resources from the Master. All tasks will be calculated in parallel by the Workers node and the results will be returned to the Master by K-V format.

### K-V format based basic fine-grained 3D array partition algorithm in Spark for parallel calculating 3D local resolution

To get the 3D local resolution map faster, 3D array partition is a key factor. K-V format based fine-grained 3D array partition algorithm theoretically confirms the maximum number of parallel execution of the algorithm. Therefore we design a K-V format based basic 3D array partition algorithm for parallel calculating 3D local FSC and estimating 3D local resolution map. The K-V format based basic 3D array partition algorithm in Spark is as follows:

Step 1: Each voxel of 3D array (3D volume density map) is transformed quadruple block (QB=(x,y,z,value)) in list.

Step 2: List containing quadruple block are transformed RDD.

Step 3: All data in RDD are added key in order to get correct partition, such as <key, QB>.

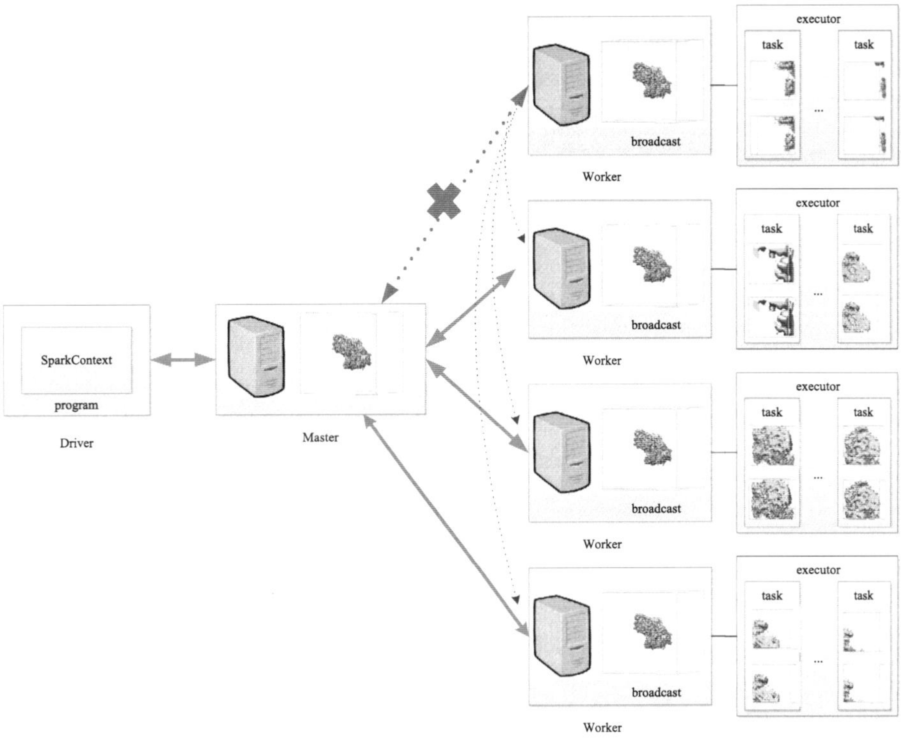


**Fig. 4** Spark-based parallel calculating 3D local FSC for 3D local resolution estimation work flow. The green arrow indicates the normal transmission and distribution of data. The red arrow indicates that the Worker node cannot connect to the Master node. The gray dotted arrow indicates that the tasks originally assigned to the crash node are reassigned to other Worker nodes

Step 4: Two RDDs are run two times operation of mapPartition and partionBy for getting the max parallel number.

Step 5: Two RDDs which should have the same number of partitions and the same number of elements in each partition, are zipped as new RDD.

Step 6: Data in each partition are transformed two 3D array, which is calculated for 3D local FSC and 3D local resolution.

Step 7: The FSC calculation results of each pair of 3D sub-volumes are filled into corresponding coordinates of the new volume, and the coordinates without filling values in the new volume is interpolated by cubic interpolation algorithm.

Here, since data in RDD is K-V pair format, in which condition all data can realize partition, 3D volume density map (3D array) must be transformed into a list of K-V format. In order to gain K-V format, all voxels in 3D volume density map must be transformed quadruple block, such as (x,y,z,v), where x, y, z is coordinate of the voxel, v is value of voxel. Then all voxels become quadruple block in list.

We show the K-V format based basic 3D array partition work flow for 3D local resolution estimation in Fig. 5. Our K-V format based basic 3D array partition algorithm in Spark requires that all values of the 3D array participate in each RDD operation, so these consume a large amount of memory space.

The details of the K-V format based basic fine-grained 3D array partition algorithm in Spark are shown in Fig. 6. Values of 3D array (voxels) transforming into quadruple block

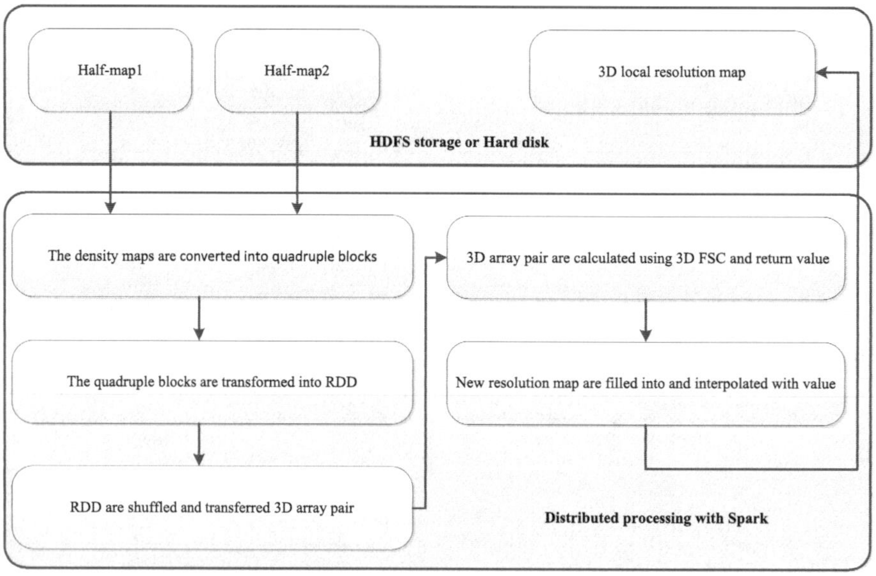


**Fig. 5** The workflow of K-V format based basic fine-grained 3D array partition algorithm in Spark for 3D local resolution estimation. Two 3D density maps are transformed RDD in Spark. K-V data in RDD are shuffled and transferred, data in each partition are transformed 3D array, which is calculated for 3D local resolution using 3D FSC. A resolution map is filled into and interpolated with calculation results according to corresponding coordinate

are a pivotal part of 3D array partition. When 3D array becomes a list of quadruple block, the 3D array partition algorithm can exploit key and quadruple block (e.g. <key, value> pair) to operate data shuffle and partition. Finally data in each partition is converted into 3D arrays again to realize 3D array partition. In each pair 3D array, the FSC value and local resolution value are calculated, and the local resolution value is filled into corresponding coordinates of 3D local resolution map.

#### K-V format based optimized fine-grained 3D array partition algorithm in Spark for parallel calculating 3D local resolution

Although the K-V format based basic fine-grained 3D array partition algorithm can achieve maximum parallelization in Spark, it incurs plenty of data shuffle and transfer. So the K-V format based basic fine-grained 3D array partition algorithm results in data redundancy and consumes a large memory space in each computing node (Fig. 7). Those actions lower parallel performance in Spark. So we conceive a K-V format based optimized fine-grained 3D array partition algorithm in Spark for calculating 3D local resolution. The K-V format based optimized fine-grained 3D array partition algorithm can reduce data redundancy and shuffle in Spark.

First, the K-V format based optimized fine-grained 3D array partition algorithm in Spark reads two density map into memory; second, it gains all start coordinates of window traversing density map, and transforms K-V pair in a list; third, it transforms the list into RDD, and partition RDD; fourth, for each partition, it acquires 3D sub-array pair from 3D array, and calculates FSC and resolution with two sub-array.

From the Algorithm 1, the optimized fine-grained 3D array partition method is more concise than the basic fine-grained 3D array partition method in K-V format of Spark. The

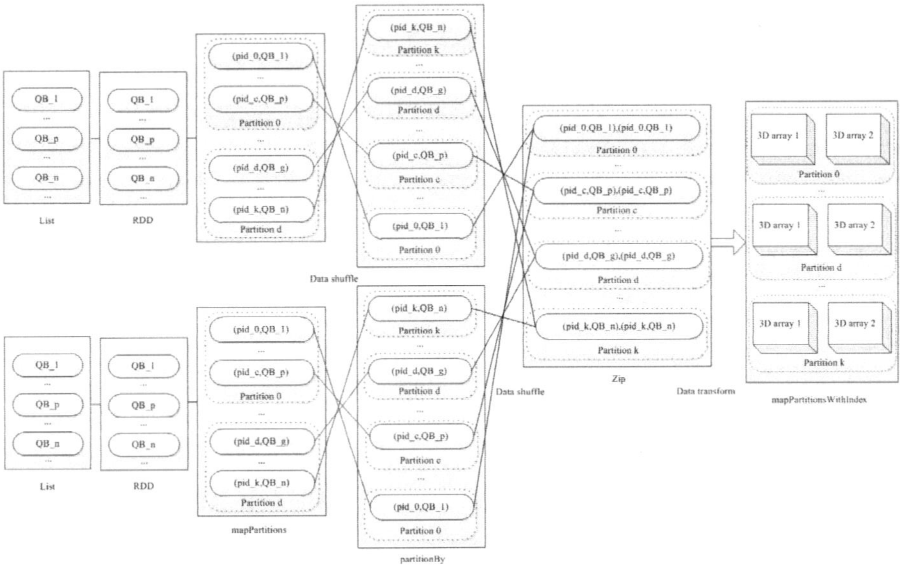


**Fig. 6** The K-V format based basic 3D array partition algorithm for 3D local resolution estimation in Spark. QB stand for quadruple block, QB=(x,y,z,v), where x, y, z is coordinate, v is voxel value. In the figure, n is volume side length, 0, c, d, k are partition number, and $0 \leq c \leq d \leq k$, k is the largest partition number. 3D array partition are realized by K-V data shuffle and transformation in Spark. Each 3D array pair are calculated for 3D local resolution estimation

K-V format optimized fine-grained 3D array partition method in Spark doesn't transform values of 3D array into RDD. Each RDD only contains start coordinates of partition array relative to original 3D array.

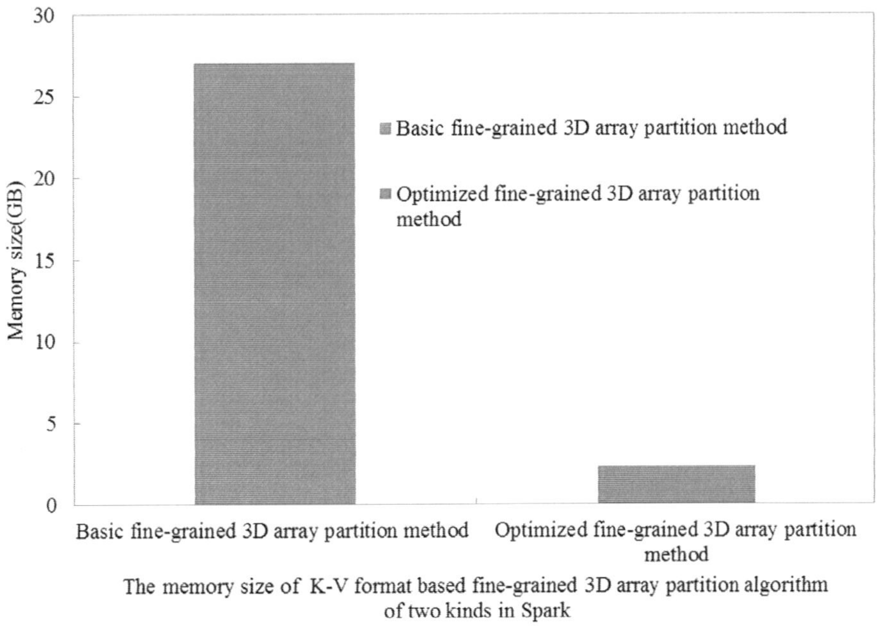


**Fig. 7** For 3D local resolution evaluation, the memory size of K-V format based basic and optimized fine-grained 3D array partition algorithm in spark

**Algorithm 1** The K-V format based Optimized fine-grained 3D array partition algorithm in Spark for parallel calculation 3D local resolution

**Input:**
$V_1 = [a_{ijk}], 1 \le i,j,k \le n \rightarrow$ **The input half density map1**
$V_2 = [a_{ijk}], 1 \le i,j,k \le n \rightarrow$ **The input half density map2**
$size \rightarrow$ **windows side length**
$inc \rightarrow$ **step size**

**Output:**
$R = [a_{ijk}], 1 \le i,j,k \le n \rightarrow$ **The output resolution map**

$pid = 0$
**for** $z = 0, z < n - size, inc$ **do**
  **for** $y = 0, y < n - size, inc$ **do**
    **for** $x = 0, x < n - size, inc$ **do**
      $L_1 = (pid, (x, y, z))$
      $pid++$
    **end for**
  **end for**
**end for**
$RDD \leftarrow L_1$
**partition** $RDD$
**for** $i, j, k \in RDD$ **do**
  $v_1, v_2 \leftarrow$ **cut 3D sub array pair from** $V_1, V_2$
  $R \leftarrow$ **calculate FSC and resolution of** $v_1$ **and** $v_2$ **for each partition**
**end for**
$R \leftarrow$ **interpolation**

The details of the K-V format based optimized fine-grained 3D array partition algorithm in Spark are shown in Fig. 8. The total operation of the K-V format based optimized fine-grained 3D array partition algorithm is much less than that of the K-V format based basic fine-grained 3D array partition algorithm, and each RDD does not include the coordinate and values of 3D array. The two 3D arrays are cut for gaining 3D sub-array in the final operation.

The K-V format based optimized fine-grained 3D array partition algorithm in Spark highlights mainly as follows:

- Reduce total memory space
- Lessen data shuffle and transfer
- Enlarge parallel number

Relative to the K-V format based basic fine-grained 3D array partition algorithm in Spark, the K-V format based optimized fine-grained 3D array partition algorithm does not require many operations of data shuffle, reduces data redundancy and only needs limited memory in Spark. Hence it calculates all the tasks quickly. So we choose the K-V format based optimized fine-grained 3D array partition algorithm in Spark as the Spark-based optimized 3D local resolution estimation algorithm.

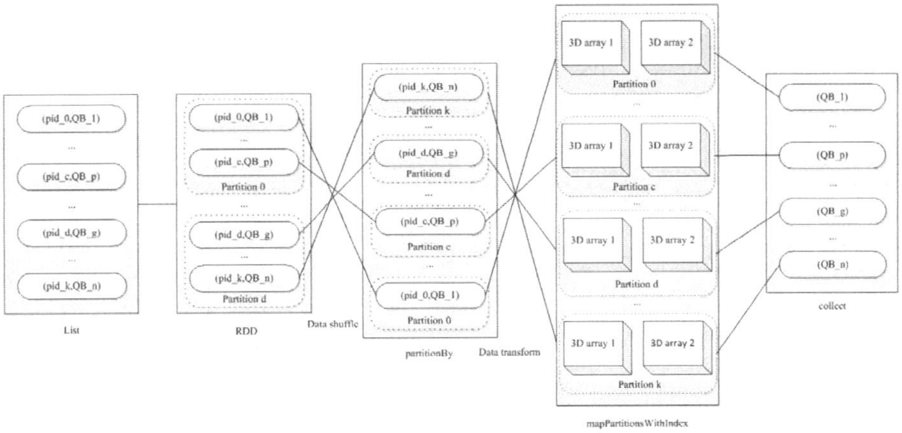


**Fig. 8** The K-V format based optimized fine-grained 3D array partition algorithm in Spark for 3D local resolution estimation. TB stand for triple block, TB=(x,y,z), where x, y, z is start coordinate of window traversing density map. 0, c, d, k are partition number, and $0 \leq c \leq d \leq k$, k is the largest partition number. List of sequence numbering and start coordinates of window traversing are transformed into RDD. 3D sub-array pair are acquired by cutting 3D array using each partition data and side length of window, and calculated for gaining 3D local FSC and resolution

## Implementation

In this paper, the 3D local resolution evaluation based on fine-grained 3D array partition method with K-V format of Spark has been implemented in Python and C++ language, and referred to PyTom [28] in part. The input parameters of 3D local resolution evaluation algorithm based on K-V fine-grained 3D array partition in Spark mainly include two half-maps, windows side length, step size, FSC threshold, and global resolution. To further reduce the computing time, a mask and no-zeros edge length can be specified, which are optional. The global resolution value is order to calibrate the average value of all local resolution on underestimating resolution. The output is a 3D local resolution map. The 3D local resolution map can be colored with final reconstruction density map to visualize representation of reconstruction density map, where subtle change of local detail of density map can be inspected during homogeneity or heterogeneity complex reconstruction. The color tool used is Chimera [29].

Our Spark-based optimized 3D local resolution estimation code can be executed in a distributed cluster containing eight compute nodes. Configurations of each compute node include, two 1.70GHz Intel Xeon Bronze 3104 CPUs with 12 physical cores, one Gigabi ethernet card and 30GB Random Access Memory (RAM).

Deployment of Spark cluster adopt stand alone mode comprising a Master node and 4 or 8 Worker nodes. Master node mainly assigns application to Work node and maintains the status of Worker node, Driver and application. Worker nodes mainly execute tasks assigned by Master node. Our code is implemented in Spark 2.0 and Python 2.7.14.

## Experimental datasets

The two experimental density maps were acquired from the Electron Microscopy Data Bank (EMDB) to calculate 3D local resolution respectively.

The first experimental density map is the structure of human transcription factor IIH (EMDB: EMD-3802) [26], which have two half maps (single particle analysis map from independently aligned half dataset 1 and 2), the resolution is 4.4Å, FSC threshold is 0.143, the number of grid points is 256 × 256 × 256, the voxel size is 1.32 × 1.32 × 1.32Å.

The second experimental density map is the structure of SA of Sar1-Sec23-Sec24 (EMDB: EMD-0044) [27], which have two half maps (average map from independently aligned half dataset 1 and 2), the resolution is 4.9Å, FSC threshold is 0.143, the number of grid points is 224 × 224 × 224, the voxel size is 1.33 × 1.33 × 1.33Å.

**Abbreviations**
cryo-EM: cryoelectron Microscopy; SPA: Single particle analysis; SA: Subtomogram averaging; cryo-ET: Electron cryotomography; FSC: Fourier shell correlation; K-V: Key-value data; SSNR: Spectral signal-to-noise ratio; OpenMP: Open multi-processing; SNR: Signal-to-noise ratio; MPI: Message passing interface; RDD: Resilient distributed datasets; API: Application programming interface; EMDB: Electron microscopy data bank; GPU: Graphics processing unit; RAM: Random access memory

**Acknowledgements**
We thank Dr. Fei Sun for fruitful discussions and suggestions. We thank Dr. Haiyang Li and Mr. Shuguang Zhao for technical assistance. We thank Wenlin Liu and Chengrui Wang for code discussions. We thank Dr. Yan Jin and Alex Singh for revising the English writing.

**About this supplement**
This article has been published as part of *BMC Bioinformatics Volume 21 Supplement 13, 2020: Selected articles from the 18th Asia Pacific Bioinformatics Conference (APBC 2020): bioinformatics*. The full contents of the supplement are available online at https://bmcbioinformatics.biomedcentral.com/articles/supplements/volume-21-supplement-13.

**Authors' contributions**
YL, XT, MX and XZ3 conceived the project, with XZ1 corresponding to Xiangrui Zeng, XZ2 corresponding Xiaohui Zheng and XZ3 corresponding to Xiaofang Zhao. YL, XZ3 and MX performed the research, designed methods, implemented methods, carried out experiments and wrote the manuscript. YL, XT, XS, HW, XZ2 and XL modified code. YL, XZ1, XZ3, MX and XG improved the paper. All the authors approved the final version of the manuscript.

**Funding**
This Publication costs are funded by the King Abdullah University of Science and Technology (KAUST) Office of Sponsored Research (OSR) under Award No. URF/1/2602-01 and URF/1/3007-01.
This work is supported by the King Abdullah University of Science and Technology (KAUST) Office of Sponsored Research (OSR) under Award No. URF/1/2602-01 and URF/1/3007-01. This work was supported by National Key R&D Program of China (Grant No. 2018YFB0904503 and 2017YFB1002703), the Key Research Program of Frontier Science of Chinese Academy of Sciences (Grant No.QYZDB-SSW-SMC004), and the National Natural Science Foundation of China (Grant No: 61379082 and 61672499). This work was supported in part by U.S. National Institutes of Health (NIH) grant P41 GM103712. This work was supported by U.S. National Science Foundation (NSF) grant DBI-1949629. XZ1 was supported by a fellowship from Carnegie Mellon University's Center for Machine Learning and Health. The founder played no role in the design of the study and collection, analysis, and interpretation of data and in writing the manuscript.

**Availability of data and materials**
Two half maps of structure of human transcription factor IIH (EMDB: EMD-3802) can be downloaded at https://www.emdataresource.org/EMD-3802
Two half maps of SA of Sar1-Sec23-Sec24 (EMDB: EMD-0044) can be downloaded at https://www.emdataresource.org/EMD-0044
The source code of the Spark-based parallel calculation of 3D Fourier shell correlation for macromolecule structure local resolution estimation algorithm is available at https://github.com/xulabs/projects.

**Ethics approval and consent to participate**
Not applicable.

**Consent for publication**
Not applicable.

**Competing interests**
The authors declare that they have no competing interests.

**Author details**
[1]Institute of Computing Technology of the Chinese Academy of Sciences, Beijing, China. [2]University of Chinese Academy of Sciences, Beijing, China. [3]Computational Biology Department, School of Computer Science, Carnegie Mellon University, Pittsburgh, USA. [4]King Abdullah University of Science and Technology (KAUST), Computational Bioscience Research Center (CBRC), Computer, Electrical and Mathematical Sciences and Engineering (CEMSE) Division, Thuwal, Saudi Arabia.

Published: 17 September 2020

**References**

1. Lučić V, Rigort A, Baumeister W. Cryo-electron tomography: the challenge of doing structural biology in situ. J Cell Biol. 2013;202(3):407–419.
2. Oikonomou CM, Jensen GJ. Cellular electron cryotomography: Toward structural biology in situ. Ann Rev Biochem. 2017;86:873–896.
3. Li R, Zeng X, Sigmund SE, Lin R, Zhou B, Liu C, Wang K, Jiang R, Freyberg Z, Lv H, et al. Automatic localization and identification of mitochondria in cellular electron cryo-tomography using faster-rcnn. BMC Bioinformatics. 2019;20(3):132.
4. Saxton W, Baumeister W. The correlation averaging of a regularly arranged bacterial cell envelope protein. J Microsc. 1982;127(2):127–138.
5. Harauz G, van Heel M. Exact filters for general geometry three dimensional reconstruction. In: Proceedings of the IEEE Computer Vision and Pattern Recognition Conf. 1986;73:146–156.
6. Cardone G, Heymann JB, Steven AC. One number does not fit all: Mapping local variations in resolution in cryo-em reconstructions. J Struct Biol. 2013;184(2):226–236.
7. Penczek PA. Three-dimensional spectral signal-to-noise ratio for a class of reconstruction algorithms. J Struct Biol. 2002;138(1-2):34–46.
8. Sousa D, Grigorieff N. Ab initio resolution measurement for single particle structures. J Struct Biol. 2007;157(1): 201–210.
9. Kucukelbir A, Sigworth FJ, Tagare HD. Quantifying the local resolution of cryo-em density maps. Nat Methods. 2013;11(1):63.
10. Vilas JL, Gómez-Blanco J, Conesa P, Melero R, de la Rosa-Trevín JM, Otón J, Cuenca J, Marabini R, Carazo JM, Vargas J, et al. Monores: Automatic and accurate estimation of local resolution for electron microscopy maps. Structure. 2018;26(2):337–344.
11. Chen Y, Pfeffer S, Fernández JJ, Sorzano COS, Förster F. Autofocused 3d classification of cryoelectron subtomograms. Structure. 2014;22(10):1528–1537.
12. Lü Y, Zeng X, Zhao X, Li S, Li H, Gao X, Xu M. Fine-grained alignment of cryo-electron subtomograms based on mpi parallel optimization. BMC Bioinformatics. 2019;20(1):1–13.
13. Jonić S. Cryo-electron microscopy analysis of structurally heterogeneous macromolecular complexes. Comput Struct Biotechnol J. 2016;14:385–390.
14. Schwander P, Fung R, Ourmazd A. Conformations of macromolecules and their complexes from heterogeneous datasets. Phil Trans R Soc B Biol Sci. 2014;369(1647):20130567.
15. Shatsky M, Hall RJ, Brenner SE, Glaeser RM. A method for the alignment of heterogeneous macromolecules from electron microscopy. J Struct Biol. 2009;166(1):67–78.
16. Zaharia M, Chowdhury M, Franklin MJ, Shenker S, Stoica I. Spark: Cluster computing with working sets. HotCloud. 2010;10(10-10):95.
17. Scheres SH, Chen S. Prevention of overfitting in cryo-em structure determination. Nat Methods. 2012;9(9):853–54.
18. The MPI Forum C. MPI: A Message-Passing Interface. In: Proceedings of the 1993 ACM/IEEE Conference on Supercomputing; 1993. p. 878–883.
19. Kumar V, Ravikumar K, Aravinth SS, Rajkumar B. A message passing interface to support fast data access in distributed cloud environment along with master and slave communication. In: Second International Conference on Current Trends In Engineering and Technology-ICCTET 2014. IEEE; 2014. p. 309–312.
20. Dean J, Ghemawat S. Mapreduce: simplified data processing on large clusters. Commun ACM. 2008;51(1):107–113.
21. Ulusoy H, Kantarcioglu M, Thuraisingham B, Khan L. Honeypot based unauthorized data access detection in mapreduce systems. In: IEEE International Conference on Intelligence & Security Informatics. Baltimore: IEEE; 2016. p. 126–31.
22. Zhang K, Chen X-W. Large-scale deep belief nets with mapreduce. IEEE Access. 2014;2:395–403.
23. Uddin MA, Joolee JB, Alam A, Lee Y-K. Human action recognition using adaptive local motion descriptor in spark. IEEE Access. 2017;5:21157–21167.
24. Yang L, Liu J, Cheng F, Ansari N. Spark-based large-scale matrix inversion for big data processing. IEEE Access. 2016;4:2166–2176.
25. Tian X, Zhan JF. Graphduo: A dual-model graph processing framework. IEEE Access. 2018;PP(99):1.
26. Greber BJ, Nguyen THD, Fang J, Afonine PV, Adams PD, Nogales E. The cryo-electron microscopy structure of human transcription factor iih. Nature. 2017;549(7672):414.
27. Hutchings J, Stancheva V, Miller EA, Zanetti G. Subtomogram averaging of copii assemblies reveals how coat organization dictates membrane shape. Nat Commun. 2018;9(1):4154.
28. Hrabe T, Chen Y, Pfeffer S, Cuellar LK, Mangold A-V, Förster F. Pytom: a python-based toolbox for localization of macromolecules in cryo-electron tomograms and subtomogram analysis. J Struct Biol. 2012;178(2):177–188.
29. Pettersen EF, Goddard TD, Huang CC, Couch GS, Greenblatt DM, Meng EC, Ferrin TE. Ucsf chimera—a visualization system for exploratory research and analysis. J Comput Chem. 2004;25(13):1605–1612.


## Publisher's Note

Acharya *et al. BMC Bioinformatics* 2020, **21**(Suppl 13):386
https://doi.org/10.1186/s12859-020-03681-5

BMC Bioinformatics

METHODOLOGY Open Access

# A consensus multi-view multi-objective gene selection approach for improved sample classification

Sudipta Acharya[1], Laizhong Cui[1*] and Yi Pan[2]

*From* The 18th Asia Pacific Bioinformatics Conference
Seoul, Korea. 18-20 August 2020

*Correspondence: cuilz@szu.edu.cn
[1] Big Data Institute, College of Computer Science and Software Engineering, Shenzhen University, Shenzhen, PR China
Full list of author information is available at the end of the article

**Abstract**

**Background:** In the field of computational biology, analyzing complex data helps to extract relevant biological information. Sample classification of gene expression data is one such popular bio-data analysis technique. However, the presence of a large number of irrelevant/redundant genes in expression data makes a sample classification algorithm working inefficiently. Feature selection is one such high-dimensionality reduction technique that helps to maximize the effectiveness of any sample classification algorithm. Recent advances in biotechnology have improved the biological data to include multi-modal or multiple views. Different *'omics'* resources capture various equally important biological properties of entities. However, most of the existing feature selection methodologies are biased towards considering only one out of multiple biological resources. Consequently, some crucial aspects of available biological knowledge may get ignored, which could further improve feature selection efficiency.

**Results:** In this present work, we have proposed a Consensus Multi-View Multi-objective Clustering-based feature selection algorithm called **CMVMC**. Three controlled genomic and proteomic resources like gene expression, Gene Ontology (GO), and protein-protein interaction network (PPIN) are utilized to build two independent views. The concept of multi-objective consensus clustering has been applied within our proposed gene selection method to satisfy both incorporated views. Gene expression data sets of *Multiple tissues* and *Yeast* from two different organisms (*Homo Sapiens* and *Saccharomyces cerevisiae*, respectively) are chosen for experimental purposes. As the end-product of CMVMC, a reduced set of relevant and non-redundant genes are found for each chosen data set. These genes finally participate in an effective sample classification.

(Continued on next page)

(Continued from previous page)

**Conclusions:** The experimental study on chosen data sets shows that our proposed feature-selection method improves the sample classification accuracy and reduces the gene-space up to a significant level. In the case of *Multiple Tissues* data set, CMVMC reduces the number of genes (features) from 5565 to 41, with 92.73% of sample classification accuracy. For *Yeast* data set, the number of genes got reduced to 10 from 2884, with 95.84% sample classification accuracy. Two internal cluster validity indices - Silhouette and Davies-Bouldin (DB) and one external validity index Classification Accuracy (CA) are chosen for comparative study. Reported results are further validated through well-known biological significance test and visualization tool.

**Keywords:** Feature selection, Sample classification, Gene ontology (GO), Protein protein interaction network, Multi-view clustering, Multi-objective optimization

## Background

The high dimensionality of gene expression data set causes the data-analysis techniques working inefficiently. Sample classification of gene expression data is one such data-analysis method that is majorly affected by large number of genes present within a data set. As a solution, in the past several years, researchers have come up with dimensionality reduction methods following different strategies [1–4]. It can be done in two ways; 1) Feature extraction: which combines different available features and creates a new feature and, 2) Feature selection: which eliminates irrelevant features and keeps a smaller subset of available features. In the current article, we are focused on developing an efficient feature selection methodology. Here, genes of expression data sets are treated as features; hence, throughout the paper, we will use the term 'gene selection' and 'feature selection' alternatively.

In gene expression data, tissue samples are associated with expression values of thousands of genes. But in reality, only a smaller subset of genes actively participates in sample classification. Therefore, it is necessary to eliminate irrelevant genes in order to 1) make data more interpretable 2) reduce the noise, which intuitively improves classification accuracy, and lastly, 3) decrease computational cost. An efficient gene selection algorithm is capable of reducing the gene-space dimension of a expression data set without losing relevant biological information. This embedded relevant biological data is essential for sample classification.

The existing feature selection algorithms can be based on different methods such as clustering-based [3, 5], graph-based [6], feature similarity-based [7], content-based [8]. The traditional clustering-based feature selection methods [3, 5] often suffer from inter-method inconsistency in assigning data points to respective clusters [9]. Therefore, obtaining consensus clusters of multiple clustering solutions can improve the overall feature selection performance.

Another point of our motivation is that biological data can be interpreted in different ways. In other word, biological data is 'multi-omics' (such as genomics, proteomics, and methylomics) or having multiple 'views' ([10–12]). For example, to identify clusters of 'similar' genes, several biological resources like gene expression data, gene annotation data from GO, protein interaction data from PPIN, protein sequence data can be used independently. Each of these controlled vocabularies interprets gene-gene similarity in different aspects or views. Therefore, each of them produces different clustering solutions. Single-omic clustering has proven invaluable for non-biological, biological and medical research [13, 14]. Hence, for the problem mentioned above, the consensus of different views certainly helps to achieve better clustering solutions of similar genes. Though in recent years, several clustering-based gene selection strategies have been developed to identify the non-redundant set of genes, most of them are single 'view' or 'omics' data-based. In articles like Acharya et al. [5], Mitra et al. [3], the gene selection task is solved through single-view clustering approaches.

### Related works and motivation

In the last few years, several efficient feature/gene selection algorithms have been proposed. Some existing works relevant to this paper is discussed here in brief. A supervised feature selection technique was proposed by Liu et al. [15], which combines recursive feature addition (RFA) - a supervised learning method and a statistical similarity measure. Their proposed method considers only gene expression data. In contrast, Acharya et al. [5] proposed an unsupervised gene selection method based on annotation data of GO. Their method does not consider gene expression data but only GO annotation data. A single-objective genetic algorithm-based feature selection method can be found in Gunavathi et al. [2]. For sample classification utilizing their obtained feature set, kNN and support vector machine (SVM) classifiers are used. In Mirta et al. [3], authors developed a gene selection algorithm named Clustering Large Applications based upon RANdomized Search (CLARANS), using GO annotation data and then performed single-objective clustering on samples of gene expression data. A graph-theoretic approach of feature selection was proposed by Mandal et al. [6], where, from input gene expression data, a weighted dissimilarity graph is created. The nodes of the graph represent genes and edges represent their dissimilarity. More of the edge weight represents the more dissimilarity, and higher node weight represents the corresponding gene is highly relevant. Finally, the feature selection problem is modeled as a dense sub-graph finding problem that is solved through multi-objective binary particle swarm optimization (MBPSO).

One point common about the above-mentioned existing works is that all of them are single-view approaches. The working principles of most of the methods are either only based on gene expression data or gene annotation data from GO.

Recently, applications of multi-view learning on feature selection has become very popular. But, most of existing multi-view feature seletion methods are designed for video or text data analysis. For example, in Xu et al. [16], authors have proposed a weighted multi-view clustering and feature selection technique and shown its application on real-life text and image data sets. Also, their proposed approach follows single-objective clustering methodology. Similarly, in Shao et al. [17], an online unsupervised multi-view feature selection algorithm has been proposed by authors, but that is also for video or text data analysis. In Liu et al. [18] and Xue et al. [19], authors have developed feature selection algorithms using multi view data for video, text, and image data sets. On the problem of gene selection, an unsupervised graph-theoretic multi-view clustering approach was proposed bt Swarnkar et al. [20]. Though their proposed algorithm is a multi-view approach, here, different 'views' are developed by considering different feature subspaces over a gene expression data set. So, multi-omics data has not been utilized by them. A good review article on available multi-view feature selection algorithms can be found in Zhang et al. [21] and Yang et al. [22].

From the existing literature, it is evident that limited work has been done on the development of multi-view feature selection methods for biological data. Hence, there is a significant scope to explore this area. Inspired by this, in the current article, we have proposed an unsupervised consensus multi-view gene selection approach called CMVMC (Consensus Multi-View Multi-objective Clustering). we have amalgamated three concepts in our proposed gene selection method, such as consensus clustering, multi-view learning, and multi-objective optimization. Three primary controlled genomic/proteomic sources, i.e., gene expression, GO, and PPIN are chosen to form two different views. To the best of our knowledge, our proposed multi-view multi-objective approach of gene selection is new in this problem domain and has not explored before.

## Methods

In this section, first, we have formulated the chosen problem mathematically. Next, we have described the building mechanism of two views for the proposed multi-view framework. Finally, we will describe different steps of CMVMC-based gene selection approach.

### The mathematical formulation of chosen problem

- Given:
  - An original gene expression data set of $n$ genes $G_{org} = \left\{g_1, g_2, \ldots, g_n\right\}$.
  - Each gene has expression values of $d$ different samples.
  - Total $v$ different views, i.e., $v$ number of $n \times n$ similarity matrices.
  - $(v + 1)$ number of objective functions

    $$OF_1, OF_2, \ldots, OF_v, AI,$$

    where $OF_i$ is a fitness function of the clustering solution corresponding to view $i$. The last objective function is Agreement Index ($AI$), which measures the agreement between clustering solutions obtained by $v$ views.

- Outcomes:
  - A consensus clustering solution ($CU$) of similar genes, which satisfies $v$ views and simultaneously optimizes $(v + 1)$ objective functions.
  - The genes of $G_{org}$, is partitioned into $K$ clusters, $\{CU_1, CU_2, \ldots, CU_K\}$
  - $CU_i = \left\{g_1^i, g_2^i, \ldots, g_{n_i}^i\right\}$; $n_i$: number of points in cluster $i$; $g_j^i$: $j^{th}$ point of consensus cluster $i$.
  - $\cup_{i=1}^{K} CU_i = G_{org}$ and $CU_i \cap CU_j = \emptyset$ for all $i \neq j$.
  - $K$ cluster medoids of consensus clustering solution, which can be declared as selected features (genes) after proper validation test.
  - A reduced gene expression data set with $K$ genes $G_{org} = \left\{g_1, g_2, \ldots, g_K\right\}$.

### Two designed views of CMVMC

The first view used in CMVMC is the gene-gene similarity network based on pairwise Euclidean distance between expression vectors of genes. It represents the similarity between genes based on their sample-specific expression values. For the second view, the gene-gene similarity network is created utilizing a very recently developed fused gene-gene similarity measure *FuSim* [23]. This measure incorporates the biological properties of genes, both from GO and the corresponding organism's PPIN. The promising performance of *FuSim* inspired us to choose it for defining the second view. The similarity network of this view signifies the semantic and functional similarity between genes, which is not specific to samples but captures a global functional relatedness among them. The formation of both views is illustrated graphically in Fig. 1.

***Building view 1:***

Let us consider $n$ = number of genes and $d$= number of samples in the input gene expression data set. The original expression data is represented as $G_{org}[n][d]$. $g_{ik}$ represents expression value of $i^{th}$ gene of data set for $k^{th}$ sample where, $i \in n$ and $k \in d$. The similarity network for *view 1* is represented as two-dimensional similarity matrix $S_{View1}[n][n]$ of dimension $n \times n$. Each entry of $S_{View1}[\,][\,]$ matrix is calculated according to Eq. 1.

$$(1 - Eucli(g_i, g_j)) = \left(1 - \sqrt{\sum_{k=1}^{d}(g_{ik} - g_{jk})^2}\right) \tag{1}$$

where, $i,j \in n$ and $i \neq j$. We have chosen Euclidean distance as it is widely accepted and chosen distance measure while identifying co-expressed genes. The $Eucli(g_i, g_j)$ is subtracted from 1, which indicates the pair-wise similarity between gene $g_i$ and $g_j$.

***Building view 2:***

For *view 2*, the corresponding similarity matrix $S_{View2}[n][n]$ is created by calculating pair-wise similarity measure as follows.

$$FuSim\left(g_i, g_j\right) = \frac{Multi\text{-}SIM(g_i, g_j) + PPI\text{-}SIM\left(g_i, g_j\right)}{2} \tag{2}$$

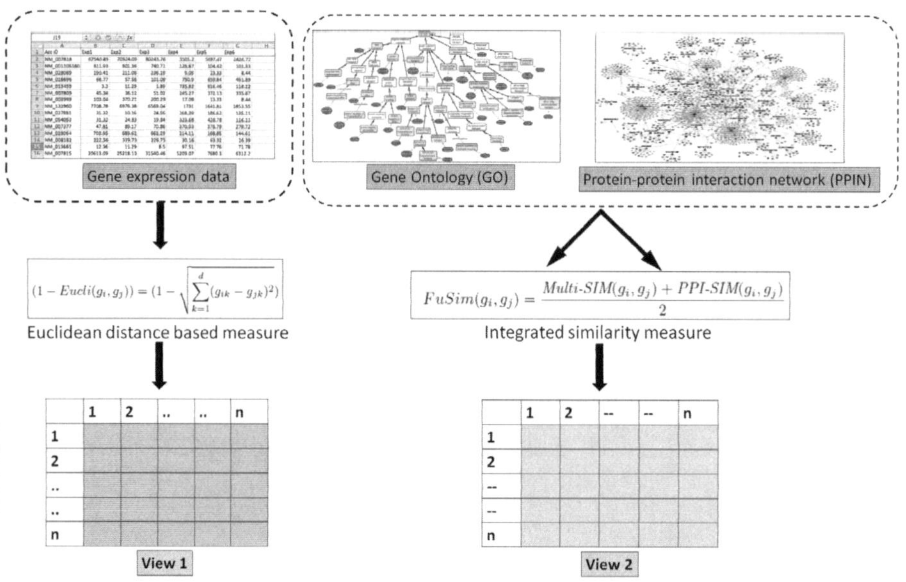


**Fig. 1** Two views developed based on multiple *'omics'* data

where, *Fusim* $(g_i, g_j)$ is GO (Acharya et al. [24]) and PPIN based hybrid similarity measure between gene $g_i$ and $g_j$ according to Acharya et al. [23]. *Multi-SIM* $(g_i, g_j)$ is GO-based multi-factored measure and *PPI-SIM* $(g_i, g_j)$ is confidence of association within PPI-based similarity measure between genes. The detailed mathematical formulation of Eq. 2 can be found in Acharya et al. [23].

The developed two views i.e., $S_{View1}[n][n]$ and $S_{View2}[n][n]$ are employed as different views in proposed CMVMC-based gene selection algorithm.

### Working strategy of proposed CMVMC-based gene selection method

In this section, each step of CMVMC-based gene selection algorithm is described in detail. The working flowchart of proposed gene selection method is shown in Fig. 2.

#### *Step 1: initializing clustering solutions*

The execution of CMVMC starts with the initialization of a number of clustering solutions with randomly selected centroid genes from the set of '*n*' genes in a given data set. The structure of a single solution is shown in Fig. 3. We can see from the figure that both for *view 1* and *view 2*, two separate strings (String 1 and String 2) within a single solution are created. String 3 within the same solution represents the consensus of String 1(*view 1*) and String 2(*view 2*). We called the clustering solution containing all three strings a '*Parent*' solution. String 3 is updated in each iteration (except the first iteration, initialized as null string). Each string (String 1, String 2, and String 3) of *parent* solution represents a complete clustering solution developed following its corresponding view. As shown in Fig. 3, $C_i$ and $C'_i$ are cluster center or centroid genes of $i^{th}$ cluster corresponding to *view 1* and *view 2*, respectively. $C_i^c$ is the $i^{th}$ consensus-cluster center of String 3. $K$ denotes

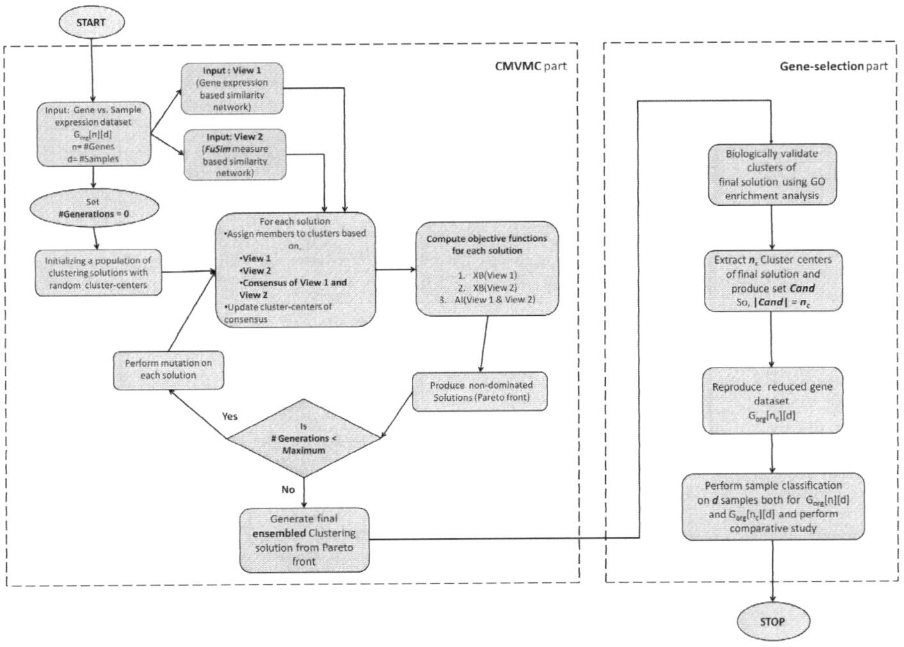


**Fig. 2** The flowchart of proposed CMVMC-based gene selection algorithm

# of clusters or **K** varies between 2 to $\sqrt{n}$

| | | | | | | | |
|---|---|---|---|---|---|---|---|
| String 1 → View 1 | $C_1$ | $C_2$ | $C_3$ | .. | .. | .. | $C_K$ |
| String 2 → View 2 | $C'_1$ | $C'_2$ | $C'_3$ | .. | .. | .. | $C'_K$ |
| String 3 → Consensus | $C^C_1$ | $C^C_2$ | $C^C_3$ | .. | .. | .. | $C^C_K$ |

**Fig. 3** Structure of each *parent* clustering solution in proposed CMVMC

the number of clusters. CMVMC detects the number of clusters in an optimized solution automatically; hence, $K$ varies between range 2 to $\sqrt{n}$ [25] for each clustering solution.

***Step 2: assigning genes to clusters and creating consensus clustering solution***

In this step, the rest of the genes (except centroid genes) are assigned to corresponding clusters of both strings (String 1 and String 2) within a *parent* solution. For a particular string, a gene is assigned to a cluster whose centroid gene is most similar to that gene concerning the corresponding view. The similarity is measured utilizing associated similarity matrices - $S_{View1}[n][n]$ and $S_{View2}[n][n]$ as developed previously.

Once individually, both strings are full of assigned genes, the consensus clustering solution combining String 1 and String 2 is created. The complete consensus solution is represented by String 3. An example of our strategy of consensus clustering is graphically illustrated in Fig. 4. First of all, the maximum overlapping clusters between String 1 and String 2 are identified. Next, the most centrally located gene among common genes between two mostly overlapping clusters is chosen as the new center of consensus cluster. The rest genes which can not be categorized are assigned to consensus clusters according to maximum-average-similarity strategy utilizing both $S_{View1}[n][n]$ and $S_{View2}[n][n]$.

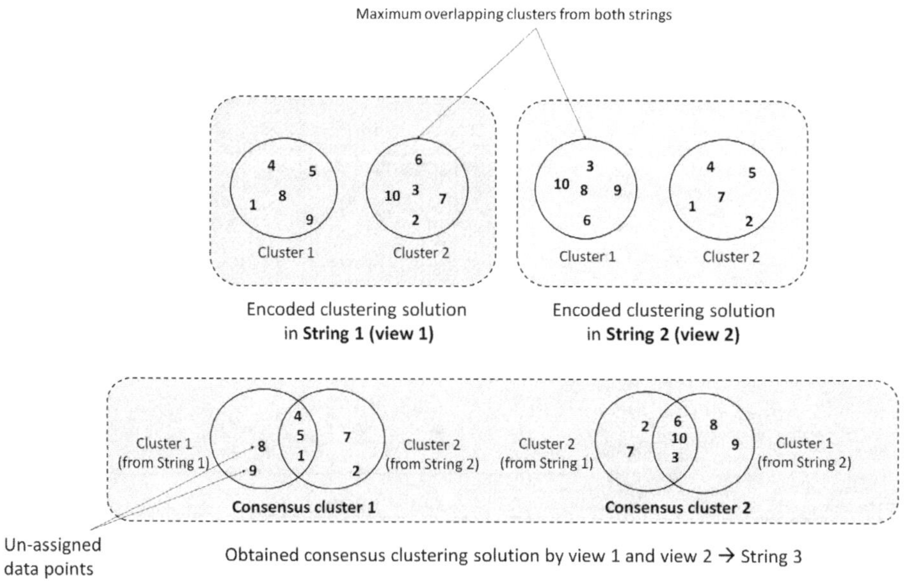


**Fig. 4** Formation of consensus clusters of *view 1* and *view 2*

### *Step 3: measuring objective functions*

CMVMC is a multi-objective clustering approach; hence, like other multi-objective optimization algorithms, it optimizes multiple fitness functions at its every iteration. For each *parent* solution, three of our chosen objective functions are Xie Beni (XB) index [26] for String 1, XB index for String 2, and Agreement Index (AI) [27] between String 1 and String 2. XB index measures cluster quality concerning better intra-cluster and inter-cluster distance and have been chosen in several past works of literature in recent years [23, 28]. For a good clustering solution, the XB index value should be minimum.

The third objective function AI [27] represents the similarity between clustering solutions obtained by two String 1 and String 2 (following both views) and higher its value signifies a better solution. The formulation of AI between two clustering solutions $CS_{view1}$ and $CS_{view2}$ based on two views, *view1* and *view2*, is described as follows.

Assume the agreement matrices corresponding to clustering solution $CS_{view1}$ and $CS_{view2}$ are $AM^{view1}$ and $AM^{view2}$ respectively. If there are $n$ genes, then the dimension of each agreement matrix is (n × n). From both obtained agreement matrices, the count of agreements ($AG$) is $AG = \sum_{i=1}^{n} \sum_{j=1}^{n} I_{AM_{ij}^{view1}, AM_{ij}^{view2}}$. where,

$$I_{AM_{ij}^{view1}, AM_{ij}^{view2}} = 1 \quad \text{if } AM_{ij}^{view1} = AM_{ij}^{view2}$$
$$= 0 \quad \text{otherwise}$$

Also the count of disagreements ($DG$) is $DG = n^2 - AG$.

Hence the AI between two clustering solutions $CS_{view1}$ and $CS_{view2}$ is formulated as follows.

$$AI_{CS_{view1}, CS_{view2}} = \frac{AG + 1}{DG + 1}$$

To avoid the *division by zero* problem, '1' is added as a normalization factor.

As the underlying multi-objective optimization strategy and generating non-dominated solutions in each iteration, we have used Archived Multi-Objective Simulated Annealing (AMOSA) [29] within CMVMC. In the field of multi-objective optimization, AMOSA has shown its superiority over several other multi-objective as well as single-objective optimization techniques. Its unique characteristic of accepting bad solution makes AMOSA not-to-get-stuck into local optima and hence provide optimal solutions.

### *Step 4: perturbation operators*

To achieve optimal clustering solution and hence to explore the search space efficiently, three types of perturbation operators are applied on both strings (String 1 and String 2) of each *parent* solution. They are described in brief as follows.

**Perturbation 1:** This perturbation operator aims to replace one existing centroid-gene from String 1 and String 2 with another gene. The gene to be selected as a new centroid is drawn from the original dataset randomly. Then it is replaced with an existing centroid-gene from a random position of the string.

**Perturbation 2:** This operation aims to increase the number of encoded clusters both in String 1 and 2 by one. For each string, a random gene from the original data set is chosen and added as the centroid of a new cluster.

**Perturbation 3:** This is the last type of perturbation operator, which aims to decrease the number of clusters by one from each string. A random position from both strings is chosen and assigned with a null value. Hence, the clusters count is decreased by one.

After performing any of the three mentioned mutation types, *Step 2* is repeated as membership has to be re-computed for the new cluster centers as well as consensus partitioning.

#### Step 5: choosing the final clustering solution through ensemble clustering

In the end, CMVMC produces multiple non-dominated Pareto optimal clustering solutions [30]. Next, all of the consensus partitions obtained from different non-dominated solutions are ensembled following the majority voting strategy. According to this strategy, if a pair of genes are members of the same cluster for the majority of produced consensus clustering solutions, then they kept together in the final solution. The rest unallocated genes are placed to corresponding clusters by following the maximum-average-similarity (average similarity based on $S_{View1}[n][n]$ and $S_{View2}[n][n]$) strategy as mentioned in *Step 2*. The ensembled clustering solution is further utilized to identify the set of relevant candidate genes.

#### Step 6: validate the obtained final clustering solution and prepare reduced gene set

Before obtaining the candidate genes from the ensembled clustering solution, to ensure that the obtained solution is biologically validated, we have performed GO enrichment analysis. GO tool (http://geneontology.org/) has been used for this significance test. After validating the clustering solution, the centers of obtained clusters embedded within the solution are extracted as candidate genes (features) of reduced gene-space. Suppose, this set of candidate genes is represented as *Cand*. Let, $|Cand|=n_c$, which represents $n_c$ number of genes are selected as relevant features/genes. Here, $n_c < n$ and also $n_c$ = # of clusters in the final solution. Next, from the original expression data set - $G_{org}[n][d]$, rows corresponding to $n_c$ candidate genes are only kept in reduced expression data set - $G_{org}[n_c][d]$. This is the end product of CMVMC-based gene selection algorithm. Finally, the sample classification on $d$ samples can be performed for obtained gene expression data set with reduced dimensionality.

## Results

### Description of chosen data sets and specifications of views

For this study, we have chosen benchmark gene expression data sets of *Multiple tissues* (http://portals.broadinstitute.org/cgi-bin/cancer/datasets.cgi) and *Yeast* (http://arep.med.harvard.edu/). Also, their corresponding PPINs, i.e., *Homo Sapiens* (*H. Sapiens*) and *Saccharomyces cerevisiae* (*S. Cerevisiae*), respectively, are utilized. The PPINs are downloaded from HitPredict [31] - an open-access resource of experimentally determined protein-protein interaction data over several organisms. Originally, *Multiple tissues* gene expression data set has 5565 genes(features) and 103 samples. Similarly, *Yeast* has 2884 genes and 17 samples. 103 samples of *Multiple tissues* are distributed among four normal tissue types, which are prostate, breast, colon, and lung. Similarly, 17 samples of *Yeast* are devided in two major classes and again, each major class is divided into four sub-classes G1, S, G2, and M [32]. The specifications, of *view 1* and *view 2* for both data sets are also noted as follows.

- *Multiple tissues*: To generate *view 2*, we have first obtained annotated GO-terms of 5565 genes using the online tool Gene Ontology Resource (http://geneontology.org/

docs/download-ontology/). The full GO-tree also downloaded from the same source. The number of mapped genes was 5205. The corresponding protein UniProt IDs of mapped 5205 genes are downloaded from 'https://www.uniprot.org/uploadlists/'. The protein IDs are used to retrieve their interacting proteins and associated statistics from downloaded *H. Sapiens* PPIN. Once these data are ready, then according to Eq. 2, *FuSim* measure [23] is calculated between each pair of 5205 genes and the gene-gene similarity network $S_{View2}[\,][\,]$ of dimension ($5205 \times 5205$) is created as *view 2*.
To generate *view 1*, the sample vectors of the same 5205 mapped genes are extracted from the original gene expression data $G_{org}[n][n]$. Then according to Eq. 1, the pair-wise similarity between all genes is calculated, and the similarity network $S_{View1}[\,][\,]$ of dimension ($5205 \times 5205$) is obtained and treated as *view 1*.

- *Yeast*: The view generation procedure for this data set is the same as the previous data set. Out of 2884 genes, 2307 genes get mapped to one or more significant GO-terms by our chosen GO tool. Following the same strategy as mentioned before, the similarity networks or views $S_{View1}[\,][\,]$ and $S_{View2}[\,][\,]$ ( of $2307 \times 2307$ dimension) are generated.

### Chosen external and internal validity measures

For that purpose of comparative analysis, two widely used internal cluster validity measures, which are Silhouette index [33] and DB index [34] have been used. A higher value of the Silhouette index and lower value of DB index indicates a better clustering solution. Also, to compare sample classification outcomes with original sample classes, an external validity index - Classification accuracy (CA) has been utilized.

## Discussion

### Discussion on results for *Multiple tissues* data

For *Multiple tissues* data set, at first, we have executed the CMVMC-based gene selection algorithm on its 5205 genes to identify gene clusters. To perform a comparative study, we have also executed single-view versions (considering *view 1* and *view 2* separately) of our proposed algorithm (Unsupervised Multi-objective Clustering (UMC)) on the chosen set of genes. We have also compared our obtained results with the approach proposed in Acharya et al. [5], which is shown in Table 1. From the table, we can see, for CMVMC, the final ensembled clustering solution has 41 clusters (indicated in bracket '()' in Table 1), and the corresponding Silhouette value is 0.446. For the dimension of reduced gene-space, the method of Acharya et al. [5] attains a slightly better dimension, i.e., 40 genes. The aim of gene selection is not only to reduce the dimension of gene-space but also to select relevant and non-redundant genes. Consequently, a better clustering solution ensures to identify more relevant genes from its outcome. Our obtained solution is better in terms of

**Table 1** Silhouette values corresponding to optimal gene clustering solution obtained by proposed CMVMC as well as other clustering approaches. In bracket () number of gene clusters is mentioned

| Data sets | CMVMC | UMC-view1 | UMC-view2 | PAM(GO-based) [5] |
|---|---|---|---|---|
| *Multiple tissues* | 0.446(**41**) | 0.434(**45**) | 0.439(**47**) | 0.4299(**40**) |
| *Yeast* | 0.4671(**10**) | 0.457(**13**) | 0.462(**10**) | 0.4531(**10**) |

Silhouette value compared to Acharya et al. [5] and other single-view approaches reported in Table 1. Hence, our obtained reduced gene-space with 41 genes seems more relevant.

Next, we have performed the biological significance test through GO enrichment analysis (http://geneontology.org/) for the obtained clustering solution having 41 clusters. The outcome of this test for the random two clusters is shown in Table 2. In the table for each of the significant GO-terms, the percentage of genes from the obtained cluster and full genome set in GO tool sharing that term is reported. It is quite evident from the table that a higher percentage of genes from obtained clusters mapped into the corresponding GO-term compared to the full genome set. This outcome indicates that genes within a same cluster of obtained solution are more involved in similar biological processes compared to other genes in the data set.

Also, to visualize the quality of obtained clusters, we have plotted expression values of genes through cluster-profile plot corresponds to one random cluster out of 41 clusters. It is shown in Fig. 5. The x-axis of the plot represents samples, and the y-axis represents expression value of a gene corresponds to a sample. As we can see in Fig. 5, the expression values of genes of the obtained cluster are mostly co-aligned. It represents genes within the same cluster are co-expressed. Therefore, it is another evidence for obtained clusters of being good quality.

Once the biological and visual validity of the obtained clustering solution is proved, next, centroid genes of 41 clusters are extracted. The IDs of these 41 genes are TMEM1, ZFR, ASH2L, RGS3, SMARCC1, TAF1C, BZRP, MAPK1, STAC, SEC22L1, PTGER1, HIPK3, PSMD12, ORM1, TPM3, HOXC5, CCL22, HIST2H2AA, CACYBP, RPL10A, KIAA0980, CRSP9, EPHX1, NPIP, CORO2B, KIAA0350, MYLK, DC12, ST5, RECK, ISGF3G, CDKN1A, LRPAP1, M6PRBP1, KIAA0792, CSTF1, MSL3L1, PDLIM7, SEC11L1, TNFRSF25, ZNF629, These genes act as candidate genes in obtained reduced gene-space. For *Multiple tissues* data set, originally, $n = 5565$ and reduced gene-space becomes $|Cand| = n_c = 41$.

Next, the classification of 103 samples is performed for both original (5565 genes) and reduced *Multiple tissues* gene expression data set. Besides our proposed one, we separately executed UMC-view1 and UMC-view2 based gene selection algorithm on the original data (5565 genes) and chosen common genes between two obtained reduced gene sets. Later, we perform sample classification with this common gene set (34 genes). We named the approach as UMC-v1-v2. Also, the reduced gene expression data set (40 genes) obtained by the approach of Acharya et al. [5] (we named it as PAM(GO)) has been chosen for comparative study. For original, as well as these three different reduced gene expression data sets, sample clustering is performed utilizing AMOSA-based multi-objective clustering algorithm.

The output sample clustering solutions are compared with each other with respect to Silhouette and DB index. Corresponding results are reported in Table 3 and plotted in Fig. 6. From the reported results in Table 3, it is evident that sample clusters obtained from reduced data set (41 genes) by our method have better quality with respect to both indices. Again, the sample clustering outcome is compared with results reported in Acharya et al. [5] and Mitra et al. [3] with respect to classification accuracy (%CA). The obtained comparative result is shown in Table 4 and plotted in Fig. 7. From Table 4, it is evident that sample classification using the reduced gene set developed by CMVMC method meets higher accuracy compared to other gene selection strategies.

**Table 2** Biological significance test outcome (under 'Biological process' ontology) for random two gene clusters obtained by CMVMC for both data sets

| *Multiple tissues* | | | | *Yeast* | | | |
|---|---|---|---|---|---|---|---|
| **Cluster** | **Significant GO-term** | **Cluster %** | **Genome %** | **Cluster** | **Significant GO-term** | **Cluster %** | **Genome %** |
| Cluster 1 | GO:0050896 | 55.38% | 40.14% | Cluster 1 | GO:0071704 | 61.2% | 53.7% |
| 134 genes | response to stimulus | | | 217 genes | organic substance metabolic process | | |
| | GO:0071840 | 40.7% | 27.86% | | GO:0044237 | 63.8% | 54.8% |
| | cellular component organization or biogenesis | | | | cellular metabolic process | | |
| | GO:0048518 | 42.9% | 29.3% | | GO:0044249 | 31.4% | 24.5% |
| | positive regulation of biological process | | | | cellular biosynthetic process | | |
| | GO:0032501 | 41.65% | 33.07% | | GO:0016043 | 43.4% | 33.4 |
| | multicellular organismal process | | | | cellular component organization | | |
| | GO:1901564 | 36.3% | 25.3% | | GO:0044260 | 42.8% | 33.5% |
| | organonitrogen compound metabolic process | | | | cellular macromolecule metabolic process | | |
| Cluster 2 | GO:0065008 | 37.39% | 19.24% | Cluster 2 | GO:0006807 | 53.9% | 47.18% |
| 124 genes | regulation of biological quality | | | 197 genes | nitrogen compound metabolic process | | |
| | GO:0051049 | 19.76% | 8.8% | | GO:0071840 | 40.09% | 37.4% |
| | regulation of transport | | | | cellular component organization or biogenesis | | |
| | GO:0032879 | 26.85% | 13.12% | | GO:0016043 | 40% | 33.4% |
| | regulation of localization | | | | cellular component organization | | |
| | GO:0009966 | 25.04% | 14.89% | | GO:0065007 | 41.7% | 34% |
| | regulation of signal transduction | | | | biological regulation | | |
| | GO:0051128 | 24.05% | 11.69% | | GO:0043170 | 50.8% | 40.4% |
| | regulation of cellular component organization | | | | macromolecule metabolic process | | |

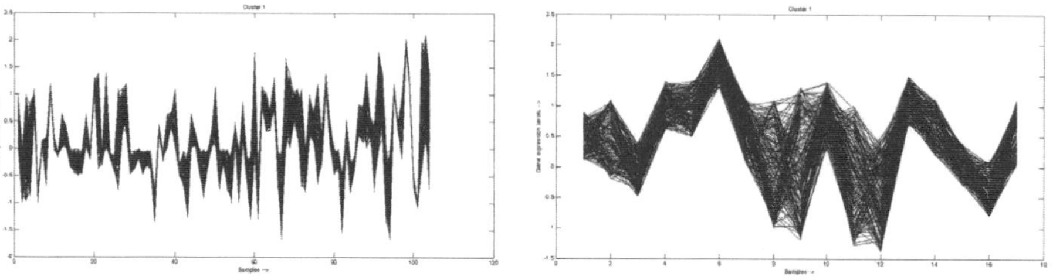

*Multiple tissues* *Yeast*

**Fig. 5** Cluster-profile plot for one random gene cluster from *Multiple tissues* (131 genes and 103 samples) and *Yeast* (180 genes and 17 samples) data set

#### Discussion on results for *Yeast* data

For *Yeast* data set, we have 2307 mapped genes on which we applied the proposed CMVMC-based gene selection technique and performed a comparative study with respect to other gene-selection algorithms. The results are reported in Table 1. Here, the reported Silhouette value corresponding CMVMC is better than the reported approach of Acharya et al. [5] by scale 0.014. According to our method, the number of genes in reduced gene-space is 10. The clustering solution obtained by UMC-view2 has Silhouette value 0.462, which is lesser than our proposed one, but it also detects a reduced set of 10 genes. For *Yeast* data set, $n = 2884$ and the size of reduced gene-space becomes $|Cand|=n_c = 10$. The clusters of obtained clustering solution by CMVMC-based gene selection method are also validated through GO enrichment analysis. The outcome of the analysis is shown in Table 2. It is evident from the reported biological significance test outcome, that, obtained clusters are biologically enriched. Again, the cluster-profile plot for one random cluster from the final gene clustering solution, as shown in Fig. 5, shows that the genes within the same cluster are co-expressed. The IDs of 10 selected genes in the reduced gene-space are YBL084C, YGR152C, YOL058W, YDR379W, YDR165W, YLR325C, YFL008W, YLR103C, YPL150W, YOL047C.

Similar to *Multiple tissues*, with the original (2884 genes) as well as reduced gene expression *Yeast* data sets by different approaches, we have executed multi-objective clustering-based sample classification. The summarized results are shown in Table 3 an plotted in Fig. 6. Table 3 shows with respect to the Silhouette index, the quality of sample clusters obtained by CMVMC is better than other methods. However, it is slightly lower than the gene selection method of Acharya et al. [5] for DB index. The comparative classification accuracy of samples is reported in Table 4 and Fig. 7. The obtained results prove

**Table 3** Comparative analysis of obtained sample clusters with respect to internal validity measures

| Data sets | # of genes (features) | # of samples | Silhouette | DB |
|---|---|---|---|---|
| *Multiple tissues* | 5565 (original) | 103 | 0.2527 | 0.998 |
| | 41(reduced by CMVMC) | | **0.4314** | **0.997** |
| | 34 (reduced by UMC-v1-v2) | | 0.3526 | 1.37 |
| | 40 (reduced by PAM(GO-based)) | | 0.4299 | 1.0065 |
| *Yeast* | 2884 (original) | 17 | 0.2365 | 0.149 |
| | 10 (reduced by CMVMC) | | **0.4568** | 0.087 |
| | 6 (reduced by UMC-v1-v2) | | 0.385 | 0.251 |
| | 10 (reduced by PAM(GO-based)) | | 0.4531 | **0.081** |

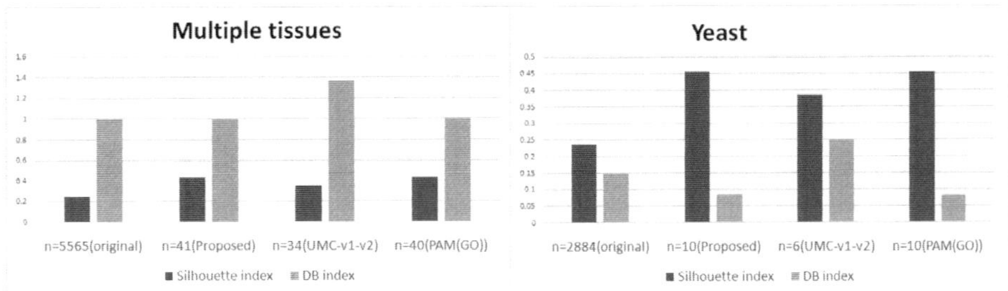


**Fig. 6** The comparative Silhouette and DB values for obtained sample clustering solutions for both data sets

that the reduced gene-space produced by our method has more biologically informative genes compared to other methods. As a result, the sample classification with our obtained gene-space attains higher accuracy compared to others.

Finally, from the obtained results and comparative analysis, it is clear that CMVMC-based gene selection approach identifies the more informative, relevant, and non-redundant set of genes from both data sets, which also assists in attaining higher accuracy in sample classification.

## Conclusions

In the current work, we have developed an unsupervised consensus multi-view multi-objective gene selection approach (CMVMC-gene selection) for efficient sample classification. Multi-faceted biological sources like gene expression data, GO, and PPIN are involved in defining two views of gene-gene similarity networks. These complimentary views are further utilized by the proposed gene selection algorithm. Experiments are performed on two widely used benchmark gene expression data sets - *Yeast* and *Multiple tissues*. Their corresponding PPINs of *S. Cerevisiae* and *H. Sapiens* are obtained from Hit-Predict [31], and also exploited for preparing one of the two developed views. Reported

**Table 4** Comparative percentage of Classification Accuracy (%CA) values of proposed CMVMC-based gene selection as well as other existing methods

| Data set | # of genes | Algorithms | %CA |
|---|---|---|---|
| *Multiple tissues* | **41** | CMVMC-gene selection | **92.73** |
| | 40 | PAM+AMOSA | 92.14 |
| | 34 | UMC-v1-v2 | 78.4 |
| | 42 | CLARANS+k-NN | 81.03 |
| | | CLARANS+RF | 76.0 |
| | | CLARANS+C4.5 | 65.0 |
| | | CLARANS+NB | 92.23 |
| | | CLARANS+MLP | 89.32 |
| *Yeast* | **10** | CMVMC-gene selection | **95.84** |
| | 10 | PAM+AMOSA | 95.63 |
| | 6 | UMC-v1-v2 | 79.5 |
| | 15 | CLARANS+k-NN | 86.78 |
| | | CLARANS+RF | 94.12 |
| | | CLARANS+C4.5 | 94.12 |
| | | CLARANS+NB | 94.12 |
| | | CLARANS+MLP | 94.12 |

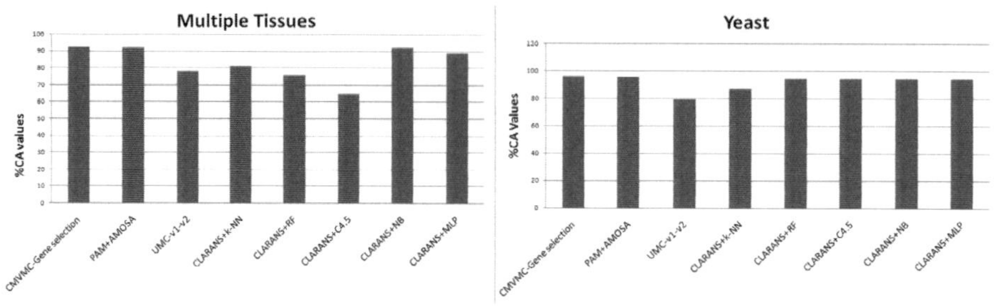


**Fig. 7** The comparative Classification Accuracy (CA) of samples by proposed and existing gene selection approaches

results are compared with state-of-the-art gene selection algorithms, and supported through proper biological significance and a visualization tests.

A thorough analysis of the reported results proves that our proposed CMVMC-based gene selection technique not only reduces the gene-space up to a reasonable scale but also ensures higher accuracy in sample classification. According to the current state of this work, one or more *'omics'* data can be further incorporated in designing a multi-view clustering algorithm to make it more efficient in finding groups of functionally similar genes. The authors are currently working in that direction.

#### Abbreviations
CMVMC: Consensus Multi-View Multi-objective Clustering; GO: Gene Ontology; PPIN: Protein-Protein Interaction Network; FuSim: Fused Similarity measure; CLARANS: Clustering Large Applications based upon RANdomized Search; UMC: Unsupervised Multi-objective Clustering; XB: Xie Beni; AI: Agreement Index; AMOSA: Archived Multi Objective Simulated annealing; CA: Classification Accuracy; DB index: Davies Bouldin index; MBPSO: Multi-objective Binary Particle Swarm Optimization; k-NN: k Nearest Neighbour; RF: Random Forest; NB: Naive Bayes; MLP: Multi-layer Perceptron

#### Acknowledgements
Not applicable.

#### About this supplement
This article has been published as part of *BMC Bioinformatics Volume 21 Supplement 13, 2020: Selected articles from the 18th Asia Pacific Bioinformatics Conference (APBC 2020): bioinformatics*. The full contents of the supplement are available online at https://bmcbioinformatics.biomedcentral.com/articles/supplements/volume-21-supplement-13.

#### Authors' contributions
The idea was given by SA and YP. SA has implemented the work. The analysis of results was jointly done by SA and LC. SA, LC and YP have participated in writing the paper. All authors have read and approved the final manuscript.

#### Funding
This work has been partially supported by National Key Research and Development Plan of China under Grant No. 2018YFB1800302, National Natural Science Foundation of China under Grant No.61772345, Major Fundamental Research Project in the Science and Technology Plan of Shenzhen Grant No. JCYJ20190808142207420, and Pearl River Young Scholars funding of Shenzhen University. Publication costs are funded by Pearl River Young Scholars funding of Shenzhen University.

#### Availability of data and materials
All data sets used in the work are publicly available and the source reference are given in main manuscript.

#### Ethics approval and consent to participate
Not applicable.

#### Consent for publication
Not applicable.

#### Competing interests
The authors declare that they have no competing interests.

#### Author details
[1]Big Data Institute, College of Computer Science and Software Engineering, Shenzhen University, Shenzhen, PR China. [2]Department of Computer Science, Georgia State University, Atlanta, USA.

Published: 17 September 2020

**References**

1. Chandra B, Gupta M. An efficient statistical feature selection approach for classification of gene expression data. J Biomed Inform. 2011;44(4):529–35.
2. Gunavathi C, Premalatha K. Performance analysis of genetic algorithm with kNN and SVM for feature selection in tumor classification. Int J Comput Electr Autom Control Inform Eng. 2014;8(8):1490–7.
3. Mitra S, Ghosh S. Feature selection and clustering of gene expression profiles using biological knowledge. IEEE Trans Syst Man Cybern Part C Appl Rev. 2012;42(6):1590–9.
4. Mudiyanselage TKB, Xiao X, Zhang Y, Pan Y. Deep fuzzy neural networks for biomarker selection for accurate cancer detection. IEEE Trans Fuzzy Syst. 2019.
5. Acharya S, Saha S, Nikhil N. Unsupervised gene selection using biological knowledge: application in sample clustering. BMC Bioinformatics. 2017;18(1):513.
6. Mandal M, Mukhopadhyay A. A graph-theoretic approach for identifying non-redundant and relevant gene markers from microarray data using multiobjective binary PSO. PloS ONE. 2014;9(3):90949.
7. Mitra P, Murthy C, Pal SK. Unsupervised feature selection using feature similarity. IEEE Trans Pattern Anal Mach Intell. 2002;24(3):301–12.
8. Dy JG, Brodley CE, Kak A, Broderick LS, Aisen AM. Unsupervised feature selection applied to content-based retrieval of lung images. IEEE Trans Pattern Anal Mach Intell. 2003;25(3):373–8.
9. Kellam P, Swift S, Tucker A, Vinciotti V, Martin NJ, Orengo CA, Liu X. Consensus clustering and functional interpretation of gene expression data. In: Spanish Bioinformatics Conference; 2004. p. 6.
10. Zeng M, Li M, Fei Z, Wu F, Li Y, Pan Y, Wang J. A deep learning framework for identifying essential proteins by integrating multiple types of biological information. IEEE/ACM Trans Comput Biol Bioinforma. 2019.
11. Hu J, Zhou X, Zhu YH, Yu DJ, Zhang G. TargetDBP: accurate DNA-binding protein prediction via sequence-based multi-view feature learning. IEEE/ACM Trans Comput Biol Bioinforma. 2019.
12. Acharya S, Cui L, Pan Y. A refined 3-in-1 fused protein similarity measure: application in threshold-free hub detection. IEEE/ACM Trans Comput Biol Bioinforma. 2020. https://doi.org/10.1109/TCBB.2020.2973563.
13. Rappoport N, Shamir R. Multi-omic and multi-view clustering algorithms: review and cancer benchmark. Nucleic Acids Res. 2018;46(20):10546–62.
14. Kumar S, Singh M. A novel clustering technique for efficient clustering of big data in Hadoop Ecosystem. Big Data Min Anal. 2019;2(4):240–7.
15. Liu Q, Sung AH, Chen Z, Liu J, Huang X, Deng Y. Feature selection and classification of MAQC-II breast cancer and multiple myeloma microarray gene expression data. PloS ONE. 2009;4(12):8250.
16. Xu YM, Wang CD, Lai JH. Weighted multi-view clustering with feature selection. Pattern Recogn. 2016;53:25–35.
17. Shao W, He L, Lu CT, Wei X, Philip SY. Online unsupervised multi-view feature selection. In: 2016 IEEE 16th International Conference on Data Mining (ICDM). IEEE; 2016. p. 1203–8.
18. Liu H, Mao H, Fu Y. Robust multi-view feature selection. In: 2016 IEEE 16th International Conference on Data Mining (ICDM). IEEE; 2016. p. 281–90.
19. Xue Y, Wang N, Niu Y, Zhong P, Niu S, Song Y. Robust re-weighted multi-view feature selection. Comput Mater Contin. 2019;60(2):741–56.
20. Swarnkar T, Mitra P. Graph-based unsupervised feature selection and multiview clustering for microarray data. J Biosci. 2015;40(4):755–67.
21. Zhang R, Nie F, Li X, Wei X. Feature selection with multi-view data: A survey. Inform Fusion. 2019;50:158–67. https://doi.org/10.1016/j.inffus.2018.11.019.
22. Yang Y, Wang H. Multi-view clustering: A survey. Big Data Min Anal. 2018;1(2):83–107.
23. Acharya S, Cui L, Pan Y. Automated hub-protein detection via a new fused similarity measure-based multi-objective clustering framework. In: International Symposium on Bioinformatics Research and Applications. Springer; 2019. p. 138–45.
24. Acharya S, Saha S, Pradhan P. Multi-factored gene-gene proximity measures exploiting biological knowledge extracted from gene ontology: application in gene clustering. IEEE/ACM Trans Comput Biol Bioinforma. 2018.
25. Bezdek JC, Pal NR. Some new indexes of cluster validity. IEEE Trans Syst Man Cybern B. 1998;28(3):301–15.
26. Xie XL, Beni G. A validity measure for fuzzy clustering. IEEE Trans Pattern Anal Mach Intell. 1991;13(8):841–7.
27. Rabbany R, Zaïane OR. A general clustering agreement index: For comparing disjoint and overlapping clusters. In: Thirty-First AAAI Conference on Artificial Intelligence; 2017.
28. Acharya S, Saha S, Thadisina Y. Multiobjective simulated annealing-based clustering of tissue samples for cancer diagnosis. IEEE J Biomed Health Inform. 2016;20(2):691–8.
29. Bandyopadhyay S, Saha S, Maulik U, Deb K. A simulated annealing-based multiobjective optimization algorithm: AMOSA. IEEE Trans Evol Comput. 2008;12(3):269–83.
30. Deb K, Pratap A, Agarwal S, Meyarivan T. A fast and elitist multiobjective genetic algorithm: NSGA-II. IEEE Trans Evol Comput. 2002;6(2):182–97.
31. López Y, Nakai K, Patil A. HitPredict version 4: comprehensive reliability scoring of physical protein–protein interactions from more than 100 species. Database. 2015;2015:.
32. Cho SB, Yoo SH. Fuzzy Bayesian validation for cluster analysis of yeast cell-cycle data. Pattern Recognit. 2006;39(12): 2405–14.
33. Rousseeuw PJ. Silhouettes: a graphical aid to the interpretation and validation of cluster analysis. J Comput Appl Math. 1987;20:53–65.
34. Davies DL, Bouldin DW. A cluster separation measure. IEEE Trans Pattern Anal Mach Intell. 1979;2:224–7.

## Publisher's Note

Zhou *et al. BMC Bioinformatics* 2020, **21**(Suppl 13):387
https://doi.org/10.1186/s12859-020-03682-4

BMC Bioinformatics

METHODOLOGY Open Access

# NEDD: a network embedding based method for predicting drug-disease associations

Renyi Zhou[1], Zhangli Lu[1], Huimin Luo[1,2], Ju Xiang[1,3], Min Zeng[1] and Min Li[1*]

*From* The 18th Asia Pacific Bioinformatics Conference
Seoul, Korea. 18-20 August 2020

* Correspondence: limin@csu.edu.cn
[1]Hunan Provincial Key Lab on Bioinformatics, School of Computer Science and Engineering, Central South University, Changsha, China
Full list of author information is available at the end of the article

## Abstract

**Background:** Drug discovery is known for the large amount of money and time it consumes and the high risk it takes. Drug repositioning has, therefore, become a popular approach to save time and cost by finding novel indications for approved drugs. In order to distinguish these novel indications accurately in a great many of latent associations between drugs and diseases, it is necessary to exploit abundant heterogeneous information about drugs and diseases.

**Results:** In this article, we propose a meta-path-based computational method called NEDD to predict novel associations between drugs and diseases using heterogeneous information. First, we construct a heterogeneous network as an undirected graph by integrating drug-drug similarity, disease-disease similarity, and known drug-disease associations. NEDD uses meta paths of different lengths to explicitly capture the indirect relationships, or high order proximity, within drugs and diseases, by which the low dimensional representation vectors of drugs and diseases are obtained. NEDD then uses a random forest classifier to predict novel associations between drugs and diseases.

**Conclusions:** The experiments on a gold standard dataset which contains 1933 validated drug-disease associations show that NEDD produces superior prediction results compared with the state-of-the-art approaches.

**Keywords:** Drug repositioning, Heterogeneous network, Network embedding, Meta path

## Background

Drug discovery is known for a large amount of money and time it consumes and the high risk it takes [1]. The investments grow continuously in recent years, but the total number of approved drugs remains constant [2]. Therefore, drug repositioning has become a popular approach to save cost by finding novel indications for approved drugs. Since these commercialized drugs have passed various clinical tests, it would save tremendous effort if we could reuse them directly. As reported, relaunching a repositioned drug can save about 80% of the cost compared with launching a reformulation of an existing drug [3].

The goal of drug repositioning is to find potential new target diseases for an existing drug and apply the newly identified drug to the treatment of diseases other than the drug's originally intended ones [4]. Historically, the discovery of finding new indications for existing drugs is mostly the result of a better understanding of a drug [5] or serendipity [6]. Then as the omics data accumulate, new bioinformatics methods emerge and play an increasingly important role. The newly proposed methods generally can be categorized as 'drug based' or 'disease based' [7]. For instance, IDMap [8] is a drug-based method and it mainly focuses on exploring the chemical structure information of drugs. Later, with the growth of drug-related data and the initiative of open data, recent studies pay more attention to integrating heterogeneous information. For example, Gottlieb et al. proposed PREDICT [9], a method that integrates various drug-drug similarity and disease-disease similarity from different sources.

These computational repositioning approaches can be roughly divided into three types: machine learning methods, text mining methods, and network-based methods [5, 10].

The aforementioned method PREDICT is an example of machine learning methods. The authors used similarities as features and applied a logistic regression classifier to predict novel indications for drugs. Moreover, Napolitano et al. proposed an approach which used a combination of drug-related data to train a multi-class SVM (Support Vector Machine) classifier to identify latent drug-disease associations [11]. Besides, there are also researches comparing traditional machine learning methods and deep learning methods [12, 13].

Electronic health records (EHR) of the patients and other literature contains a vast amount of information about drugs and diseases that can be explored using the text mining technique. For instance, Zhu et al. explored pharmacogenomics studies and modelled FDA-approved breast cancer drugs by using Semantic Web notions which support automated semantic inference [14]. Chen et al. integrated and annotated data from public datasets and developed a statistical model called Semantic Link Association Prediction (SLAP) to assess drug–target associations based on semantic links [15].

Network-based methods have been wildly used for computational drug repositioning. Martínez V et al. proposed DrugNet [16], which is based on a heterogeneous network prioritization approach that can utilize heterogeneous information. Luo, Y et al. developed a pipeline called DTINet [17]. It originally aims to find interactions between drugs and targets, but can also be applied in drug repositioning. DTINet uses a matrix completion method to calculate the low dimensional feature vectors which capture the topological properties of nodes in the network and uses these features to predict novel associations. Luo, H et al. proposed MBiRW [18], which used the Bi-Random Walk algorithm to predict potential novel indications of a drug. However, current network-based methods often show a certain preference for drugs that have more known drug-disease associations. Therefore, they are not good at finding novel indications for drugs that are less explored or new drugs.

In this work, we propose NEDD, a network embedding based method for predicting novel interactions between drugs and diseases using heterogeneous information. NEDD tries to solve the above problems by adopting the concept of inductive learning and meta path. The results of experiments show that NEDD outperforms other methods in drug repositioning.

## Methods

In this section, we will introduce our method NEDD. Generally, the whole procedure of NEDD consists of three steps. First, based on the heterogeneous information related to drugs and diseases, an undirected graph with weighted edges is constructed. Second, we train a meta-path-based representation learning model to learn the embedding vectors of each entity. Last, using the vectors learned, we train a classifier to identify potential associations between drugs and diseases.

### Construction of the drug-disease network

We construct the network as an undirected graph that consists of two node types (drug node $n_{dr}$ and disease node $n_{di}$) and three edge types (drug-disease edge $r_{dr\text{-}di}$, drug-drug edge $r_{dr\text{-}dr}$, and disease-disease edge $r_{di\text{-}di}$). An edge of $r_{dr\text{-}di}$ represents the association between a drug and a disease. An edge of $r_{dr\text{-}dr}$ represents the connection between two drugs which have a high similarity and an edge of $r_{di\text{-}di}$ denotes the connection between two diseases which have a high similarity.

After calculating all the similarities among drugs and that among diseases, we filter the edges with a certain threshold. In particular, the threshold of drug similarity is 0.8 and the threshold of disease similarity is 0.7. The thresholds are determined by experiments. Each similarity edge below its threshold is removed unless it has the greatest weight (similarity) among other homogeneous edges for a drug or a disease and removing this edge may cause one node to be isolated. The insight of filtering is that drug pairs with low similarity have an insignificant probability in indicating common diseases while drug pairs with high similarity have a strong probability to indicate common diseases [18], and the same is also true for disease pairs. This step is illustrated in Fig. 1.

Though here we only make use of three types of relationships within drugs and diseases, other types of information, like disease-target interactions and so on, can be used to expand the graph.

### Network embedding

Most existing network representation learning methods can be summarized into two steps: proximity matrix construction and dimension reduction [19]. NEDD uses

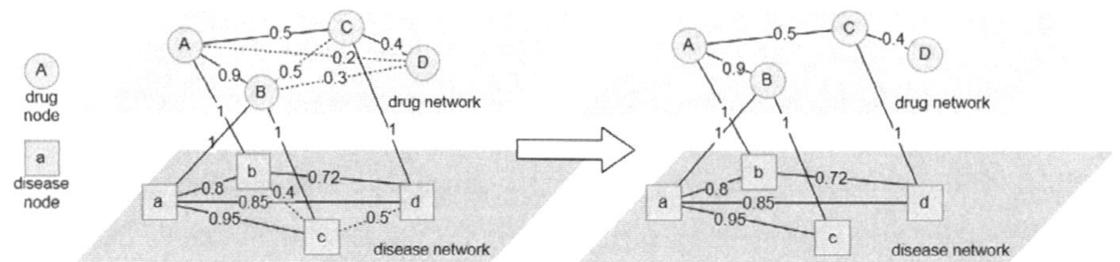


**Fig. 1** Network Construction. A demo graph constructed by integrating drug similarity network and disease similarity network. Dotted lines, like $Edge_{bc}$, represent removed edges whose weights under thresholds. Though weights of $Edge_{AC}$ is under the threshold, this edge is not removed because firstly all similarity edges of $Node_C$ are below the threshold and secondly it is the edge with the most weight among them. And it is the same with $Edge_{CD}$

random walk to perform the first step. One of the strengths of random walk is that it is efficient in both time and space. In the next step, NEDD adopts a neural-network-based method, HIN2vec [20], to learn network embedding vectors.

The key concept of the first step is the meta path. Given a drug-disease network $G(V, E)$, the node type set $T_V$ and the edge type set $T_E$, a meta path can be defined as a triplet $(n_h, n_t, m)$, where $n_h, n_t \in T_V$ are the head node type and the end node type, and m is a sequence of edge types $r_1 \rightarrow r_2 \rightarrow \ldots \rightarrow r_l$ $(r_1, r_2, \ldots, r_l \in T_E)$ which indicates a composite relation between the two node types. For example, a meta path $(n_{di}, n_{di}, r_{dr\text{-}di} \rightarrow r_{dr\text{-}di})$ indicates the relationship that two diseases share the same treatment, while a meta path $(n_{dr}, n_{dr}, r_{dr\text{-}di} \rightarrow r_{di\text{-}di} \rightarrow r_{dr\text{-}di})$ describes the relationship of two drugs that could treat two similar diseases.

NEDD first uses a random walk algorithm to generate long possible paths in the graph. In this process, node numbers and node types are recorded. Then, these paths are cut into shorter paths with lengths from 1 to W. These shorter paths represent the network proximity between two nodes of the $1^{st}$ to the $W^{th}$ order. In our experiments, W is set to 6. Besides, negative sampling [21] is used to generate negative data entries.

Then, NEDD adopts HIN2vec [20] to learn network embedding vectors. HIN2vec is a representation learning method. It assigns a low dimensional embedding vector for each entity in the graph, including drug nodes and disease nodes, and each meta path. It trains a neural network classifier with one hidden layer to identify if two nodes have a certain relationship and take weights of the hidden layer as embedding vectors like word2vec [21]. The prediction which node $v_i$ and node $v_j$ have an association that matches a particular meta path $R$ is given as below:

$$P(R|v_i, v_j) = sigmoid\left(\sum e_{v_i} \odot e_{v_j} \odot f_{01}(e_R)\right)$$

where $e_{v_i}, e_{v_j}, e_R \in R^D$ is the embedding vector of the nodes $v_i$, $v_j$, and the meta path $R$, $f_{01}$ is a regularization function that regularizes values in $e_R$ within 0 and 1, and $\odot$ is the element-wise product. HIN2vec uses cross-entropy loss to measure the prediction error and uses stochastic gradient descent to update embedding vectors.

A visualization using T-SNE [22] of meta path embedding vectors is shown in Fig. 2.

### Using embedding vectors to predict novel associations

In this step, we use a random forest classifier to make final predictions. We apply the element-wise product to aggregate a drug node embedding vector and a disease node embedding vector together as the input of the random forest classifier. The output of the random forest classifier is the predicted probability that a drug and a disease have associations:

$$P(durg_i\ treats\ disease_j) = random\_forest\_classifier(e_{drug_i} \odot e_{disease_j})$$

We use scikit-learn [23] to implement the random forest classifier. While optimizing the parameters of the random forest, we evaluated the forest on the same training set and validation set. Because random forests are less likely to overfit, we started with a large classifier and gradually cut down the scale. In the end, we set the max depth of trees to 25, the number of estimators to 300, the minimum number of samples to split to 2, and use the Gini coefficient as the criterion.

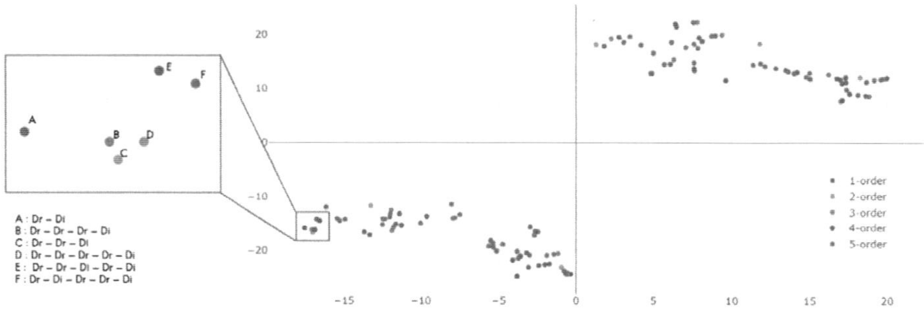


**Fig. 2** Visualization of meta path embedding vectors. Visualization of meta path embedding vectors using T-SNE [22]. The visualization result is very intuitive. Since the graph we created is an undirected one, the embeddings of paths are naturally symmetric. The lower-left group is meta paths which start from drug nodes and the upper-right group represent meta paths which start from disease nodes. Each point represents a meta path vector and different color represents different order of the relationships. For instance, point C on the left is a meta path of "Drug-Drug-Disease", which represents the second order relationship that a drug might cure a disease that a similar drug can treat

## Results

In this section, we evaluate the performance of NEDD on the gold standard dataset. First, we introduce the evaluation metrics. Then, by performing these measurements, we compare NEDD with five other methods: MBiRW [18], DTINet [17], HGBI [24], NBI [25], and JUST [26].

### Dataset

In the dataset, drug-disease associations are collected from multiple data sources. This gold standard dataset which has been used in reference [9] includes 593 drugs from DrugBank [27], 313 diseases from the Online Mendelian Inheritance in Man (OMIM) [28] and 1933 validated drug–disease associations.

The drug similarity data is calculated by the Chemical Development Kit (CDK) [29] based on SMILES [30] chemical structures and the disease similarity data is obtained from MimMiner [31] which is based on disease phenotype similarity using text mining analysis of their medical descriptions information in the OMIM database.

### Evaluation metrics

In order to evaluate the ability of NEDD in finding new possible target diseases of a specific drug, we conduct 10-fold cross-validation and perform the top-ranked candidate disease analysis.

In 10-fold cross-validation, all 1933 known drug-disease associations in gold standard datasets are randomly divided into 10 partitions with each roughly equal in size. Then, 1 of 10 partitions in turn serves as the test set, while the remaining as the training set. After the whole process, each possible association is given a score representing the confidence of the association. Then these associations are sorted in descending order according to their score. Next, for each ranking threshold, true positive rate (TPR), which measures the proportion of known associations that are correctly identified, and false positive rate (FPR), which measures the proportion of unverified associations that are predicted as real associations, are calculated based on the ranking results. By changing settings of the threshold, we can get various pairs of TPR and FPR. Based on these pairs, the receiver operating characteristic curve (ROC curve) can be drawn with FPR as the x-axis and TPR as the y-axis. The area under ROC (AUC) is then calculated to measure the performance.

Besides, as top ranked predicting results may be important in practice, we also test our method in terms of top ranked results. We count the number of true drug-disease associations proved by other sources, e.g. other datasets or literatures to evaluate NEDD and other methods.

## Comparison with other methods

NEDD is compared with five state-of-the-art methods: MBiRW [18], DTINet [17], HGBI [24], NBI [25], and JUST [26]. These five methods are all computational methods which are based on network and can utilize the heterogeneous network of drugs and diseases. NBI is a method based on a two-state diffusion model in a bipartite graph. HGBI is a method based on the guilt-by-association principle and an intuitive interpretation of information flow on a heterogeneous graph. MBiRW and DTINet have been introduced in the above. MBiRW is based on random walk and DTINet is based on matrix completion. Moreover, similar to NEDD, DTINet also adopts the concept of inductive learning. HGBI and NBI are also originally developed for drug-target association prediction but they have also been used in drug-disease association prediction [32]. JUST improves the random walk sampling method on the heterogeneous network which avoids using meta paths and use Skip-Gram model [21] to learn network embedding vectors. In the study, the parameter $\alpha$ is set to 0.3, and $l$, $r$ to 2 for MBiRW according to the default parameter setting in [18]. The parameter $\alpha$ of HGBI is set to 0.4 as suggested in [24]. We change the lengths of vectors to 100 in DTINet, and we use the default settings for other parameters. Besides, we use additional information to train DTINet model because the method mainly focuses on integrating various information. The additional information includes drug-drug interactions and drug-protein interactions collected from DrugBank [27], associations between drugs and side-effects from SIDE R [33], and disease-gene associations from CTD [34]. For JUST, we set $\alpha$ to 0.4, $m$ to 2 as suggested in the original paper [26]. The length of the embedding vectors is set to 128 and the window size of the Skip-Gram model is set to 10.

Through repeating the 10-fold cross-validation experiment specified in the above 1 hundred times with different random seeds, we calculate average AUC to estimate the performance of each method. In the 10-fold cross-validation test, the experiment results show that NEDD outperforms the other methods. NEDD achieves the AUC value of 0.923, while MBiRW, DTINet, HGBI, NBI, and JUST obtain inferior results of 0.912, 0.869, 0.820, 0.584 and 0.828, respectively. The results are illustrated in Fig. 3.

## Case study

After confirming the ability of NEDD to predict potential drug-disease associations based on 10-fold cross-validation, we further conduct a case study to search evidences in other sources. In this step, all known associations in the gold standard dataset are used as the training set, and then the possible associations are ranked according to NEDD's prediction. Next, top-ranked predictions for each drug are verified based on CTD [34]. The top-N results of NEDD are summarized in Fig. 4. In this step, it shows

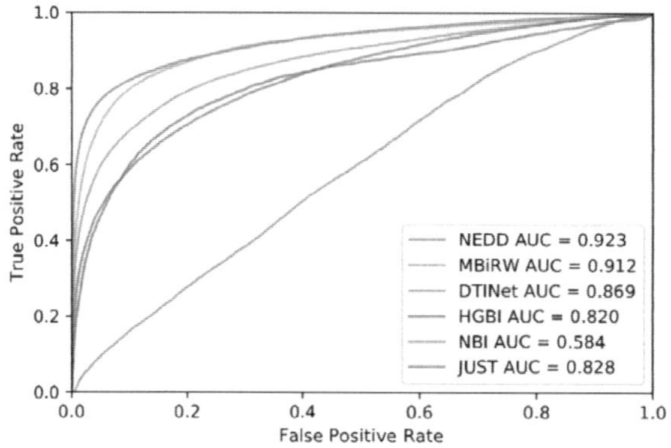


**Fig. 3** Results of ten-fold cross-validation. AUC and ROC curve of ten-fold cross-validation

that 204 novel indications in the top 5 predictions are verified by CTD; 438 novel indications are verified in the top 10 and 849 novel indications are verified in the top 20.

Fifteen drug-disease associations with the highest prediction scores of all are listed in Table 1. None of these associations are verified by CTD. So, we conduct a further study to find supporting evidences by literature searching. The result shows that 7 out of 15 predictions are verified by literature. In predictions ranked 1 and 6, drugs are related to corresponding diseases, but may target another subtype of the diseases. Predictions ranked 2, 3, 5, 7, 11, and 12 have not been adequately investigated. Some details are provided below.

Levetiracetam can treat epilepsy but one of its side-effects is ataxia [35]. So, it may not be helpful in treating ataxia with myoclonic epilepsy and presenile dementia. Ifosfamide is predicted to treat reticulum cell sarcomaand it has been used in treating soft tissue sarcoma [43]. Enalapril is an orally-active antihypertensive agent that can suppress the renin-angiotensin-aldosterone system to lower blood pressure. NEDD predicts Enalapril can treat renal failure, which has been tested in [36]. Teniposide is used for the treatment of refractory acute lymphoblastic leukaemia. The prediction result suggests that it may be applied in the treatment of mismatch repair cancer syndrome (MMRCS) as well. Dihydrotachysterol is predicted to treat vitamin D-dependent rickets, type 2a, which has been studied in [37]. NEDD suggests using Ibandronate in the treatment of inclusion body myopathy with early-onset Paget disease with or without frontotemporal dementia 1 and the treatment of Paget

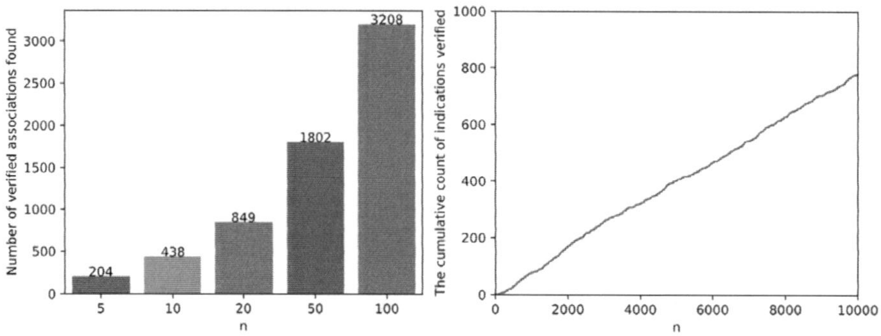


**Fig. 4** Result of top n test. Number of verified novel drug–disease associations found by NEDD. On the left is the sum of verified associations which rank in the top *n* results for each drug. On the right is the number of verified associations found in the top *n* results of all

**Table 1** Associations with the highest prediction scores

| Rank | Drug | | Disease | | References |
|---|---|---|---|---|---|
| | ID (DrugBank) | Name | ID (OMIM) | Name | |
| **1** | DB01202 | Levetiracetam | 208,700 | ATAXIA WITH MYOCLONIC EPILEPSY AND PRESENILE DEMENTIA | [35] |
| **2** | DB01181 | Ifosfamide | 267,730 | RETICULUM CELL SARCOMA | – |
| **3** | DB00937 | Diethylpropion | 303,110 | CHOROIDEREMIA, DEAFNESS, AND MENTAL RETARDATION | – |
| **4** | DB00584 | Enalapril | 161,900 | RENAL FAILURE, PROGRESSIVE, WITH HYPERTEN SION; RFH1 | [36] |
| **5** | DB00444 | Teniposide | 276,300 | MISMATCH REPAIR CANCER SYNDROME; MMRCS | – |
| **6** | DB01070 | Dihydrotachysterol | 277,440 | VITAMIN D-DEPENDENT RICKETS, TYPE 2A; VDDR2A | [37] |
| **7** | DB00176 | Fluvoxamine | 131,300 | CAMURATI-ENGELMANN DISEASE; CAEND | – |
| **8** | DB00710 | Ibandronate | 167,320 | INCLUSION BODY MYOPATHY WITH EARLY-ONSET PAGET DISEASE WITH OR WITHOUT FRONTOTEMPORAL DEMENTIA 1; IBMPFD1 | [38] |
| **9** | DB00710 | Ibandronate | 602,080 | PAGET DISEASE OF BONE 2, EARLY-ONSET; PDB2 | [38] |
| **10** | DB00282 | Pamidronic acid | 602,080 | PAGET DISEASE OF BONE 2, EARLY-ONSET; PDB2 | [39] |
| **11** | DB01551 | Dihydrocodeine | 147,530 | INSENSITIVITY TO PAIN WITH HYPERPLASTIC MYELINOPATHY | – |
| **12** | DB00500 | Tolmetin | 147,530 | INSENSITIVITY TO PAIN WITH HYPERPLASTIC MYELINOPATHY | – |
| **13** | DB00136 | Calcitriol | 241,519 | HYPOPHOSPHATEMIA, RENAL, WITH INTRACER EBRAL CALCIFICATIONS | [40] |
| **14** | DB00214 | Torasemide | 256,370 | NEPHROTIC SYNDROME, TYPE 4; NPHS4 | [41] |
| **15** | DB01120 | Gliclazide | 600,496 | MATURITY-ONSET DIABETES OF THE YOUNG, TYPE 3; MODY3 | [42] |

disease of bone 2, early-onset. And Ibandronate has long been used in treating Paget disease [38]. Pamidronic acid is used to prevent bone loss and to strengthen the bone in Paget disease [39], which verified the prediction of our method. Calcitriol, an active metabolite of vitamin D is predicted to treat renal hypophosphatemia with intracerebral calcifications. In real life, it is used in the treatment of hypophosphatemia [40]. Despite concerns that the use of calcitriol may contribute to vascular calcification, there is no clear evidence [44]. Torasemide is a high-ceiling loop diuretic [45]. It is predicted to treat nephrotic syndrome, type 4; this prediction is verified in [41]. For Gliclazide, the prediction to treat maturity-onset diabetes of the young of type 3 has been tested in clinical trials [42].

#### Parameter sensitivity

In this section, we investigate the parameter sensitivity. We change the thresholds for drug similarities and disease similarities and the window size to see how these parameters affect the result. We conduct 10-fold cross-validation five times for each parameter setting and evaluate the performance using AUC values. In each experiment, we only change one corresponding parameter and set the others as default—i.e. 0.7 for disease similarity threshold, 0.8 for drug similarity threshold, and 6 for window size.

The results are illustrated in Fig. 5. From Fig. 5, we can recognize a similar pattern. The performance rises initially when the values of the corresponding parameters rise.

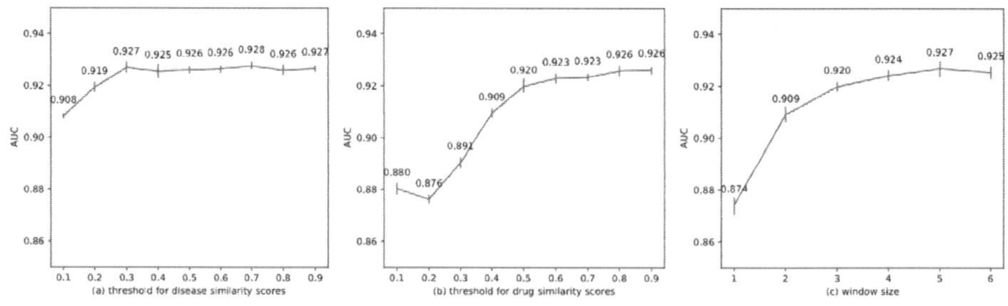


**Fig. 5** Result of parameter sensitivity test. AUC of ten-fold cross-validation on different parameter settings

However, after a certain point, NEDD becomes insensitive towards that parameter. For window size, this is because most of the useful information is already encoded in the embedding vectors. For thresholds of similarity scores, this is because most of the noise has already been ruled out at the beginning when we raise the thresholds.

### Model robustness

In this section, we investigate model robustness over different similarity measures.

We evaluate NEDD's performance using three different disease similarity measures, i.e. MimMiner [31] which is in the golden test dataset and utilizes disease phenotype information, NetSim [46] which employs the protein interaction network, and RADAR [47] which we used to get similarity scores based on pathways. Since we use the disease similarity scores provided by the authors and they used other types of ID for disease and some IDs do not have any mapping information, the experiments are done on a subset of the original dataset, which consists of 196 diseases, 593 drugs, and 1052 drug-disease associations.

We also evaluate NEDD's performance using three different drug similarity measures which utilize the information from the chemical structure, the corresponding side effects of each drug and the drug-related genes respectively. The three different types of drug similarity are calculated according to [9].

We repeat 10-fold cross-validation 10 times on each type of similarity scores. The results are illustrated in Fig. 6, in which NEDD produces similar AUC over different disease similarity measures.

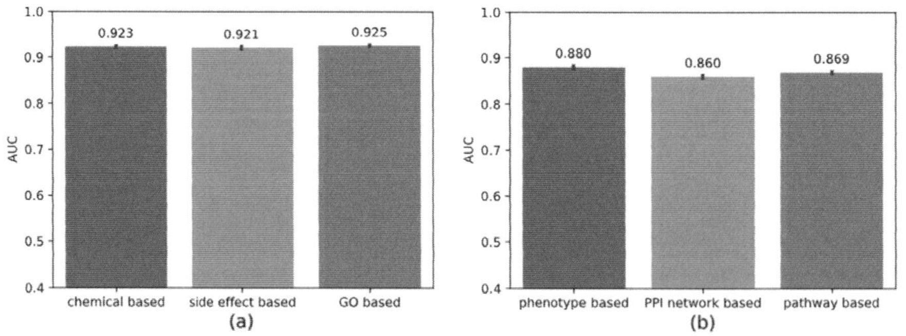


**Fig. 6** Result of model robustness test. **a** AUC of ten-fold cross-validation tests over different drug similarity measures; **b** AUC of ten-fold cross-validation tests over different disease similarity measures on a subset of the original dataset

## Discussion

We think that the superior performance of NEDD in finding indications for new drugs stems from two aspects: inductive learning and meta path. Inductive learning methods can be applied to entities not seen at train time. Compared with MBiRW, HGBI and NBI, DTINet and NEDD, which used inductive methods, yields relatively higher AUC score in the test. And with the concept of meta path, NEDD is able to explicitly capture high order proximities. This is especially important in tasks like drug repositioning where many latent links between drugs and diseases are unknown. If the associations between two nodes are missing, their first-order proximity is zero, so it is essential to exploit high order information.

Besides, NEDD can be easily adopted in larger datasets with more types of biological entities such as target, gene, side-effect, etc.

However, the limitations of NEDD should also be acknowledged. First, in order to use various information like drug-target interactions, the maximum length of the meta path should be increased, which might significantly increase the computational cost. Second, because the trained embeddings are not specifically fine-tuned for association prediction between drugs and diseases, the difficulty in training the vectors is increased when adding more information to the network.

## Conclusion

In this work, we present NEDD, a new computational approach for drug repositioning. NEDD uses a meta-path-based representation method to inductively learn node embedding vectors of drugs and diseases on a given graph. The graph is constructed by integrating heterogeneous biological information related to drugs and diseases. After learning the network embedding vectors, a random forest classifier is trained to predict the probabilities of drugs and diseases being associated.

NEDD shows competitive results in the 10-fold cross-validation test. The case study of NEDD gives fair results that remain to be further explored. In summary, the results prove that NEDD is practical in drug repositioning tasks toward existing drugs.

### Abbreviations

SVM: Support vector machine; HER: Electronic health records; OMIM: Online Mendelian Inheritance in Man; CDK: Chemical Development Kit; TPR: True positive rate; FPR: False positive rate; AUC: Area under curve; ROC: Receiver operating characteristic

### Acknowledgements

Not applicable.

### About this supplement

This article has been published as part of *BMC Bioinformatics Volume 21 Supplement 13, 2020: Selected articles from the 18th Asia Pacific Bioinformatics Conference (APBC 2020): bioinformatics*. The full contents of the supplement are available online at https://bmcbioinformatics.biomedcentral.com/articles/supplements/volume-21-supplement-13 .

### Authors' contributions

Method design: RZ, ZL; overall study design: ML, HL; coding: RZ, ZL; literature and database search: JX, MZ; comparing with other methods: JX, MZ; analysis of results: ML, MZ, HL; paper writing: RZ, JX. All authors have read and approved the final manuscript.

### Funding

Publication costs are funded by the National Natural Science Foundation of China under Grants (No. 61832019).

### Availability of data and materials

The used dataset can be downloaded from bioinformatics.csu.edu.cn/resources/softs/DrugRepositioning/n_Web/NEDD.html.

**Ethics approval and consent to participate**
Not applicable.

**Consent for publication**
Not applicable.

**Competing interests**
The authors declare that they have no competing interests.

**Author details**
[1]Hunan Provincial Key Lab on Bioinformatics, School of Computer Science and Engineering, Central South University, Changsha, China. [2]School of Computer and Information Engineering, Henan University, Kaifeng 475001, China. [3]Neuroscience Research Center & School of Basic Medical Sciences, Changsha Medical University, Changsha, China.

Published: 17 September 2020

**References**

1. Dudley JT, Deshpande T, Butte AJ. Exploiting drug–disease relationships for computational drug repositioning. Brief Bioinform. 2011;12(4):303–11.
2. Ashburn TT, Thor KB. Drug repositioning: identifying and developing new uses for existing drugs. Nat Rev Drug Discov. 2004;3(8):673.
3. Persidis A. The benefits of drug repositioning. Drug Discov World. 2011;12:9–12.
4. Shim JS, Liu JO. Recent advances in drug repositioning for the discovery of new anticancer drugs. Int J Biol Sci. 2014; 10(7):654.
5. Yella J, Yaddanapudi S, Wang Y, Jegga A. Changing trends in computational drug repositioning. Pharmaceuticals. 2018; 11(2):57.
6. Bolgár B, Arany A, Temesi G, Balogh B, Antal P, Matyus P. Drug repositioning for treatment of movement disorders: from serendipity to rational discovery strategies. Curr Top Med Chem. 2013;13(18):2337–63.
7. Shaughnessy AF. Old drugs, new tricks. BMJ. 2011;342:d741.
8. Ha S, Seo YJ, Kwon MS, Chang BH, Han CK, Yoon JH. IDMap: facilitating the detection of potential leads with therapeutic targets. Bioinformatics. 2008;24(11):1413–5.
9. Gottlieb A, Stein GY, Ruppin E, Sharan R. PREDICT: a method for inferring novel drug indications with application to personalized medicine. Mol Syst Biol. 2011;7(1):496.
10. Li J, Zheng S, Chen B, Butte AJ, Swamidass SJ, Lu Z. A survey of current trends in computational drug repositioning. Brief Bioinform. 2015;17(1):2–12.
11. Napolitano F, Zhao Y, Moreira VM, Tagliaferri R, Kere J, D'Amato M, Greco D. Drug repositioning: a machine-learning approach through data integration. J Cheminform. 2013;5(1):30.
12. Aliper A, Plis S, Artemov A, Ulloa A, Mamoshina P, Zhavoronkov A. Deep learning applications for predicting pharmacological properties of drugs and drug repurposing using transcriptomic data. Mol Pharm. 2016;13(7):2524–30.
13. Ramsundar B, Kearnes S, Riley P, Webster D, Konerding D, Pande V. Massively multitask networks for drug discovery. arXiv preprint arXiv. 2015:1502.02072.
14. Zhu Q, Tao C, Shen F, Chute CG. Exploring the pharmacogenomics knowledge base (PharmGKB) for repositioning breast cancer drugs by leveraging web ontology language (OWL) and cheminformatics approaches. Biocomput. 2014; 2014:172–82.
15. Chen B, Ding Y, Wild DJ. Assessing drug target association using semantic linked data. PLoS Comput Biol. 2012;8(7): e1002574.
16. Martínez V, Navarro C, Cano C, Fajardo W, Blanco A. DrugNet: network-based drug–disease prioritization by integrating heterogeneous data. Artif Intell Med. 2015;63(1):41–9.
17. Luo Y, Zhao X, Zhou J, Yang J, Zhang Y, Kuang W, et al. A network integration approach for drug-target interaction prediction and computational drug repositioning from heterogeneous information. Nat Commun. 2017;8(1):573.
18. Luo H, Wang J, Li M, Luo J, Peng X, Wu FX, Pan Y. Drug repositioning based on comprehensive similarity measures and bi-random walk algorithm. Bioinformatics. 2016;32(17):2664–71.
19. Yang C, Sun M, Liu Z, Tu C. Fast network embedding enhancement via high order proximity approximation. In: Proceedings of the 26th International Joint Conference on Artificial Intelligence. United States: AAAI Press; 2017:3894–900.
20. Fu TY, Lee WC, Lei Z. Hin2vec: explore meta-paths in heterogeneous information networks for representation learning. In: Proceedings of the 2017 ACM on conference on information and knowledge management. United States: Association for Computing Machinery; 2017. 1797–806.
21. Mikolov T, Sutskever I, Chen K, Corrado GS, Dean J. Distributed representations of words and phrases and their compositionality. In: Advances in neural information processing systems; 2013. p. 3111–9.
22. Maaten LVD, Hinton G. Visualizing data using t-SNE. J Mach Learn Res. 2008;9:2579–605.
23. Pedregosa F, Varoquaux G, Gramfort A, Michel V, Thirion B, Grisel O, et al. Scikit-learn: machine learning in python. J Mach Learn Res. 2011;12:2825–30.
24. Wang W, Yang S, Li JING. Drug target predictions based on heterogeneous graph inference. Biocomput. 2013;2013:53–64.
25. Cheng F, Liu C, Jiang J, Lu W, Li W, Liu G, et al. Prediction of drug-target interactions and drug repositioning via network-based inference. PLoS Comput Biol. 2012;8(5):e1002503.
26. Hussein R, Yang D, Cudré-Mauroux P. Are Meta-paths necessary?: revisiting heterogeneous graph Embeddings. In: Proceedings of the 27th ACM international conference on information and knowledge management. United States: Association for Computing Machinery; 2018. p. 437–46.

27. Wishart DS, Knox C, Guo AC, Shrivastava S, Hassanali M, Stothard P, et al. DrugBank: a comprehensive resource for in silico drug discovery and exploration. Nucleic Acids Res. 2006;34(suppl_1):D668–72.
28. Hamosh A, Scott AF, Amberger JS, Bocchini CA, McKusick VA. Online Mendelian Inheritance in Man (OMIM), a knowledgebase of human genes and genetic disorders. Nucleic Acids Res. 2005;33(suppl_1):D514–7.
29. Steinbeck C, Han Y, Kuhn S, Horlacher O, Luttmann E, Willighagen E. The chemistry development kit (CDK): an open-source Java library for chemo-and bioinformatics. J Chem Inf Comput Sci. 2003;43(2):493–500.
30. Weininger D, Weininger A, Weininger JL. SMILES. 2. Algorithm for generation of unique SMILES notation. J Chem Inf Comput Sci. 1989;29(2):97–101.
31. Van Driel MA, Bruggeman J, Vriend G, Brunner HG, Leunissen JA. A text-mining analysis of the human phenome. Eur J Hum Genet. 2006;14(5):535.
32. Wang W, Yang S, Zhang X, Li J. Drug repositioning by integrating target information through a heterogeneous network model. Bioinformatics. 2014;30(20):2923–30.
33. Kuhn M, Letunic I, Jensen LJ, Bork P. The SIDER database of drugs and side effects. Nucleic Acids Res. 2015;44(D1):D1075–9.
34. Davis AP, Grondin CJ, Johnson RJ, Sciaky D, King BL, McMorran R, et al. The comparative toxicogenomics database: update 2017. Nucleic Acids Res. 2016;45(D1):D972–8.
35. Farooq MU, Bhatt A, Majid A, Gupta R, Khasnis A, Kassab MY. Levetiracetam for managing neurologic and psychiatric disorders. Am J Health Syst Pharm. 2009;66(6):541–61.
36. Abraham PA, Opsahl JA, Halstenson CE, Keane WF. Efficacy and renal effects of enalapril therapy for hypertensive patients with chronic renal insufficiency. Arch Intern Med. 1988;148(11):2358–62.
37. Rosen JF, Finberg L. Vitamin D-dependent rickets: actions of parathyroid hormone and 25-hydroxycholecalciferol. Pediatr Res. 1972;6(6):552.
38. Grauer A, Heichel S, Knaus J, Dosch E, Ziegler R. Ibandronate treatment in Paget's disease of bone. Bone. 1999;24(5 Suppl):87S–9S.
39. Seifi M, Amdjadi P, Tayebi L. Pharmacological agents for bone remodeling: an experimental approach. In: Biomaterials for oral and dental tissue engineering. United Kingdom: Woodhead Publishing; 2017. p. 503–23.
40. Imel EA, Econs MJ. Approach to the hypophosphatemic patient. J Clin Endocrinol Metabol. 2012;97(3):696–706.
41. Qavi AH, Kamal R, Schrier RW. Clinical use of diuretics in heart failure, cirrhosis, and nephrotic syndrome. Int J Nephrol. 2015;2015:975934.
42. Habeb AM, George ET, Mathew V, Hattersley AL. Response to oral gliclazide in a pre-pubertal child with hepatic nuclear factor-1 alpha maturity onset diabetes of the young. Ann Saudi Med. 2011;31(2):190–3.
43. Suppiah R, Wood L, Elson P, Budd GT. Phase I/II study of docetaxel, ifosfamide, and doxorubicin in advanced, recurrent, or metastatic soft tissue sarcoma (STS). Investig New Drugs. 2006;24(6):509–14.
44. Wolisi, G. O., & Moe, S. M. (2005). Vitamin D in health and disease: the role of vitamin D in vascular calcification in chronic kidney disease. In Seminars in dialysis. Oxford: Blackwell Science Inc.; 18(4):307–314.
45. Lopez B, González A, Hermida N, Laviades C, Díez J. Myocardial fibrosis in chronic kidney disease: potential benefits of torasemide: new strategies to prevent cardiovascular risk in chronic kidney disease. Kidney Int. 2008;74:S19–23.
46. Li P, Nie Y, Yu J. Fusing literature and full network data improves disease similarity computation. BMC Bioinform. 2016;17(1):326.
47. Qin R, Duan L, Zheng H, Li-Ling J, Song K, Zhang Y. An ontology-independent representation learning for similar disease detection based on multi-layer similarity network. In: IEEE/ACM transactions on computational biology and bioinformatics; 2019.

## Publisher's Note

Springer Nature remains neutral with regard to jurisdictional claims in published maps and institutional affiliations.

Li et al. BMC Bioinformatics 2020, 21(Suppl 13):381
https://doi.org/10.1186/s12859-020-03683-3

BMC Bioinformatics

METHODOLOGY

Open Access

# Prediction of hot spots in protein–DNA binding interfaces based on supervised isometric feature mapping and extreme gradient boosting

Ke Li[1,2†], Sijia Zhang[2†], Di Yan[2,3], Yannan Bin[2] and Junfeng Xia[2*]

*From* The 18th Asia Pacific Bioinformatics Conference
Seoul, Korea. 18-20 August 2020

* Correspondence: jfxia@ahu.edu.cn
†Ke Li and Sijia Zhang contributed equally to this work.
[2]Key Laboratory of Intelligent Computing and Signal Processing of Ministry of Education, Institutes of Physical Science and Information Technology, Anhui University, Hefei 230601, Anhui, China
Full list of author information is available at the end of the article

## Abstract

**Background:** Identification of hot spots in protein-DNA interfaces provides crucial information for the research on protein-DNA interaction and drug design. As experimental methods for determining hot spots are time-consuming, labor-intensive and expensive, there is a need for developing reliable computational method to predict hot spots on a large scale.

**Results:** Here, we proposed a new method named sxPDH based on supervised isometric feature mapping (S-ISOMAP) and extreme gradient boosting (XGBoost) to predict hot spots in protein-DNA complexes. We obtained 114 features from a combination of the protein sequence, structure, network and solvent accessible information, and systematically assessed various feature selection methods and feature dimensionality reduction methods based on manifold learning. The results show that the S-ISOMAP method is superior to other feature selection or manifold learning methods. XGBoost was then used to develop hot spots prediction model sxPDH based on the three dimensionality-reduced features obtained from S-ISOMAP.

**Conclusion:** Our method sxPDH boosts prediction performance using S-ISOMAP and XGBoost. The AUC of the model is 0.773, and the F1 score is 0.713. Experimental results on benchmark dataset indicate that sxPDH can achieve generally better performance in predicting hot spots compared to the state-of-the-art methods.

**Keywords:** Protein–DNA complexes, Hot spot, Supervised isometric feature mapping, Extreme gradient boosting

## Background

Protein-DNA interactions play a crucial role in many biological processes, such as gene transcription and translation, DNA repair and assembly [1, 2]. In pioneering research work on the binding of human growth hormone to its receptor, a small number of interface residues, known as hot spots, were found to contribute more affinity compared with other amino acid residues [3]. In the experiments, alanine scanning mutation technology has been used to identify hot spots when their free energy changes exceed a certain threshold [4]. This experimental method was also used to explore the mechanism of protein-DNA recognition. As the experimental method is high-cost and time-consuming, the computational method provides another way for studying hot spots prediction.

A series of methods based on feature selection have been used to study the hot spots in protein binding interfaces. Xia et al. selected the three optimal features with the largest contribution through a two-step feature selection approach including maximum relevance minimum redundancy (mRMR) and exhaustive search [5]. Pan et al. used gradient tree boosting algorithm to find the smallest optimal features from 125 candidates [6]. Qiao et al. proposed a hybrid feature selection strategy, combining the feature subsets selected by decision tree and mRMR respectively, and finally obtained six features using pseudo sequential forward selection [7]. Deng et al. adopted a two-step feature selection method consisting of mRMR and sequential forward selection (SFS) to select the best 6 features from a group of 156 features [8]. Hot spots identification is of great significance for exploring the potential binding mechanism and the stability of protein-DNA interactions [9]. So far, many studies have focused on the prediction of binding sites in protein-DNA complexes [10]. However, there is little research on the prediction of hot spots in protein-DNA complexes. Recently, Zhang et al. used a computational approach to predict the hot spots in protein-DNA binding interfaces [11].

The above methods have some disadvantages. For example, the mRMR-based method has good time performance, but its classification accuracy is general and it cannot eliminate redundancy completely [12]. Although the SFS-based method has good feature resolution, it has high computational complexity and is easy to over-fit [13]. Manifold learning is a nonlinear dimensionality reduction method appeared in recent years. It can map the high-dimensional input data to the low-dimensional manifold and preserve the topological structure of the data while reducing the dimension. The classical manifold learning methods include isometric feature mapping (ISOMAP) [14], local linear embedding (LLE) [15], etc. However, these are unsupervised dimensionality reduction methods, which cannot make full use of the class label information of samples. Here, we propose a new method based on supervised manifold learning to predict the hot spots in protein-DNA binding interfaces. We extracted 64 DNA-binding proteins and collected 114 features based on our previous work [11]. In order to improve prediction performance, supervised isometric feature mapping (S-ISOMAP) [16] algorithm considering the class label information was used to implement dimensionality reduction. Finally, we employed an improved version of the Gradient Boosting algorithm, extreme gradient boosting (XGBoost) [17], to build the prediction model. Experimental results show that compared with the state-of-the-art prediction methods, our method sxPDH (S-ISOMAP and XGBoost based model for prediction of protein-DNA binding hot spots) has higher prediction performance.

## Methods

### Dataset and features used in this work

In this study, we used the same dataset and features as our previous work [11]. Among 64 protein-DNA complexes, 40 complexes were selected randomly as the training dataset including 62 hot spots and 88 non-hot spots and the other 24 complexes were used as the test dataset with 26 hot spots and 38 non-hot spots. We obtained 114 features from four feature groups, namely, solvent accessible surface area, sequence, structure and network. For details, the interested readers can refer to our previous work [11].

### Feature dimensionality reduction

If the dimension of the features is too high, the classifier will over-fit. Therefore, in order to improve the prediction performance of classifiers, reducing the feature dimension is essential. Here, we used S-ISOMAP algorithm, which can make the data of the same category close to and different categories distant from each other in the dimension reduction space, thus achieve dimensionality reduction. The framework of manifold learning algorithm based on S-ISOMAP is as follows [16].

Step 1: Define the dissimilarity distance:

Assuming that the given data are $(x_i, y_i)$, where $x_i \in R^D (i = 1, 2, \ldots, N)$, $y_i$ is the category label for $x_i$, we define the dissimilarity between two points $x_i$ and $x_j$ as [16]:

$$D(x_i, x_j) = \begin{cases} \sqrt{1 - \exp(-d^2(x_i, x_j)/\beta)} y_i = y_i \\ \sqrt{\exp(d^2(x_i, x_j)/\beta)} - \alpha y_i \neq y_i \end{cases} \quad (1)$$

where $d(x_i, x_j)$ represents the Euclidean distance between $x_i$ and $x_j$, the parameter $\beta$ is used to control the growth rate of $D(x_i, x_j)$, and the parameter $\alpha$ is used to control the distance between different classes [16].

Step 2: Construct the neighborhood graph:

Firstly the dissimilarity distance between the sample point $x_i \in R^D$ and sample points $x_j \in R^D$ is calculated [16]. When $x_j$ is one of the nearest $K$ points of $x_i$, they are adjacent, that is, there is edge $x_i x_j$ in the graph $G$ (k-neighborhood). If $x_j$ is not the nearest $K$ points of $x_i$, and the Euclidean distance between $x_i$ and $x_j$ is less than the fixed value $\varepsilon$, it is considered that there is edge $x_i x_j$ in the graph $G$ ($\varepsilon$-neighborhood). Here, the weight of the edge is set to dissimilarity distance $D(x_i, x_j)$ [16].

Step 3: Compute the shortest paths:

We initialize the shortest path $d_G(x_i, x_j) = D(x_i, x_j)$, if there's an edge $x_i x_j$ in graph $G$; Otherwise $d_G(x_i, x_j) = \infty$. Then we calculate $d_G(x_i, x_j)$ for each data $(x_i, y_i)$ [16]:

$$d_G(x_i, x_j) = \min\{d_G(x_i, x_j), d_G(x_i, x_l) + d_G(x_l, x_j)\} \quad (2)$$

where $l = 1, 2, \ldots, N$.

In this way, the shortest path distance matrix $\boldsymbol{D}_G = \{d_G(x_i, x_j)\}$ can be obtained. This process is called Floyd algorithm [16].

Step 4: Construct $d$-dimensional embedding:

Multidimensional scaling (MDS) [18] is applied to the distance matrix $\boldsymbol{D}_G$. The global low-dimensional coordinates are obtained by minimizing the cost function $E$:

$$E = \|\tau(\boldsymbol{D}_G) - \tau(\boldsymbol{D}_Y)\|_{L^2} \quad (3)$$

where the operator $\tau$ is defined by $\tau(\boldsymbol{D}) = -\boldsymbol{HSH}/2$, in which $H = \{H_{ij}\} = \{\delta_{ij} - 1/N\}$ is the "centering matrix", and $\boldsymbol{S} = \{S_{ij}\} = \{D^2(x_i, x_j)\}$ is the square distance matrix. The eigenvector corresponding to the maximum $d$ eigenvalues $\lambda_1, \lambda_2, \cdots, \lambda_d$ of $\tau(\boldsymbol{D}_G)$ is $u_1$, $u_2, \cdots, u_d$ [16]. Then $Y = diag(\lambda_1^{1/2}, \lambda_2^{1/2}, \cdots, \lambda_d^{1/2})[u_1, u_2, \cdots, u_d]^T$ is the $d$-dimensional embedding result [16].

### Model construction

XGBoost has achieved the most advanced results in many machine learning challenges based on the idea of continuously reducing the residual of the previous model in the gradient direction to obtain a new model. As an improved version of the Gradient Boosting algorithm, XGBoost performs a second-order Taylor expansion on the loss function to obtain the optimal solution for the regular term outside the loss function. The advantages of multi-core CPU parallel computing is fully utilized to improve the accuracy and speed. Therefore, we established a prediction model for hot spots in protein-DNA binding interfaces based on XGBoost. In order to achieve good experimental results, the XGBoost was tuned using a grid search method, and obtained the optimal parameters with n_estimators = 500, learning_rate = 0.1, and max_depth = 30.

### Evaluation criteria

The computer model used in the simulation is an ASUS FX503VD, the CPU is a dual-core processor i7-7700HQ model with a main frequency of 2.8 GHz, and its memory is 8G. In order to improve the robustness of the prediction model, we used 10-fold cross validation and performed 20 experiments to obtain average results. To evaluate the classification performance of our model, we adopted some commonly used evaluation metrics, including sensitivity (SEN), specificity (SPE), precision (PRE), F1 score (F1), accuracy (ACC), and Matthews correlation coefficient (MCC) [19–23]:

$$SEN = TP/(TP + FN) \tag{4}$$

$$SPE = TN/(TN + FP) \tag{5}$$

$$PRE = TP/(TP + FP) \tag{6}$$

$$F1 = \frac{2 \times SEN \times PRE}{SEN + PRE} \tag{7}$$

$$ACC = \frac{TP + TN}{TP + TN + FP + FN} \tag{8}$$

$$MCC = \frac{TP \times TN - FP \times FN}{\sqrt{(TP + FP)(TP + FN)(TN + FP)(TN + FN)}} \tag{9}$$

where TP, FP, TN, FN represent the number of true positive (correctly predicted hot spot residues), false positive (non-hot spot residues incorrectly predicted as hot spots), true negative (correctly predicted non-hot spot residues) and false negative (hot spot residues incorrectly predicted as non- hot spots), respectively. We also adopted the ROC curve as the assessment criteria in this work. From the ROC curve, we calculated the area under the ROC curve (AUC).

## Results and discussion

### Overview of sxPDH

Figure 1 shows the workflow of our method sxPDH. First, a benchmark dataset consisting of 88 hot spots and 126 non-hot spots from 64 protein-DNA complexes was constructed. Then, four types of features were generated, namely, solvent-accessible surface area, sequence features, structural features and network features. S-ISOMAP algorithm was then used to reduce the dimension of these feature. On this basis, XGBoost was applied to construct a prediction model of hotspots in protein-DNA binding interface. Finally, according to the feature set after dimensionality reduction, the prediction results are output through the XGBoost model.

### Evaluation of different manifold learning methods

In this study, we reduce feature dimension based on the S-ISOMAP. In order to evaluate the practicability of the S-ISOMAP method, it is compared with three other manifold learning-based methods, including LLE, ISOMAP and supervised locally linear embedding (SLLE) [24], with the XGboost is used as the classification model. LLE method is to obtain low-dimensional embedded coordinates by linear reconstruction of local neighborhood in high-dimensional data, thereby keeping the neighborhood relationship of high-dimensional data unchanged. The goal of ISOMAP method is to maintain the geodesic distance between the points in the original data set to the greatest extent. Both methods are based on unsupervised dimensionality reduction. SLLE introduces class labels by calculating the maximum Euclidean distance between classes, which is based on supervised dimensionality reduction. Table 1 shows the performance of the model using S-ISOMAP compared with the other three manifold learning methods on the test set. From these

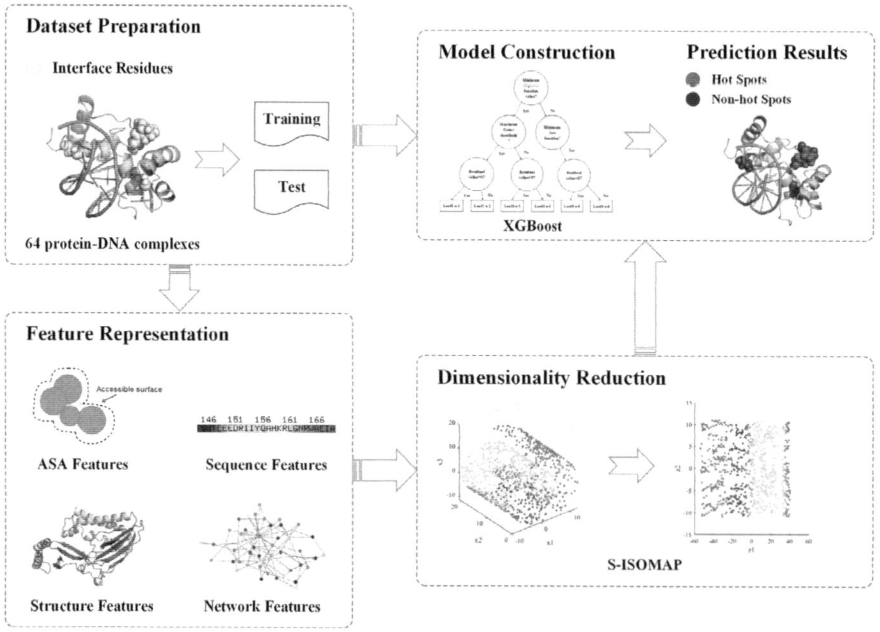


**Fig. 1** The workflow of sxPDH

**Table 1** Performance of different manifold learning methods on the test set

| Method | SEN | SPE | PRE | F1 | ACC | MCC | AUC |
|---|---|---|---|---|---|---|---|
| LLE (10) | 0.653 | 0.711 | 0.607 | 0.629 | 0.687 | 0.361 | 0.693 |
| ISOMAP (10) | 0.687 | 0.766 | 0.692 | 0.695 | 0.709 | 0.476 | 0.738 |
| SLLE (3) | 0.671 | 0.732 | 0.648 | 0.656 | 0.691 | 0.381 | 0.703 |
| S-ISOMAP (3) | **0.707** | **0.819** | **0.721** | **0.713** | **0.768** | **0.508** | **0.773** |

The highest value in each column is shown in bold. The numbers in parentheses represent the feature dimensions after dimensionality reduction

evaluation criteria, it can be seen that the model prediction effect using S-ISOMAP is the best (PRE = 0.707, F1 = 0.713, MCC = 0.508 and ACC = 0.768).

Figure 2 shows the runtime comparison of our method with the other three manifold learning methods. The dimensionality reduction time of S-ISOMAP is slightly higher than that of SLLE, but lower than those of LLE and ISOMAP.

## Compared with the feature selection methods

To further verify the performance of our model, we also compared its performance with four commonly used feature selection methods with the classification model XGboost. These methods are RF-based on sequential forward selection (RF-SFS) [25], mRMR [26], SVM-based recursive feature elimination (SVM-RFE) [27] and variable selection using random forests (VSURF) [28]. RF-SFS uses RF to rank the importance of features and then performs feature selection using sequential forward selection strategy. The mRMR method analyzes and evaluates features by producing a feature list based on the maximum relevance and minimum redundancy criteria. SVM-RFE is an application of RFE using the weight magnitude as the ranking standard. VSURF adopts a two-stage strategy. It first uses the importance score based on the random forest to sort features, and then uses a stepwise forward strategy to return a smaller subset that tries to avoid redundancy.

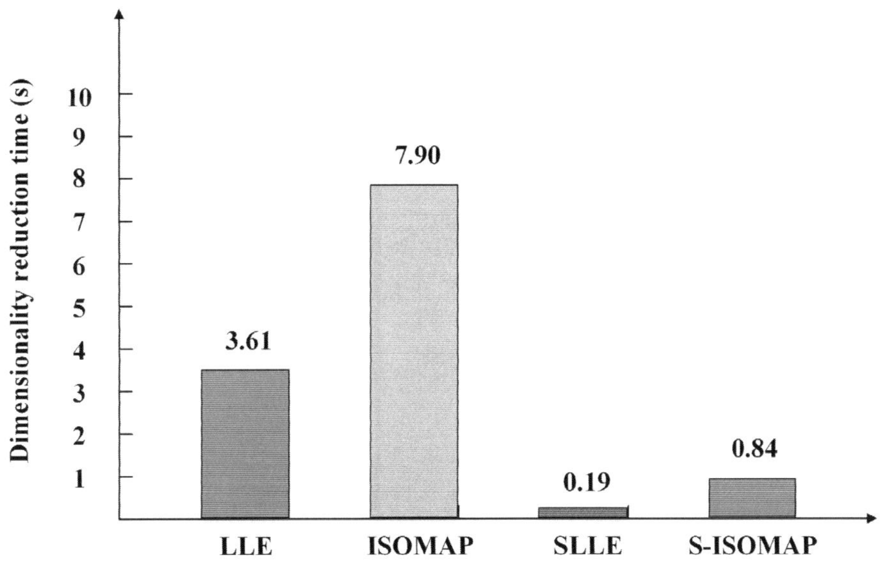


**Fig. 2** Running time of different manifold learning methods

**Table 2** Performance of S-ISOMAP compared with other feature selection methods on the test set

| Method | SEN | SPE | PRE | F1 | ACC | MCC | AUC |
|---|---|---|---|---|---|---|---|
| SVM-RFE (19) | 0.423 | 0.763 | 0.555 | 0.478 | 0.625 | 0.197 | 0.635 |
| mRMR (30) | 0.538 | 0.711 | 0.569 | 0.549 | 0.642 | 0.251 | 0.696 |
| RF-SFS (17) | 0.654 | 0.737 | 0.629 | 0.642 | 0.703 | 0.388 | 0.709 |
| VSURF (10) | 0.678 | 0.776 | 0.672 | 0.669 | 0.736 | 0.431 | 0.704 |
| S-ISOMAP (3) | **0.707** | **0.819** | **0.721** | **0.713** | **0.768** | **0.508** | **0.773** |

The highest value in each column is shown in bold. The numbers in parentheses represent the feature dimensions after dimensionality reduction

The prediction performance of the five algorithms on the test set is shown in Table 2. Our model produced the best performance with an AUC score of 0.773 on test set. In addition, the number of features after dimensionality reduction is the smallest. In contrast, the other four feature selection methods produced a relatively lower AUC score and more selected features.

Figure 3 shows the runtime comparison of S-ISOMAP with the other four feature selection methods. The dimensionality reduction time of mRMR is less than 0.01 (0.000001). The dimensionality reduction time of our method is only higher than that of mRMR, but lower than those of RF-SFS, SVM-RFE and VSURF.

### Compared with other methods

SAMPDI [29] and PremPDI [30] are two molecular mechanics-based approaches which can predict protein-DNA binding free energy changes, while mCSM-NA [31] uses the concept of graph-based signatures to quantitatively predict the influences of single mutation on protein-DNA or protein-RNA binding affinities. Recently, we proposed a computational methods called PrPDH [11] to predict DNA-binding hot spots, which uses VSURF method for feature selection and SVM as the classifier model. The comparison of our method sxPDH with these four methods is shown in Table 3. Our method sxPDH shows similar success rate in comparison with PrPDH. On the test set, the F1 score, MCC, ACC and AUC of our model sxPDH were 0.713, 0.508, 0.768 and 0.773 respectively, while PrPDH could correctly identify DNA-binding hot spots with F1 score = 0.706, MCC = 0.511, ACC = 0.766 and

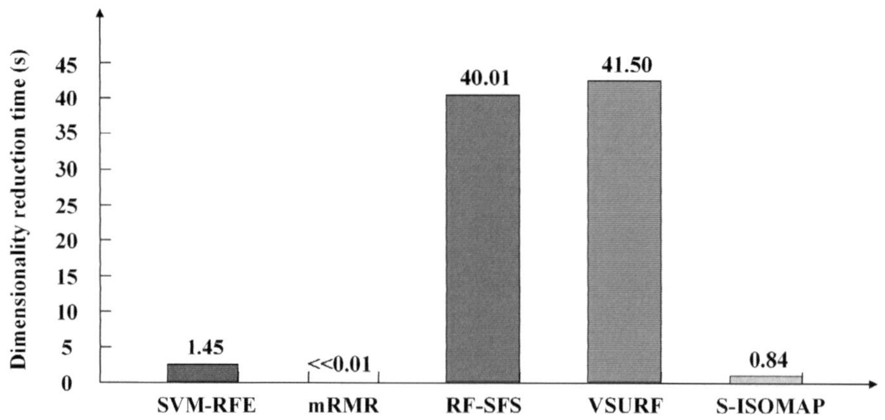


**Fig. 3** Running time of S-ISOMAP compared with other feature selection

**Table 3** Performance of different methods on the test set

| Method | SEN | SPE | PRE | F1 | ACC | MCC | AUC |
|---|---|---|---|---|---|---|---|
| SAMPDI | 0.654 | 0.658 | 0.567 | 0.607 | 0.656 | 0.307 | 0.690 |
| PremPDI | 0.577 | 0.737 | 0.600 | 0.588 | 0.672 | 0.316 | 0.708 |
| mCSM-NA | 0.538 | 0.737 | 0.583 | 0.560 | 0.656 | 0.279 | 0.661 |
| PrPDH | 0.692 | 0.816 | 0.720 | 0.706 | 0.766 | **0.511** | 0.764 |
| sxPDH | **0.707** | **0.819** | **0.721** | **0.713** | **0.768** | 0.508 | **0.773** |

The highest value in each column is shown in bold

AUC = 0.764. Since the experiments of SAMPDI, PremPDI and mCSM-NA were performed on their webserver, we only compared the time performance of sxPDH and PrPDH. Our method sxPDH is far less than PrPDH in terms of optimal feature number (Table 3) and running time (Fig. 4). Overall, our method sxPDH exerts impressive predictive and time efficiency in detecting hot spots in protein–DNA interaction interfaces.

## Conclusion

In this work, we proposed a method called sxPDH based on S-ISOMAP and XGBoost to distinguish hot spots and non-hot spots at protein-DNA interfaces. Based on our previous work [11], 64 complexes were selected as the benchmark dataset, and 114 features were calculated from four types of feature groups. Then the feature dimension was reduced to three by S-ISOMAP method. The XGBoost was used to build the final prediction model. The prediction results show that the proposed method sxPDH has better prediction performance and lower time complexity. However, there is still room to improve our method. Because most used features in this study are related to proteins and amino acids, we will explore more DNA-related features to make our model more robust in the future work.

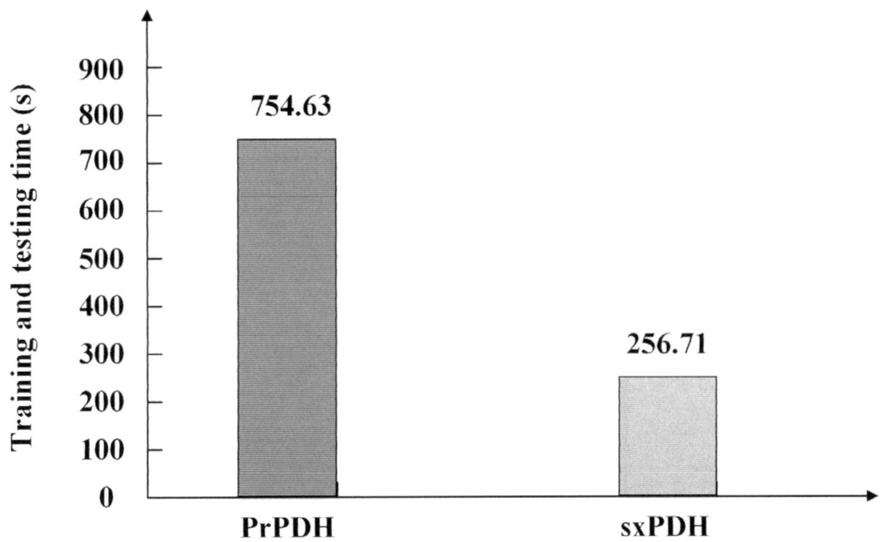


**Fig. 4** Running time of sxPDH compared with PrPDH

### Abbreviations
S-ISOMAP: Supervised isometric feature mapping; XGBoost: Extreme gradient boosting; mRMR: Maximum relevance minimum redundancy; SFS: Sequential forward selection; ISOMAP: Isometric feature mapping; LLE: Local linear embedding; SLLE: Supervised locally linear embedding; RF-SFS: RF based on sequential forward selection; SVM-RFE: SVM-based recursive feature elimination; VSURF: Variable selection using random forests; SEN: Sensitivity; SPE: Specificity; PRE: Precision; F1: F1 score; ACC: Accuracy; MCC: Matthews correlation coefficient; AUC: The area under the ROC curve

### Acknowledgments
The authors thank all members of our laboratory for their valuable discussions.

### About this supplement
This article has been published as part of *BMC Bioinformatics Volume 21 Supplement 13, 2020: Selected articles from the 18th Asia Pacific Bioinformatics Conference (APBC 2020): bioinformatics*. The full contents of the supplement are available online at https://bmcbioinformatics.biomedcentral.com/articles/supplements/volume-21-supplement-13 .

### Authors' contributions
KL performed the analysis and drafted the manuscript. SZ collected the datasets and performed the analysis. DY and YB performed the analysis. JX designed the study and performed the analysis. All authors read and approved the final manuscript.

### Funding
Publication costs are funded by the National Natural Science Foundation of China (61672037, 11835014, U19A2064, 21601001, and 31301101) and in part by the Anhui Provincial Outstanding Young Talent Support Plan (gxyqZD2017005), the Young Wanjiang Scholar Program of Anhui Province, the Recruitment Program for Leading Talent Team of Anhui Province (2019–16), the China Postdoctoral Science Foundation Grant (2018 M630699),the Anhui Provincial Postdoctoral Science Foundation Grant (2017B325), and the Key Project of Anhui Provincial Education Department (KJ2017ZD01). Funding agencies have no role in design of the study and collection, analysis, and interpretation of data and in writing the manuscript.

### Availability of data and materials
The data and python code of sxPDH are freely available via GitHub: https://github.com/xialab-ahu/sxPDH.

### Ethics approval and consent to participate
Not applicable.

### Consent for publication
Not applicable.

### Competing interests
The authors declared that they have no competing interests exist.

### Author details
[1]School of Information and Computer, Anhui Agricultural University, Hefei 230036, Anhui, China. [2]Key Laboratory of Intelligent Computing and Signal Processing of Ministry of Education, Institutes of Physical Science and Information Technology, Anhui University, Hefei 230601, Anhui, China. [3]School of Life Sciences, Anhui University, Hefei 230601, Anhui, China.

Published: 17 September 2020

### References
1. Zhang J, Zhang Z, Chen Z, Deng L. Integrating multiple heterogeneous networks for novel lncRNA-disease association inference. IEEE/ACM Trans Comput Biol Bioinform. 2017;16(2):396–406.
2. König J, Zarnack K, Luscombe NM, Ule J. Protein–RNA interactions: new genomic technologies and perspectives. Nat Rev Genet. 2012;13(2):77–83.
3. Clackson T, Wells JA. A hot spot of binding energy in a hormone-receptor interface. Science. 1995;267(5196):383–6.
4. Moreira IS, Fernandes PA, Ramos MJ. Hot spots—a review of the protein–protein interface determinant amino-acid residues. Proteins. 2007;68(4):803–12.
5. Xia J, Yue Z, Di Y, Zhu X, Zheng C-H. Predicting hot spots in protein interfaces based on protrusion index, pseudo hydrophobicity and electron-ion interaction pseudopotential features. Oncotarget. 2016;7(14):18065–75.
6. Pan Y, Wang Z, Zhan W, Deng L. Computational identification of binding energy hot spots in protein–RNA complexes using an ensemble approach. Bioinformatics. 2017;34(9):1473–80.
7. Qiao Y, Xiong Y, Gao H, Zhu X, Chen P. Protein-protein interface hot spots prediction based on a hybrid feature selection strategy. BMC Bioinformatics. 2018;19(1):14. https://doi.org/10.1186/s12859-018-2009-5.
8. Deng L, Sui Y, Zhang J. XGBPRH: prediction of binding hot spots at protein–RNA interfaces utilizing extreme gradient boosting. Genes. 2019;10(3):242. https://doi.org/10.3390/genes10030242.
9. Wang L, Liu Z-P, Zhang X-S, Chen L. Prediction of hot spots in protein interfaces using a random forest model with hybrid features. Protein Eng Des Sel. 2012;25(3):119–26.
10. Xiong Y, Zhu X, Dai H, Wei DQ. Survey of computational approaches for prediction of DNA-binding residues on protein surfaces. Methods Mol Biol. 2018;1754:223–34.

11. Zhang S, Zhao L, Zheng C-H, Xia J. A feature-based approach to predict hot spots in protein–DNA binding interfaces. Brief Bioinform. 2019. https://doi.org/10.1093/bib/bbz037.
12. Li J, Cheng K, Wang S, Morstatter F, Trevino RP, Tang J, Liu H. Feature selection: a data perspective. ACM Comput Surv. 2018;50(6):94. https://doi.org/10.1145/3136625.
13. Cai J, Luo J, Wang S, Yang S. Feature selection in machine learning: a new perspective. Neurocomputing. 2018;300:70–9.
14. Tenenbaum JB, De Silva V, Langford JC. A global geometric framework for nonlinear dimensionality reduction. Science. 2000;290(5500):2319–23.
15. Roweis ST, Saul LK. Nonlinear dimensionality reduction by locally linear embedding. Science. 2000;290(5500):2323–6.
16. Geng X, Zhan D-C, Zhou Z-H. Supervised nonlinear dimensionality reduction for visualization and classification. IEEE Trans Syst Man Cybern B Cybern. 2005;35(6):1098–107.
17. Chen T, Guestrin C. Xgboost: A scalable tree boosting system. In: Proceedings of the 22nd ACM Sigkdd International Conference on Knowledge Discovery and Data Mining; 2016. p. 785–94.
18. Borg I, Groenen P. Modern multidimensional scaling: theory and applications. J Educ Meas. 2003;40(3):277–80.
19. Chen Z, Liu X, Li F, et al. Large-scale comparative assessment of computational predictors for lysine post-translational modification sites. Brief Bioinform. 2018. https://doi.org/10.1093/bib/bby089.
20. Li F, Li C, Marquez-Lago TT, et al. Quokka: a comprehensive tool for rapid and accurate prediction of kinase family-specific phosphorylation sites in the human proteome. Bioinformatics. 2018;34(24):4223–31.
21. Li F, Wang Y, Li C, et al. Twenty years of bioinformatics research for protease-specific substrate and cleavage site prediction: a comprehensive revisit and benchmarking of existing methods. Brief Bioinform. 2018. https://doi.org/10.1093/bib/bby077.
22. Song J, Wang Y, Li F, et al. iProt-sub: a comprehensive package for accurately mapping and predicting protease-specific substrates and cleavage sites. Brief Bioinform. 2018;20(2):638–58.
23. Song J, Li F, Leier A, et al. PROSPERous: high-throughput prediction of substrate cleavage sites for 90 proteases with improved accuracy. Bioinformatics. 2017;34(4):684–7.
24. De Ridder D, Kouropteva O, Okun O, et al. Supervised locally linear embedding. In: Artificial Neural Networks and Neural Information Processing—ICANN/ICONIP: Springer; 2003. p. 333–41.
25. Lou W, Wang X, Chen F, Chen Y, Jiang B, Zhang H. Sequence based prediction of DNA-binding proteins based on hybrid feature selection using random forest and Gaussian naive Bayes. PLoS One. 2014;9(1):e86703.
26. Peng H, Long F, Ding C. Feature selection based on mutual information criteria of max-dependency, max-relevance, and min-redundancy. IEEE Trans Pattern Anal Mach Intell. 2005;27(8):1226–38.
27. Guyon I, Weston J, Barnhill S, Vapnik V. Gene selection for cancer classification using support vector machines. Mach Learn. 2002;46(1–3):389–422.
28. Genuer R, Poggi J-M, Tuleau-Malot C. VSURF: an R package for variable selection using random forests, vol. 7; 2015. p. 19–33.
29. Peng Y, Sun L, Jia Z, Li L, Alexov E. Predicting protein–DNA binding free energy change upon missense mutations using modified MM/PBSA approach: SAMPDI webserver. Bioinformatics. 2017;34(5):779–86.
30. Zhang N, Chen Y, Zhao F, et al. PremPDI estimates and interprets the effects of missense mutations on protein-DNA interactions. PLoS Comput Biol. 2018;14:e1006615.
31. Pires DEV, Ascher DB. mCSM-NA: predicting the effects of mutations on protein-nucleic acids interactions. Nucleic Acids Res. 2017;45:W241–6.

## Publisher's Note

Springer Nature remains neutral with regard to jurisdictional claims in published maps and institutional affiliations.

Albrecht *et al. BMC Bioinformatics* 2020, **21**(Suppl 13):390
https://doi.org/10.1186/s12859-020-03684-2

BMC Bioinformatics

RESEARCH

Open Access

# MAIRA- real-time taxonomic and functional analysis of long reads on a laptop

Benjamin Albrecht[1†], Caner Bağcı[1,2†] and Daniel H. Huson[1*]

*From* The 18th Asia Pacific Bioinformatics Conference
Seoul, Korea. 18–20 August 2020

*Correspondence:
daniel.huson@uni-tuebingen.de
†Benjamin Albrecht and Caner Bağcı contributed equally to this work.
[1]Department of Computer Science, University of Tübingen, Sand 14, Tübingen, Germany
Full list of author information is available at the end of the article

## Abstract

**Background:** Advances in mobile sequencing devices and laptop performance make metagenomic sequencing and analysis in the field a technologically feasible prospect. However, metagenomic analysis pipelines are usually designed to run on servers and in the cloud.

**Results:** MAIRA is a new standalone program for interactive taxonomic and functional analysis of long read metagenomic sequencing data on a laptop, without requiring external resources. The program performs fast, online, genus-level analysis, and on-demand, detailed taxonomic and functional analysis. It uses two levels of frame-shift-aware alignment of DNA reads against protein reference sequences, and then performs detailed analysis using a protein synteny graph.

**Conclusions:** We envision this software being used by researchers in the field, when access to servers or cloud facilities is difficult, or by individuals that do not routinely access such facilities, such as medical researchers, crop scientists, or teachers.

**Keywords:** Metagenomics, Microbiome, Long read sequencing, Taxonomic analysis, Functional analysis, Mobile computing, Open source software, Antibiotic resistance, Virulence factors

## Background

The Oxford Nanopore MinION USB sequencing device, the MinIT USB base-calling device and advances in lab-on-a-chip technologies allow sequencing to be taken into field [1]. With ever rising sequencing yield and continuously growing reference databases, the computational analysis of such data is challenging, and much work is being done to address this efficiently, usually on a server or using cloud computing [2, 3]. Modern laptops provide a lot of computational power, main memory and fast access to SSD storage, and thus it should be possible to perform detailed analysis of microbiome sequencing data in the field, on a laptop.

In this paper, we present a new program called MAIRA (mobile analysis of long reads) for analyzing long metagenomic reads on a laptop, without requiring external resources. The program first performs fast genus-level analysis in real time, and then also performs detailed species-level taxonomic and functional analysis. The latter refers to basic analysis of antibiotic resistance potential [4] and virulence factors [5].

We envision this software being used by researchers in the field, when access to servers or cloud facilities is difficult [6], or by individuals that do not routinely use such facilities, such as medical researchers, crop scientists, or teachers, say. The program is able to analyze sequencing data in real-time and thus may be applicable in time-critical applications.

Following the approach established in [7, 8], we base the taxonomic and functional alignment of metagenomics sequencing reads on the alignment of the reads against a protein reference database; here we use all bacterial proteins in RefSeq [9].

There are three main challenges.

1. Current long reads contain many erroneous insertions and deletions, causing frame shifts that disrupt translated alignments and result in very poor performance of methods such as BLASTX [10].
2. The alignment of gigabases of sequencing reads against a reference database containing on the order of one hundred million reference proteins requires substantial computational resources, despite major advances [11, 12].
3. For many applications, such as the assessment of antibiotic resistance or virulence, say, it is crucial to link the associated genes to specific organisms.

Here we describe a new open source program called MAIRA that addresses these three main challenges.

## Implementation

Challenge 1 can be solved using a frame-shift aware alignment tool [13], and MAIRA supports the use of LAST [11], but also contains a new, built-in frame-shift aligner called ELLA (manuscript in preparation). To address Challenge 2, the program performs a fast online genus-level analysis, whereas a more detailed analysis can be computed on-demand, for selected genera of interest. To address Challenge 3, we introduce the concept of a protein synteny graph (manuscript in preparation) whose nodes represent gene families and whose edges reflect synteny.

MAIRA is a standalone computer program for analyzing sequencing reads from a mobile sequencing device or long reads from other sources. MAIRA is written in Java and runs on all three major operating systems. The program is designed to be used interactively, however a command-line mode is also provided.

### Online and on-demand

Initial analysis is performed "online" in the sense that sequencing reads are processed incrementally in batches, in real-time, as they become available. The displayed results are updated after completion of genus-level analysis of each batch. Detailed analysis is performed "on-demand" in the sense that the user may select or prioritize the genera to be analyzed in detail, so as to focus the laptop's computational resources on the analysis of specific organisms of interest. The displayed results are updated after completion of the analysis of each batch.

We will now discuss the main workflow of MAIRA, summarized in Fig. 1.

### Input

The input consists of files of sequences in FastA or FastQ format. Input can be specified as a single file or a directory of input files. All reads are then loaded and processed by the program in batches. Alternatively, one can specify a "monitor directory" that the program will periodically inspect to determine whether any new files of reads are available, which are then processed in batches. In particular, when sequencing with a MinION and running a MinIT to perform base-calling, the MinIT can write FastQ files onto its local hard drive, which can be accessed remotely from a laptop and MAIRA will load its input from there.

### Genus-level alignment and analysis

As indicated in Fig. 1a, MAIRA processes incoming reads in batches (by default, 10,000 reads per batch). These reads are aligned against a precomputed database of genus-marker genes, currently representing 2,418 genera, based on the bacterial RefSeq protein database.

Frame-shift aware DNA-to-protein alignment is performed either using LAST [11] or the built-in alignment tool ELLA (manuscript in preparation). For aligning reads against a database both aligners use a frame-shift penalty of 15 and a multiplicity value of 10 (for the marker database) and 50 (for genus-specific databases, see below), respectively. ELLA is written in Java and uses the same algorithmic approach as LAST. It has similar specificity and sensitivity, but is about 2.5 times slower. We have implemented and incorporated ELLA so as to make MAIRA completely self-contained and platform-independent.

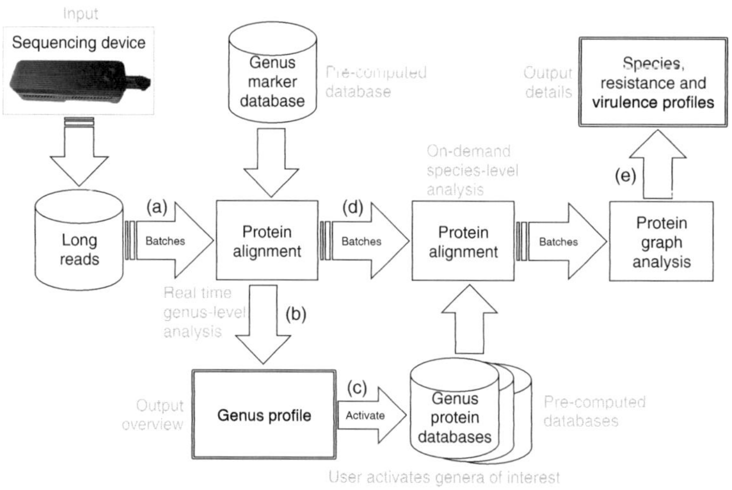


**Fig. 1** MAIRA workflow. In an online manner, **a** batches of long reads obtained from a sequencing device are subjected to protein alignment against a precomputed database of genus-specific marker genes. Based on this, **b** the list of detected genera is updated in real-time. **c** Substantial evidence for the presence of a specific genus triggers **d** batch-wise protein alignment against a full database of proteins associated with the genus. Based on this, a protein graph is updated and used to perform species and-strain level analysis of the long reads, including the identification of antibiotic resistance and virulence factors

The marker database that we provide for download contains 9,434,634 marker genes. All results presented in this paper were obtaining using this database.

The database was precomputed from data downloaded from RefSeq in February 2019, comprising 142,092 bacterial genomes in 2,420 different genera. For each genus, we first constructed a graph whose nodes are proteins and any two nodes $v$ and $w$ are connected by an edge from $v$ to $w$, if the two protein sequences have a similarity of $\geq$ 90% and if $v$ covers at least 90% of $w$, using DIAMOND [12] in BLASTP mode.

A set of candidate marker genes was then computed by greedily determining a minimum *dominating set* of nodes, that is, a subset of nodes $D$ such that every node of the graph is either contained in $D$, or is adjacent to some node in $D$. We then turned each candidate set of proteins into a set of genus representative proteins. This was done by aligning all pairs of candidate sets from different genera against each other using DIAMOND, applying an identity threshold of 80% and a coverage threshold of 80%, so as to detect and remove proteins that are not unique to a single genus.

Based on these sets of representative proteins, we then computed the marker database. Using the graph described above, we first removed all nodes that are not associated with a representative protein. Next, we greedily selected a minimum set of nodes so that all genomes are covered by a fixed minimum number of proteins, namely 1000, 500 or 100, for each of the large, medium or small database, respectively. Here, a genome is considered covered by one of its proteins $v$, if $v$ is selected, or if $v$ is adjacent to a selected node.

This resulted in 9,434,634 genus-specific *marker proteins* that represent 2,418 genera, with a median of 1,121 (mean 3,903) marker proteins (s.d. 10,588). There are 4 genera that contain less than 100 proteins and there are 6 genera that contain more than 100,000 proteins, with Bacillus (205,864), Streptomyces (194,850), and Lactobacillus (158,640) having the largest counts.

While aligning reads against the genus-marker database, for each genus $G$, we maintain and report a *presence score* $p$ that heuristically seeks to approximate the probability of genus $G$ being present in the sample. This is calculated as follows: First, each marker protein $v$ associated with $G$ is given a *weight* $\omega(v) = \frac{1}{\bar{n}}$, where $\bar{n}$ is the average number of marker proteins over all genomes that either contain the marker protein $v$ itself or a similar one (both, similarity and coverage $\geq$ 90%). The presence score for $G$ is then given by the sum of weights of all marker proteins for $G$ to which a read has been aligned to.

The genus-level analysis of the content of the sample being sequenced is summarized on a tree in the main window of the program. Within minutes, the program will provide a first crude estimation of the genus-level content of the sample being sequenced (or read from an input file). With the completion of genus-level analysis of each batch, this summary is updated.

#### Species-level alignment

The above described genus-level analysis is designed to provide a very quick taxonomic analysis of long read sequencing data, akin to traditional 16S rRNA taxonomic profiling.

In addition, the program is able to perform an in-depth, protein-alignment-based taxonomic and functional analysis of the incoming reads, in batches. To save computational resources, the in-depth analysis is performed on demand, for a specified subset of taxa.

In more detail, using the bacterial RefSeq protein database, for each genus we pooled all proteins that either correspond to the genus itself, or to an ancestor or descendant taxon. We obtained a set of 2,420 genus-specific databases, each containing an average of 43,849 proteins (median 6,310, s.d. 251,520).

Batches of incoming reads (Fig. 1d) are aligned against the reference databases for all genera that have been "activated" (Fig. 1c), either explicitly by the user, or automatically, because their presence score has exceeded the activation threshold (80%, by default). The program tracks all batches of reads and ensures that each batch of reads will eventually be compared against every activated genus-specific database, independent of when the batch of reads was loaded into the program or when the genus was activated.

### Protein synteny graph

The alignments computed during species-level analysis are used to build a *protein synteny graph PSG*. In this graph, each node represents a protein gene family. Gene families are defined as follows: The alignments produced from the genus specific databases for each long-read are pooled together, and filtered by the coverage of the reference protein (default: 80%), the percentage of positives in the alignments (default: 60%), and the raw-score of the alignment (default: 100). The alignments that stack (at least 2/3 of the smaller alignment overlapping with the longer alignment) at different loci are then binned together so as to obtain gene families. These bins are represented as *protein nodes* in the graph. Each such *protein node* $v$ is annotated by the set $\tau(v)$ of all taxa and functional classes associated with members of the gene family.

A protein node is placed into the graph, if there exists a sequencing read that aligns to a gene sequence in the corresponding family. Two such nodes $v$ and $w$ in the graph are joined by an undirected edge $\{v, w\}$ if the two corresponding gene families appear next to each other in some sequencing read. Each edge $e$ is annotated by the set $\rho(e)$ of all such reads.

Let $t$ be a taxon. The taxon-specific *induced* protein synteny graph $PSG|_t$ is given by the set of all protein nodes $v$ whose annotation contains the taxon $t$, that is, for which $t \in \tau(v)$ holds, together with all other nodes that lie between them in the graph along some read. Any two nodes in the resulting graph are connected by an edge, if and only if they were in the original graph.

### Taxonomic analysis

Using these concepts, we declare a taxon $t$ to be *present* in the sample, if the corresponding induced graph $PSG|_t$ contains enough nodes. In more detail, we define the naïve *completeness score* $k(t)$ as the number of nodes in the induced graph divided by the median number of proteins present for the same or (a similar) taxon in the RefSeq database. A taxon is considered present, if its completeness score exceeds a specified threshold (80%, by default).

We further compare all induced graphs to all others that are above the detection criteria, and eliminate those that are largely contained within the induced graph of another taxon (default: 85% of the nodes), without containing the same percentage of the nodes from the other itself. We do this in order to eliminate false-positive taxa that can still produce highly connected and complete subgraphs due to partial but not complete similarity to the true positive taxon.

### Antibiotic resistance and virulence analysis

To identify potential antibiotic resistance genes, we use information from the *comprehensive antibiotic resistance database* (CARD) [4] to annotate nodes that represent resistance-associated genes. Similarly, to identify potential virulence factors, we use information from the *virulence factor database* (VFDB) [5] to annotate nodes that are considered virulence factors. For a given taxon $t$, we report all CARD and VFDB annotations present in the induced graph $PSG|_t$.

## Results

We demonstrate the use of the program using three datasets.

### Nanopore mock community

The Nanopore mock community published in [14] consists of 3,491,078 long reads with an average length of 4,012 bases, sequenced on a GridION. The source consists of eight bacterial strains and two yeast strains.

We ran our pipeline on 25 batches of 10,000 reads each, using both LAST and ELLA. In both cases, MAIRA correctly determined all 8 present genera of bacteria after the third batch (requiring 5 minutes using LAST and 11 minutes using ELLA). Both aligners also resulted in false-positive identification of 2 genera after the third batch: *Enterobacter* and *Shigella*. Another false-positive *Citrobacter* also passed the default genus detection threshold of 0.8 for the 15-th batch for ELLA and 11-th batch for LAST (see Fig. 2a).

Running the species-level analysis to completion on 25 batches required $9\frac{1}{2}$ hours using LAST and nearly 26 hours using ELLA. All eight true-positive species were identified with strong signal (completeness >90%) using both aligners after the end of the 4-th batch, after one hour for LAST and three hours for ELLA; and their scores thereafter remained constant.

MAIRA identified one false-positive species, *Bacillus amyloliquefaciens*, using both aligners. Its completeness score, however, stayed below 0.75 using ELLA and 0.85 using LAST (see Fig. 2b). We report the coverage of genomes in Fig. 2c.

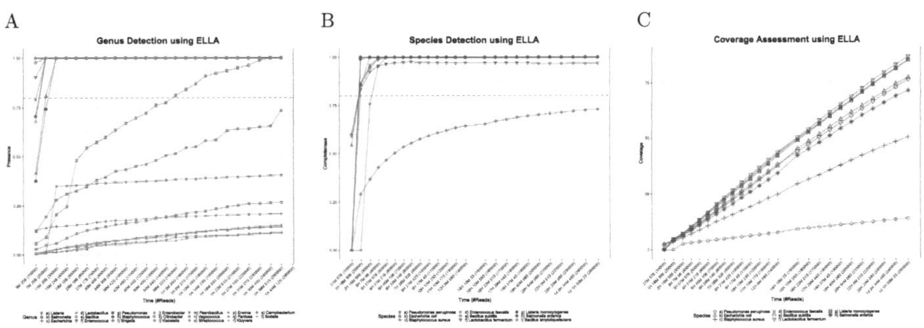


**Fig. 2** Performance on mock community. **a** The presence score for true positive (green) and false positive (red) genera as a function of computation time and number of reads processed. The blue line indicates the threshold used to deem a genus "present". **b** The completeness of true positive (green) and false positive (red) species as a function of computation time and number of reads processed. The blue line indicates the threshold used to deem a species "present". **c** The coverage of species, as a function of time and number of reads processed

Running the software on further batches eventually produces one more false positive genus *Klebsiella*.

### Simulated pathogen mock community

We now consider the simulation of a pathogen mock community. We simulated an even community consisting of ten different bacterial species, each represented by 10,000 simulated Nanopore reads generated using NanoSim [15]. The ten source genomes correspond to the components of a commercially available metagenomic mock community for pathogen detection called ATCC® MSA-4000™. This consists of 11 strains of 10 species in eight different genera. The genome sequence of one of the strains (ATCC BAA-1718) is very fragmented (174 contigs) and so we excluded it from our study.

Complete analysis of all 100,000 simulated long reads took slightly over six hours using MAIRA and LAST. The approach correctly determined all eight genera at the end of the 7-th batch, while indicating two incorrect genera, *Shigella* and *Enterobacter*. The species-level analysis resulted in identification of all 10 species present, while indicating three incorrect species, *Staphylococcus mitis*, *Streptococcus pseudopneumoniae* and *Streptococcus oralis*. Note that the dataset contains multiple strains of both *Staphylococcus* and *Streptococcus*.

To assess the ability of MAIRA to correctly determine the presence of CARD or VFDB genes, for both functional classifications, we aligned the set of their reference proteins against all coding sequences (CDSs) in the source genomes, using DIAMOND [12]. Then, for both classifications, any protein that aligned to a CDS with a percent positives of $\geq$ 80% and a reference coverage above $\geq$ 90% was considered an actual positive.

For each species, MAIRA recovered almost all actual positives for both CARD and VFDB, with mostly low numbers of false positives, see Table 1.

### Carbapenemase-producing gram-negative bacteria isolates

The long read dataset published in [16] consists of 110 MinION long read sequencing datasets, of which we were able to download 109 (sample ERR2797062 could not be found). These datasets were sequenced from 58 *Klebsiella pneumoniae*, 28 *Escherichia coli*, 13 *Pseudomonas aeruginosa*, and 10 *Acinetobacter baumannii* isolates. The authors report that 63 isolates are suspected to have carbapenemase resistance, 34 have other types of antibiotic resistance, and indicate that they failed to produce evidence for antibiotic resistance in 13 isolates.

We ran MAIRA on the first 10,000 reads of each sample (or all reads, where there were less than 10,000). It took about one minute per sample to complete genus-level analysis using MAIRA and LAST.

For all 109 samples, MAIRA reports the correct species and no false positives. However, the initial, fast genus-level analysis did report some false positive genera. For example, in the case of *Klebsiella pneumoniae*, three false positives were listed. None of the false positive genera were confirmed by the species-level analysis. The full running time varied between five and 20 minutes per sample, depending on the organism.

In order to determine carpabanem-resistance of isolates, we checked for CARD ontology terms with annotation *confers_resistance_to: ARO:0000020* and found homology-based evidence of resistance in 102 isolates with varying ranges of support. The number of reads covering resistance genes ranged from 1 to 35.

**Table 1** Assessment of CARD and VFDB hits reported by MAIRA

| Organism | Classif. | TP (CDS) | FN (CDS) | TP (Term) | FP (Term) | FN (Term) |
|---|---|---|---|---|---|---|
| *Acinetobacter baumannii* | CARD | 25 | 3 | 145 | 6 | 65 |
| | VFDB | 31 | 9 | 31 | 0 | 10 |
| *Enterococcus faecalis* | CARD | 22 | 5 | 24 | 2 | 26 |
| | VFDB | 23 | 3 | 23 | 0 | 5 |
| *Escherichia coli* | CARD | 69 | 6 | 81 | 22 | 22 |
| | VFDB | 102 | 27 | 123 | 0 | 36 |
| *Klebsiella pneumoniae* | CARD | 63 | 7 | 85 | 20 | 282* |
| | VFDB | 33 | 8 | 34 | 0 | 8 |
| *Neisseria meningitidis* | CARD | 15 | 1 | 16 | 1 | 5 |
| | VFDB | 31 | 8 | 30 | 0 | 12 |
| *Pseudomonas aeruginosa* | CARD | 55 | 15 | 87 | 7 | 38 |
| | VFDB | 119 | 16 | 119 | 0 | 22 |
| *Staphylococcus aureus* | CARD | 27 | 2 | 32 | 2 | 7 |
| | VFDB | 57 | 5 | 55 | 0 | 5 |
| *Streptococcus agalactiae* | CARD | 4 | 7 | 4 | 4 | 20 |
| | VFDB | 31 | 0 | 32 | 0 | 1 |
| *Streptococcus pneumoniae* | CARD | 17 | 1 | 19 | 8 | 14 |
| | VFDB | 11 | 2 | 11 | 0 | 4 |
| *Streptococcus pyogenes* | CARD | 4 | 8 | 4 | 4 | 16 |
| | VFDB | 29 | 1 | 29 | 0 | 3 |

For all ten bacterial species present in a simulated pathogen mock community, we report the number of true postives (TP), false positives (FP) and false negatives (FN), based on the coding sequences (CDS) in the source genomes, and based on the terms reported (Term). (*) The large false negative value for CARD terms reported for *Klebsiella pneumoniae* is based on a single CDS that is annotated with over 250 different CARD terms (WP_004176269.1) and thus gives an exaggerated impression

## Discussion

In Fig. 3 we show a screen shot of the user interface. The user sets up an analysis using the dialog shown at the bottom left. The program then processes the given data in batches, frequently updating the tables and trees that represent the taxonomic and functional analysis of the data. The user can interactively explore the taxonomic and functional analyses (CARD and VFDB), can export all tables and trees in different formats, save and reopen the analysis.

With MAIRA, we intend to provide a powerful and accurate, standalone long read analysis software that runs on a laptop and can be used in the field. The use of a comprehensive protein reference database ensures wide applicability of the software. The two step approach, light-weight genus-level analysis, followed by on-demand species-level analysis of selected genera, allows the user to interactively control how the computational resources of the laptop are used.

The presented results suggest that the software is able to perform analysis of real data in acceptable running time and with very good accuracy. Alternative approaches currently rely on the use of servers or cloud resources to perform detailed analysis.

MAIRA is a general purpose tool. Moving forward, we intend to add special purpose modules that are dedicated to specific applications, such as the detection of specific pathogens. Moreover, antibiotic resistance is a complex question that is not adequately addressed by the tool at present. In future work, we intend to also consider DNA alignments of resistance-related genes to perform a more detailed analysis. Future versions

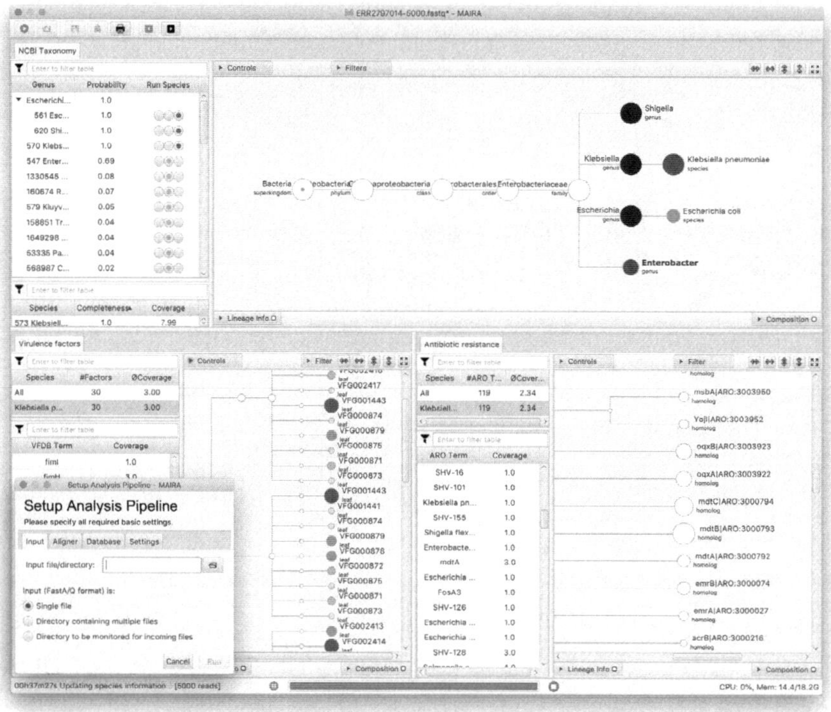


**Fig. 3** MAIRA user interface. During processing of an input file or directory, MAIRA provides an overview of the genera, species and functional classes discovered so far. The user can explore and download results while they are being produced, so as to allow insights into the processed sequences

of the program will use locality of resistance genes and virulence factors in the protein synteny graph to provide more meaningful results.

Initial genus-level analysis performed by MAIRA is very fast and can be applied online during sequencing. Detailed, species-level analysis on individual genera is also fast enough to keep pace with a current MinION+MinIT sequencing setup. Complete analysis of one Gb of long read data requires about 10 hours using LAST and can be considered comfortably doable on a laptop.

One current limitation is that MAIRA focuses on identifying genera and species. Moving forward, we plan to lift this restriction so that assignments can be made to higher level taxonomic ranks, as well. Another limitation is given by the choice of reference databases. In this paper we describe the use of bacterial RefSeq, and the use of this database will avoid false positive identifications. If false negative results are to be avoided, then the use of the more comprehensive, but less curated, NR database may be more appropriate.

The data analysis presented here suggests the current algorithm for performing a fast genus-level analysis tends to yield false positive genera that are then not subsequently confirmed by the detailed species-level analysis. This indicates that there is room for improving the accuracy of the fast genus-level analysis and we will address this in future work.

### Comparison with centrifuge

In order to further evaluate the performance of MAIRA, we compared it to Centrifuge [17], which is the backend for the real-time analysis software WIMP [18], developed by ONT. Centrifuge is designed to work with nucleotide databases on a server. We designed our evaluation so as to simulate a common obstacle in metagenomics, namely that the organism to be identified is not present in the reference databases. For this, we extracted genome assemblies from RefSeq that were published before 2016, and built MAIRA databases for them, as explained above. For Centrifuge, we were only able to build an index for the subset of complete assemblies, as using all assemblies produced an out-of-memory error on a server with 600GB of RAM. Also, the pre-built NCBI-nt index obtained from the Centrifuge website produced the same error.

We used NanoSim to collect 20x long reads on each of 100 different genomes available after 2016. These genomes were selected so as to equally cover four different categories, reflecting what information is available for the corresponding MAIRA and/or Centrifuge indices.

The 4 categories are: (a) no information at species level for MAIRA or Centrifuge; (b) no information at species level for Centrifuge, some for information for MAIRA from incomplete assemblies; (c) only one genome from the species for both MAIRA and Centrifuge; and (d) more than 15 genomes from the species for both MAIRA and Centrifuge. Species assignment accuracy was based on the top 10% of the scores of each tool (top 10% completeness for MAIRA, top 10% number of reads assigned for Centrifuge).

For category (a), MAIRA reported a false species in 5 cases, and did not report any assignment in the other 20; whereas Centrifuge reported a false assignment for all 25 genomes. For category (b), MAIRA reported the true assignment in 22 cases, failed to assign a species in 3 cases, and reported an additional species in 5 cases; whereas Centrifuge again reported a false assignment in all 25 cases. For category (c), MAIRA failed to identify the species in 3 cases, while Centrifuge failed in one case. MAIRA reported 4 false positives, while Centrifuge reported one false positive. For category (d), Centrifuge reported only the true positive species in all cases, whereas MAIRA failed to identify 4, and reported a false positive for one case.

In summary, Centrifuge shows better performance in the ideal case that the sequenced genomes are present in the reference DNA database. However, in the more realistic scenario in which the sequenced genomes are not present in verbatim in the reference databases, MAIRA shows better performance. Moreover, an added benefit of MAIRA is that it also provides a functional analysis, which Centrifuge does not attempt.

## Conclusions

This paper demonstrates that laptop-based analysis of sequencing reads from mobile sequencing devices is possible and this should extend the range of mobile sequencing. As an easy-to-install, standalone application that runs on all three major operating systems, MAIRA provides a complete analysis solution that does not require access to additional computational resources.

Beyond the practical uses of the software, this work illustrates that protein-alignment-based analysis can be performed in real-time on a laptop, that an on-demand design allows a user to direct their computational resources to species of interest, and it shows that the protein synteny graph is a useful concept for the analysis of long reads. Moreover, while designed for running on a laptop, the command-line version can also be run on a server, where the user can also benefit from its efficient design.

We believe that this type of software may play an important role in practical pathogen detection in the future [19].

### Abbreviations
PSG: protein synteny graph; CDS: coding sequence; Gb: gigabase

### Acknowledgements
Not applicable.

### About this supplement
This article has been published as part of *BMC Bioinformatics Volume 21 Supplement 13, 2020: Selected articles from the 18th Asia Pacific Bioinformatics Conference (APBC 2020): bioinformatics*. The full contents of the supplement are available online at https://bmcbioinformatics.biomedcentral.com/articles/supplements/volume-21-supplement-13.

### Authors' contributions
DHH proposed and guided the project and wrote the manuscript. BA and CB contributed to the writing. BA and CB wrote the software framework. BA designed and implemented the built-in alignment tool ELLA. CB designed and implemented the protein-graph approach. All authors read and approved the final manuscript.

### Funding
The authors acknowledge hardware support by the High Performance and Cloud Computing Group at the Zentrum für Datenverarbeitung of the University of Tübingen, the state of Baden-Württemberg through bwHPC and the German Research Foundation (DFG) through grant no INST 37/935-1 FUGG. BA and CB salaries were supported the German Research Foundation (DFG) through grant no HU 566/12-1. Publication costs were supported by the Open Access Publishing Fund of University of Tübingen.

### Availability of data and materials
The publicly available datasets used in the study are available from following BioProject accessions: PRJEB29504 and PRJEB28660.
Project name: MAIRA - mobile analysis of long reads.
Project home page: https://ab.inf.uni-tuebingen.de/software/maira
Operating system(s): Platform independent.
Programming language: Java (OpenJDK 12 and OpenJFX).
Other requirements: High-end laptop, $\geq$32 Gb of memory, 500G of SSD
License: GNU GPL
Any restrictions to use by non-academics: None

### Ethics approval and consent to participate
Not applicable.

### Consent for publication
Not applicable.

### Competing interests
The authors declare that they have no competing interests.

### Author details
[1]Department of Computer Science, University of Tübingen, Sand 14, Tübingen, Germany. [2]International Max Planck Research School From Molecules to Organisms, Max Planck Institute for Developmental Biology and Eberhard Karls University Tübingen, Max-Planck-Ring 5, 72076 Tübingen, Germany.

Published: 17 September 2020

### References
1. Tyler AD, Mataseje L, Urfano CJ, Schmidt L, Antonation KS, Mulvey MR, Corbett CR. Evaluation of Oxford Nanopore's MinION sequencing device for microbial whole genome sequencing applications. Sci Rep. 2018;8(1):1–12.
2. Huson DH, Beier S, Flade I, Górska A, El-Hadidi M, Mitra S, Ruscheweyh H-J, Tappu R. MEGAN community edition - interactive exploration and analysis of large-scale microbiome sequencing data. PLoS Comput Biol. 2016;12(6): 1004957.

3. Weber N, Liou D, Dommer J, MacMenamin P, Quinones M, Misner I, Oler AJ, Wan J, Kim L, Coakley McCarthy M, Ezeji S, Noble K, Hurt DE. Nephele: a cloud platform for simplified, standardized and reproducible microbiome data analysis. Bioinformatics. 2018;34(8):1411–3.
4. Jia B, Raphenya AR, Alcock B, Waglechner N, Guo P, Tsang KK, Lago BA, Dave BM, Pereira S, Sharma AN, Doshi S, Courtot M, Lo R, Williams LE, Frye JG, Elsayegh T, Sardar D, Westman EL, Pawlowski AC, Johnson TA, Brinkman FSL, Wright GD, McArthur AG. CARD 2017: expansion and model-centric curation of the comprehensive antibiotic resistance database. Nucleic Acids Res. 2017;45(D1):566–73.
5. Chen L, Yang J, Yu J, Yao Z, Sun L, Shen Y, Jin Q. VFDB: a reference database for bacterial virulence factors. Nucleic Acids Res. 2005;33(suppl_1):325–8.
6. Quick J, Loman NJ, Duraffour S, Simpson JT, Severi E, Cowley L, Bore JA, Koundouno R, Dudas G, Mikhail A, Ouédraogo N, Afrough B, Bah A, Baum JHJ, Becker-Ziaja B, Boettcher JP, Cabeza-Cabrerizo M, Camino-Sánchez Á, Carter LL, Doerrbecker J, Enkirch T, Dorival IG, Hetzelt N, Hinzmann J, Holm T, Kafetzopoulou LE, Koropogui M, Kosgey A, Kuisma E, Logue CH, Mazzarelli A, Meisel S, Mertens M, Michel J, Ngabo D, Nitzsche K, Pallasch E, Patrono LV, Portmann J, Repits JG, Rickett NY, Sachse A, Singethan K, Vitoriano I, Yemanaberhan RL, Zekeng EG, Racine T, Bello A, Sall AA, Faye O, Faye O, Magassouba N, Williams CV, Amburgey V, Winona L, Davis E, Gerlach J, Washington F, Monteil V, Jourdain M, Bererd M, Camara A, Somlare H, Camara A, Gerard M, Bado G, Baillet B, Delaune D, Nebie KY, Diarra A, Savane Y, Pallawo RB, Gutierrez GJ, Milhano N, Roger I, Williams CJ, Yattara F, Lewandowski K, Taylor J, Rachwal P, Turner DJ, Pollakis G, Hiscox JA, Matthews DA, Shea MKO, Johnston AM, Wilson D, Hutley E, Smit E, Di Caro A, Wölfel R, Stoecker K, Fleischmann E, Gabriel M, Weller SA, Koivogui L, Diallo B, Keïta S, Rambaut A, Formenty P, Günther S, Carroll MW. Real-time, portable genome sequencing for Ebola surveillance. Nature. 2016;530:228–32.
7. Huson DH, Auch AF, Qi J, Schuster SC. MEGAN analysis of metagenomic data. Genome Res. 2007;17(3):377–86. https://doi.org/10.1101/gr.5969107.
8. Huson DH, Albrecht B, Bağcı C, Bessarab I, Górska A, Jolic D, Williams RBH. MEGAN-LR: new algorithms allow accurate binning and easy interactive exploration of metagenomic long reads and contigs. Biol Direct. 2018;13(1):6.
9. Pruitt KD, Tatusova T, Klimke W, Maglott DR. NCBI Reference Sequences: current status, policy and new initiatives. Nucleic Acids Res. 2009;37:32–6.
10. Altschul SF, Madden TL, Schäffer AA, Zhang J, Zhang Z, Miller W, Lipman DJ. Gapped BLAST and PSI-BLAST: a new generation of protein database search programs. Nucleic Acids Res. 1997;25:3389–402.
11. Sheetlin SL, Park Y, Frith MC, Spouge JL. Frameshift alignment: statistics and post-genomic applications. Bioinformatics. 2014;30(24):3575–82.
12. Buchfink B, Xie C, Huson DH. Fast and sensitive protein alignment using DIAMOND. Nat Methods. 2015;12:59–60.
13. Arumugam K, Bağcı C, Bessarab I, Beier S, Buchfink B, Gorska A, Qiu G, Huson DH, Williams RBH. Annotated bacterial chromosomes from frame-shift-corrected long read metagenomic data. Microbiome. 2019;7(1):1–13.
14. Nicholls SM, Quick JC, Tang S, Loman NJ. Ultra-deep, long-read nanopore sequencing of mock microbial community standards. GigaScience. 2019;8(5):043.
15. Yang C, Chu J, Warren R, Birol I. NanoSim: nanopore sequence read simulator based on statistical characterization. GigaScience. 2017;6(4):1–6.
16. Noll N, Urich E, Wüthrich D, Hinic V, Egli A, Neher R. Resolving structural diversity of Carbapenemase-producing gram-negative bacteria using single molecule sequencing. bioRxiv. 2018456897.
17. Kim D, Song L, Breitwieser FP, Salzberg SL. Centrifuge: rapid and sensitive classification of metagenomic sequences. Genome Res. 2016;26(12):1721–9.
18. Juul S, Izquierdo F, Hurst A, Dai X, Wright A, Kulesha E, Pettett R, Turner DJ. What's in my pot? Real-time species identification on the MinION. bioRxiv. 2015030742.
19. Smolinski MS, Hamburg MA, Lederberg J. Microbial Threats to Health: Emergence, Detection, and Response. Washington, DC: The National Academies Press; 2003. https://doi.org/10.17226/10636.


## Publisher's Note

Ahmed *et al. BMC Bioinformatics* 2020, **21**(Suppl 13):388
https://doi.org/10.1186/s12859-020-03685-1

BMC Bioinformatics

SOFTWARE Open Access

# GPU acceleration of Darwin read overlapper for de novo assembly of long DNA reads

Nauman Ahmed[1,2*], Tong Dong Qiu[1], Koen Bertels[1] and Zaid Al-Ars[1]

*From* The 18th Asia Pacific Bioinformatics Conference
Seoul, Korea. 18-20 August 2020

*Correspondence:
n.ahmed@tudelft.nl
[1]Delft University of Technology, Delft, Netherlands
[2]University of Engineering and Technology Lahore, Lahore, Pakistan

## Abstract

**Background:** In Overlap-Layout-Consensus (OLC) based de novo assembly, all reads must be compared with every other read to find overlaps. This makes the process rather slow and limits the practicality of using de novo assembly methods at a large scale in the field. Darwin is a fast and accurate read overlapper that can be used for de novo assembly of state-of-the-art third generation long DNA reads. Darwin is designed to be hardware-friendly and can be accelerated on specialized computer system hardware to achieve higher performance.

**Results:** This work accelerates Darwin on GPUs. Using real Pacbio data, our GPU implementation on Tesla K40 has shown a speedup of 109x vs 8 CPU threads of an Intel Xeon machine and 24x vs 64 threads of IBM Power8 machine. The GPU implementation supports both linear and affine gap, scoring model. The results show that the GPU implementation can achieve the same high speedup for different scoring schemes.

**Conclusions:** The GPU implementation proposed in this work shows significant improvement in performance compared to the CPU version, thereby making it accessible for utilization as a practical read overlapper in a DNA assembly pipeline. Furthermore, our GPU acceleration can also be used for performing fast Smith-Waterman alignment between long DNA reads. GPU hardware has become commonly available in the field today, making the proposed acceleration accessible to a larger public. The implementation is available at https://github.com/Tongdongq/darwin-gpu.

**Keywords:** Genomics, Read overlapper, De novo assembly, Long DNA reads, GPU acceleration

## Background

DNA sequencing techniques used today produce short pieces of data (called reads) that represent parts of the sampled DNA, possibly containing some errors. The length and error rate of these reads depends on the sequencing technique used. DNA assembly tries to combine the reads into larger, more accurate DNA segments. For these DNA reads, graph-based assemblers are used for the assembly process, which comes in two flavors: Overlap-Layout-Consensus (OLC) and de Bruijn Graph (dBG).

The OLC assemblers [1] first find overlaps and build an overlap graph. Each node represents a read, and each edge represents an overlap between two reads. During the layout phase, the graph is analyzed to find paths, corresponding to segments of the original genome. The perfect graph contains one path that visits each node exactly once. This problem can be described as finding a Hamiltonian Path. A Hamiltonian Path includes all vertices of a graph exactly once. Examples of assemblers that use the OLC approach are Dazzler [2] and SGA [3].

In dBG based assemblers [4], each read is divided into $K$-*mers*. A $K$-*mer* is a substring of a read having a length $K$. Each $K$-*mer* represents a directed edge between two vertices, where the source vertex represents the first $K-1$ bases of the $K$-*mer*, and the destination vertex the last $K-1$ bases of the $K$-*mer*. When a particular $K-1$ vertex does not exist, it is created, otherwise, the existing one is reused. The weights of the edges indicate how many times a particular $K$-*mer* is encountered. The next step is to find an Eulerian Path, which is a path that includes all edges of a graph exactly once. Examples of dBG assemblers are Velvet [5], ABySS [6] and SOAPdenovo2 [7].

So called Next Generation Sequencing (NGS) techniques produce reads with lengths anywhere from 50 to 500 base pairs. They can be produced at high throughput but at the expense of a smaller read length. However, DNA can contain repeat regions, where a certain piece of DNA is repeated many times back-to-back, or a repeat could appear in many different places in the genome. Since these repeats can be longer than the produced short reads, this means the reads cannot be used to resolve these repeat regions. Third generation sequencing produces much longer reads, of up to 60K base pairs. Due to their length, these reads are more likely to contain a whole repeat region, which makes them suitable for accurately reconstructing the repeat regions. A major drawback of longer DNA reads is their higher error rate, ranging from 15-30%, depending on the exact sequencing technology. dBG based assembly is the more preferred approach for NGS reads, which are much shorter and have much lower error rates. However, de Bruijn Graphs are quite susceptible to sequencing errors, since one substituted base pair causes $K$ incorrect $K$-*mers*. Pair this with an often-used values of $K$ above 50 and third generation sequencing error rate of about 15%, it is clear that the graph will contain a lot of incorrect edges. Therefore, OLC based assemblers are more suitable for state-of-the-art third generation long DNA reads produced by Pacbio and Oxford Nanopore sequencers.

Darwin [8] is a read overlapper for the assembly of long DNA reads. Darwin is designed to be highly accurate, achieving a sensitivity of 99.89% and a precision of 88.30% for simulated Pacbio reads. This is higher than other commonly used read overlappers such as Daligner [2]. The ASIC (Application-Specific Integrated Circuit) implementation of Darwin is shown to be hundreds of times faster than other software based overlappers. However, ASIC implementation requires bulk volume production to be economically feasible. Moreover, DNA analysis using high-throughput DNA sequencing is an evolving field, and any major improvement in the algorithm will require a new ASIC implementation which costs both time and money.

Heterogeneous systems with GPU accelerators have become easily accessible due to their widespread use. They have shown convincing speedups in many high performance computing applications. Therefore, GPU acceleration of various genome analysis algorithms has been the topic of many research works, like in [9, 10] and [11]. In this paper, we present a GPU accelerated version of Darwin. We identified the computational bottleneck in the Darwin software and replaced it with the GPU accelerated version. The accelerated implementation proposed in this paper is orders of magnitude faster than its software counterpart. The contributions of the paper are as follows:

- The paper shows the GPU implementation of the Darwin read overlapper used in the de novo assembly of long DNA reads.
- The paper shows that the GPU acceleration of Darwin is orders of magnitude faster than the multithreaded software version on both IBM Power8 and Intel Xeon machines using a real Pacbio dataset.
- The results in the paper show that the GPU implementation of Darwin can also be applied for accelerating Smith-Waterman alignment of long DNA reads.

## Background

Smith-Waterman (SW) [12] algorithm finds local alignment between a pair of sequences. Smith-Waterman is exact, producing the optimal local alignment. It can be implemented using dynamic programming which computes a 2D matrix $S$. Let $V$ and $W$ be the two sequences to be aligned. Let $V_0, V1, \ldots, V_{|V|-1}$ and $W_0, W1, \ldots, W_{|W|-1}$ be the bases of $V$ and $W$, respectively. $|V|$ and $|W|$ are the lengths of $V$ and $W$. $S(i,-1) = S(-1,j) = 0$ for $i = 0, 1, 2 \ldots, |V|-1$ and $j = 0, 1, 2 \ldots, |W|-1$. The cells in the matrix are computed using the following recurrence relation:

$$S(i,j) = max \begin{cases} S(i-1,j) + gap \\ S(i,j-1) + gap \\ S(i-1,j-1) + subt\left(V_i, W_j\right) \\ 0 \end{cases} \tag{1}$$

$$m_{i,j} = \begin{cases} (i,j) & S(i,j) > m \\ m_{i,j} & S(i,j) \leq m \end{cases} \tag{2}$$

$$m = max \begin{cases} m \\ S(i,j) \end{cases} \tag{3}$$

$$D(i,j) = \begin{cases} 0 & S(i,j) = 0 \\ \uparrow & S(i,j) = S(i-1,j) \\ \leftarrow & S(i,j) = S(i,j-1) \\ \nwarrow & S(i,j) = S(i-1,j-1) + subt\left(V_i, W_j\right) \end{cases} \tag{4}$$

Here, $S$ and $D$ are the score and traceback matrices, respectively. *match*, *mismatch* and *gap* are numeric parameters. $subt\left(V_i, W_j\right)$ is equal to *match* if $V_i = W_j$, and is equal to *mismatch* otherwise. *gap* is the penalty for inserting a gap. $m$ is the alignment score, which is initialized to zero, and $m_{i,j}$ is the corresponding position on $V$ and $W$. The traceback matrix is required to compute the actual alignment. Traceback starts from the highest scoring cell and follows the arrows in $D$ until a zero or a boundary of the matrix is encountered. Equations 1 and 3 indicate that for computing the alignment score $m$ there is no need to store the whole $S$ matrix as all cells of the $S$ matrix are computed using only the values of three other cells $S(i-1,j)$, $S(i,j-1)$ and $S(i-1,j-1)$. Hence, to compute the alignment score, storing only the values in the previous row and column are sufficient to compute $m$. The above equations are for calculating the alignment with a linear-gap scoring model. However, Darwin and our GPU implementation also support the more commonly used affine gap penalty model in which there are separate penalties for opening a gap (*gapo*) and extending a gap (*gape*).

A straightforward way of finding all overlaps is performing an alignment algorithm, such as Smith-Waterman, on every pair of reads. The number of alignments is quadratic with the number of reads, and the runtime of one alignment is quadratic with the lengths of the involved reads, making this method not feasible. Many heuristic algorithms have been developed to perform this alignment, for different lengths and error rates. Seed-and-extend is one heuristic, which dramatically reduces the amount of computation needed [13]. A seed is a $K$-*mer* made up of $K$ consecutive bases of a read. Instead of performing Smith-Waterman on each read pair, only read pairs that have one or more common $K$-*mers* are aligned. A common $K$-*mer* between two or more reads is known as a "seed hit". Darwin also uses the seed-and-extend approach, which reduces the amount of computation needed, without compromising the output by much. Other algorithms, like BLAST [14], also use the seed-and-extend approach, but give sub-optimal alignments. Results in [8] show that Darwin provides optimal Smith-Waterman alignments between long DNA sequences with error rates up to 40%.

### Darwin

Darwin is read overlapping algorithm for de novo assembly of third-generation long DNA reads. It is based on the seed-and-extend. It consists of a filter called D-SOFT (Diagonal-band Seed Overlapping based Filtration Technique), which finds seed hits, and GACT (Genome Alignment using Constant memory Traceback), which extends the seed hit by performing sequence alignment between the sequences on the left and right of the seed hit. Figure 1 shows the seed-and-extend technique employed in Darwin to find the overlap between *Read A* and *Read B*. To compute the overlap, a seed hit is extended on both sides by aligning *R_left* with *Q_left* and *R_right* with *Q_right*. This speedups the computation by avoiding the computation of a large number of dynamic programming matrix cells (grey cells in Fig. 1).The dynamic programming matrix computed to align *R_left* with *Q_left* is known as *left extension matrix*. Similarly, the dynamic programming matrix computed to align *R_right* with *Q_right* is known as *right extension matrix*.

#### *D-SOFT*

D-SOFT is the seeding stage of Darwin, also known as the *filtering* stage. Darwin uses minimizers [15] as seeds which are $K$-*mers* extracted from all the reads to be overlapped. The position of a seed is stored in a minimizer table which records the location of the seed in a read along with the identifier of the read. The window size $w$ is the most important parameter for building a minimizer table and must be smaller than the seed length ($K$). To obtain seed hits, '$N$' $K$-*mers* of a read are used as seeds that are located in other reads using the minimizer table. Two reads are considered for alignment in the extension phase

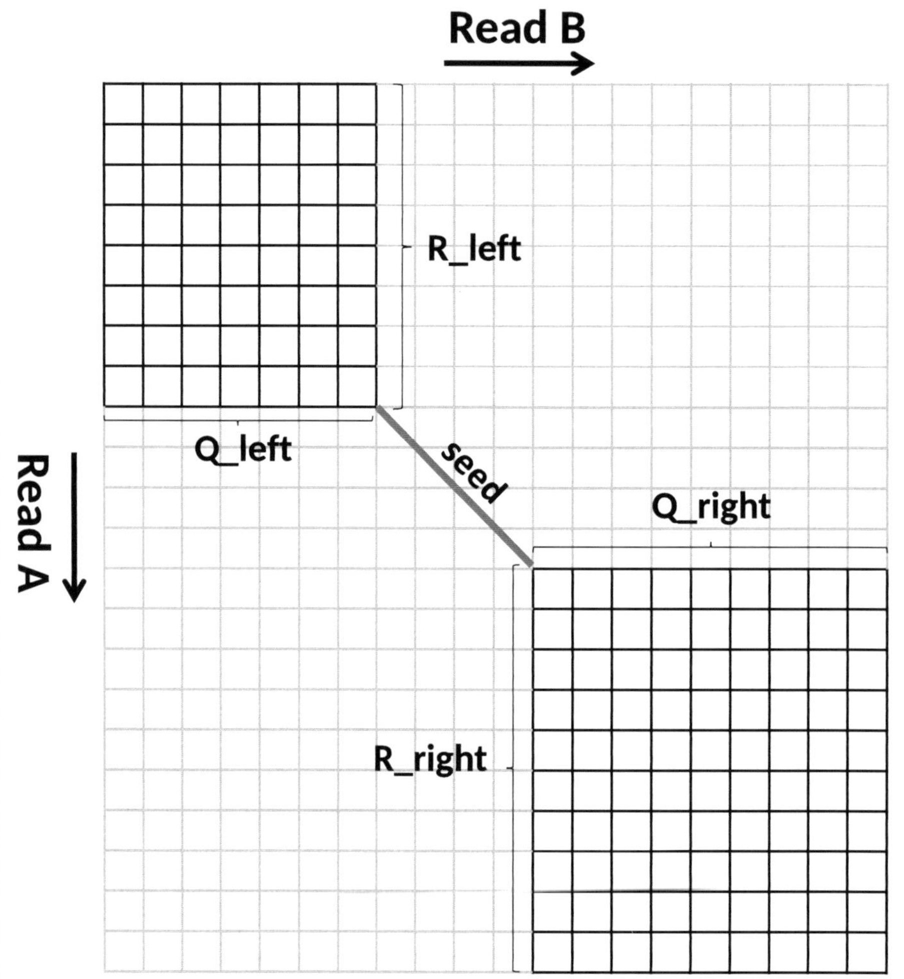


**Fig. 1** The seed-and-extend method used in Darwin to find the overlap between *Read A* and *Read B*. *R_left* and *R_right* are the substrings of Read A. *Q_left* and *Q_right* are the substrings of Read B

if they have at least $h$ unique bases in common. The pair of reads passing this filter are aligned using a modified Smith-Waterman algorithm described below.

### *GACT*

The seed-and-extend approach to find overlap between two reads is much faster than performing the complete Smith-Waterman alignment algorithm. However, the reduced dynamic programming matrices could still be quite large. State-of-the-art third generation sequencers produce reads having lengths in megabases and the dynamic programming matrix in the seed extension may have around 1 Tera cells to compute. For example, if the seed hit lies at the beginning of the two reads, i.e. near the top left corner in Fig. 1, the right extension matrix is nearly as large as the full matrix.

Numerous efforts to accelerate Smith-Waterman have been made, both by using hardware like [16] and software [17]. But the memory required to store the traceback matrix $D$ is still an issue. One can apply the Hirschberg's algorithm described in [18] to reduce the RAM storage but at the cost of increase in computation time. Therefore, Darwin proposed the GACT algorithm for seed extension. It has two advantages: 1) All the cells of the right and left extension matrix are not computed reducing the computation time. 2) The traceback matrix is very small. GACT performs normal Smith-Waterman on a submatrix of the extension matrix, known as *tiles* of size $T$x$T$. After computing a tile it computes the next tile, which overlaps the previous tile with at least $O$ cells on both axes. For reasonable values for $T$ and $O$, GACT has shown to produce the same result as normal Smith-Waterman [8]. Figure 2 shows an example of computing the extension matrix with the GACT algorithm. In the example $T = 8$ and $O = 3$. The tiles are computed in the order $T1, T2, T3$ and $T4$. The example in Fig. 2 can be used to explain both the computation of left and right extension matrices. The only difference is $R = \overline{R\_right}$ and $Q = \overline{Q\_right}$ in case of right extension, where $\overline{R\_right}$ and $\overline{Q\_right}$ is the reverse of $R\_right$ and $Q\_right$ sequences, respectively. Listing 1, shows the algorithm for left extension. Positions $i\_curr$ and $j\_curr$ are produced by D-SOFT. The start and end position of the current tile are stored in $(i\_start, j\_start)$ and $(i\_curr, j\_curr)$, respectively. The traceback path of the whole left extension is kept in *tb_left*. The function *Align*() uses Smith-Waterman to compute traceback matrix $D$ between subsequences *R_tile* and *Q_tile*. Once the traceback matrix is filled, traceback is performed starting from the bottom-right cell, except for the first tile, where traceback starts from the highest-scoring cell. The starting cells of the traceback are coloured yellow in Fig. 2. *Align*() returns the number of bases in $R$ and $Q$ aligned by this tile (*i_off,j_off*), the traceback arrows/pointers (*tb*) and the position of the highest-scoring cell (*i_max,j_max*). *Align*() also limits *i_off* and *j_off* to at most

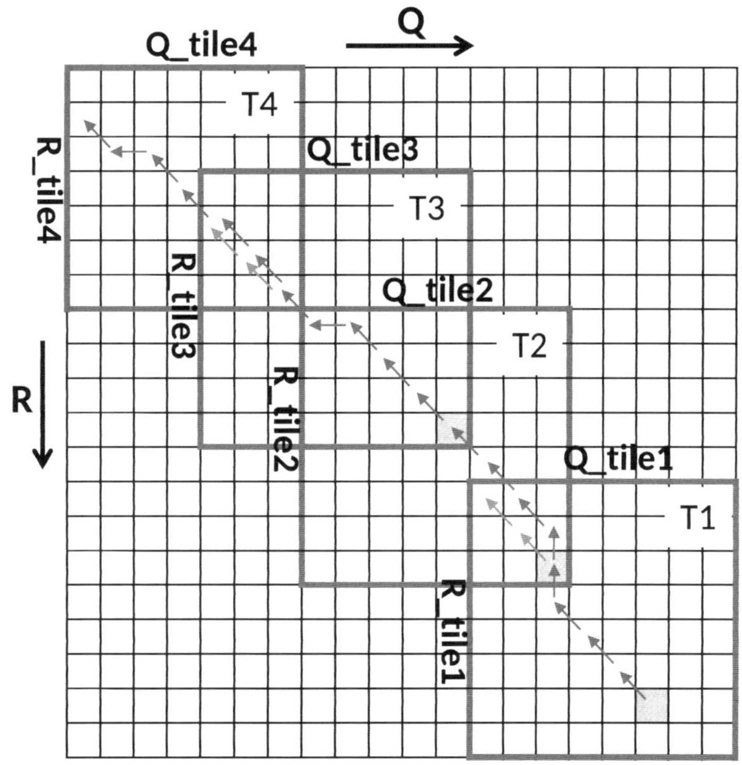


**Fig. 2** An example of the GACT algorithm

**Listing 1** GACT algorithm

```
tb_left = []
(i_curr, j_curr) = (i_seed, j_seed)
t = 1
 while (i_curr > 0 and j_curr > 0)
{
 (i_start, j_start) = (Max(0, i_curr - T), Max(0, j_curr - T
    ))
 (R_tile, Q_tile) = (R[i_start : i_curr], Q[i_start : i_curr
    ])
 (i_off, j_off, i_max, j_max, tb) = Align(R_tile, Q_tile, t,
    T-O)
 tb_left.Prepend(tb)
 if (t == 1){
 (i_curr, j_curr) = (i_max, j_max)
 t = 0
 }
 if (i_off == 0 and j_off == 0) {
 break;
 }
 else {
 (i_curr, j_curr) = (i_curr - i_off, j_curr - j_off)
 }
}
return (i_max, j_max, tb_left)
```

$T - O$ bases, to ensure the next tile overlaps by at least $O$ bases on both $R$ and $Q$. The green arrows shows the path taken by the traceback in a tile if there is no limit and the traceback is allowed to complete. The left extension finishes when it hits the end of $R$ or $Q$, or when traceback cannot add any bases to the existing alignment. The memory needed for the traceback is $\mathcal{O}\left(T^2\right)$, which is constant since $T$ is chosen upfront. The whole alignment of the extension is contained in *tb_left* and is equivalent to the path traced by the red arrows in Fig. 2. The alignment score of the extension can also be computed with the help of *tb_left* The right extension operates on the reverse of $R$ and $Q$.

The performance of GACT is linear ($O(max\{|ReadA|, |ReadB|\} \cdot T)$) where $|ReadA|$ and $|ReadB|$ are the lengths of *Read A* and *Read B*, respectively. It is more suited for long reads than banded alignment [19] because banded alignment uses a static band around the main diagonal. GACT allows for flexible bands since the position of the new tile depends on the traceback path, this is useful for long reads that have high indel rates.

### GPU processing

A GPU is a Graphics Processing Unit, which is a processor that is mainly used to perform video processing. GPUs contain many cores that allow them to perform parallelizable tasks very quickly. A GPGPU, or General Purpose GPU, can be programmed to perform tasks that are different from video processing. GPUs cannot operate on their own, they must be guided by a CPU. The functions that run on a GPU are called *kernels* and are usually launched by a CPU.

CUDA is a parallel programming platform that allows people to use Nvidia GPUs for their applications. Developers can write kernels, launched from a CPU function. The GPU is referred to as *device* and the CPU as *host*. Kernels can be launched from the CPU with a certain number of *thread blocks* with each block containing many GPU threads. The number of blocks and the number of threads in a block are the kernel launch parameters. Each thread executes the kernel code, although they usually operate on different data.

On a hardware level, an NVIDIA GPU is divided into Streaming Multiprocessors (SM). Each SM contains several cores, or Streaming Processors (SP), these are the basic building blocks and perform the actual calculations. Each block is assigned to at most one SM. This block's threads are then executed as warps, with 32 threads per warp. Each SM has multiple warp schedulers, so multiple warps can run in parallel on an SM. All threads in a warp must execute the same instruction, if a thread is the only to take a branch, the other threads must wait until the branch is completed, this is called *thread divergence*.

GPUs have several different memory types and levels. It has its own DRAM known as the *global memory* and a cache shared by all SM's. Accesses to the global memory are also executed in parallel, this means that all threads try to read/write to the memory in parallel. If the addresses are next to each other, only one memory transaction is needed, since a transaction processes a whole memory line. This is known as coalescing. Non-coalesced memory accesses cause multiple memory transactions.

A general workflow using a GPU is:

1 Data is copied from the main memory to the GPU global memory.
2 The CPU launches the GPU kernel.
3 The GPU executes the kernel.
4 Results are copied from the GPU to the CPU memory.

### Previous research

Multiple efforts to accelerate the DNA alignment algorithm on GPU have been made. MUMmerGPU [20] is one of the first GPU accelerated algorithms, it stores a suffix tree of the reference sequence on the GPU, and aligns it with queries. Its newest GPU implementation shows a 13x speedup over the CPU implementation. CUDAlign [21] accelerates the exact Smith-Waterman algorithm and allows an affine gap. The input sequence length is only restricted by the available global memory. It uses linear space and boasts a 702x and 19.5x speedup compared to 1 core and 64 cores, respectively. CUSHAW2-GPU [22] is an accelerated short read aligner. Other work has been done on accelerating BWA-MEM [23] and Protein database search [24].

## Implementation

### Profiling

We measured the runtime of various elements of the Darwin algorithm on the CPU. Two notable parts are D-SOFT (which consists of building the minimizer table, and finding the seeds) and aligning using GACT. Of those two, the *Align*() function in GACT (Listing 1) takes the most time, namely 99.9% for Pacbio reads. Therefore, we accelerated the *Align*() function on GPU. We selected a tile size $T$ of 320 as it gives optimal Smith-Waterman alignment scores [8]. With this setting, 63% of the tiles are exactly 320x320. The remaining tiles are smaller as they occur near the edges of *R*-*Q* matrix (Fig. 2). If some GPU threads have a smaller tile size, it will cause some divergence, because they will have to wait until the threads with larger tile sizes are finished.

### Acceleration

It is possible to run the whole GACT kernel on the GPU, for both left and right extension. However, since it is not known how long the resulting alignment will be, and all GPU threads have to wait until all threads are done, this will cause lots of idle time. Instead, it is chosen to only have a single tile of size $T$x$T$ aligned per GPU-thread per GPU-invocation as shown in Fig. 3. The *Align*() function for many different $R, Q$ pairs are executed in parallel on the GPU. Figure 4 shows the flow graph of GPU accelerated Darwin. It has two GPU kernels shown by green processes in the flow graph. All other tasks are performed on the CPU. The CPU builds the minimizer table using all the reads for which the overlaps have to be computed. The accelerated algorithm processes a set of reads to exploit the massive parallelism of the GPU. The CPU first computes the seed hits for all the reads in the set using the D-SOFT algorithm. With the help of the seed hit location the sequences for the left and right extension matrices are determined. i.e. (*R_left,Q_left*) and (*R_right,Q_right*). One tile (*R_tile, Q_tile* pair) from each extension matrix is assigned to a GPU thread for alignment. All the tile alignments are computed in parallel on the GPU. There are enough seed hits, and hence sufficient extension matrices in the set of reads to fully utilize all the GPU resources. In the post-processing step, full alignment of the extension using *tb_left* (and *tb_right* for right extension) is constructed on the CPU. As described in the "Profiling" section all the tasks other than computing the alignment between *R_tile* and *Q_tile* takes a negligible amount of time on the CPU.

To reduce the GPU memory accesses, the alignment is preceded by a packing step as indicated in the flow graph of Fig. 4. The bases of both sequences are packed in a 4-bit format, where 8 bases are packed into a 32-bit integer. This packing is performed on the GPU and it is hundreds of times faster than packing bases on CPU [25]. To align *R_tile* with *Q_tile*, we extended the local alignment kernel of GASAL (GPU Accelerated Sequence Alignment Library) library [25]. The tile is subdivided into submatrices of size 8x8. Since there are 8 bases in one integer, only two global memory accesses are required to compute a single submatrix. The layout of a tile computed on the GPU is shown in Fig. 5. Each green box contains an 8x8 submatrix. The submatrices are computed in the order

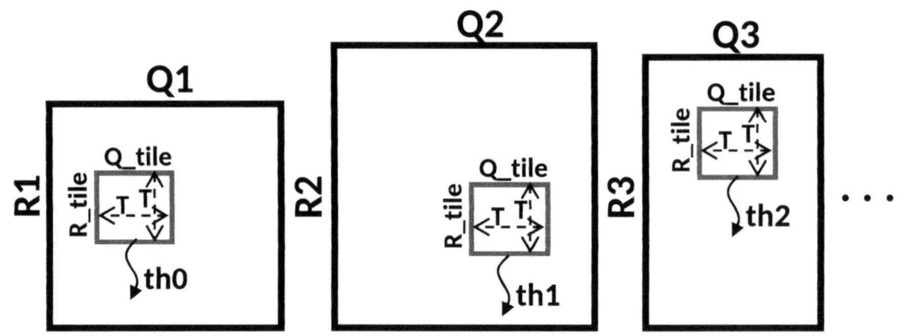


**Fig. 3** GPU thread assignment for parallelization of the tile computation in GACT. *th* is a GPU thread

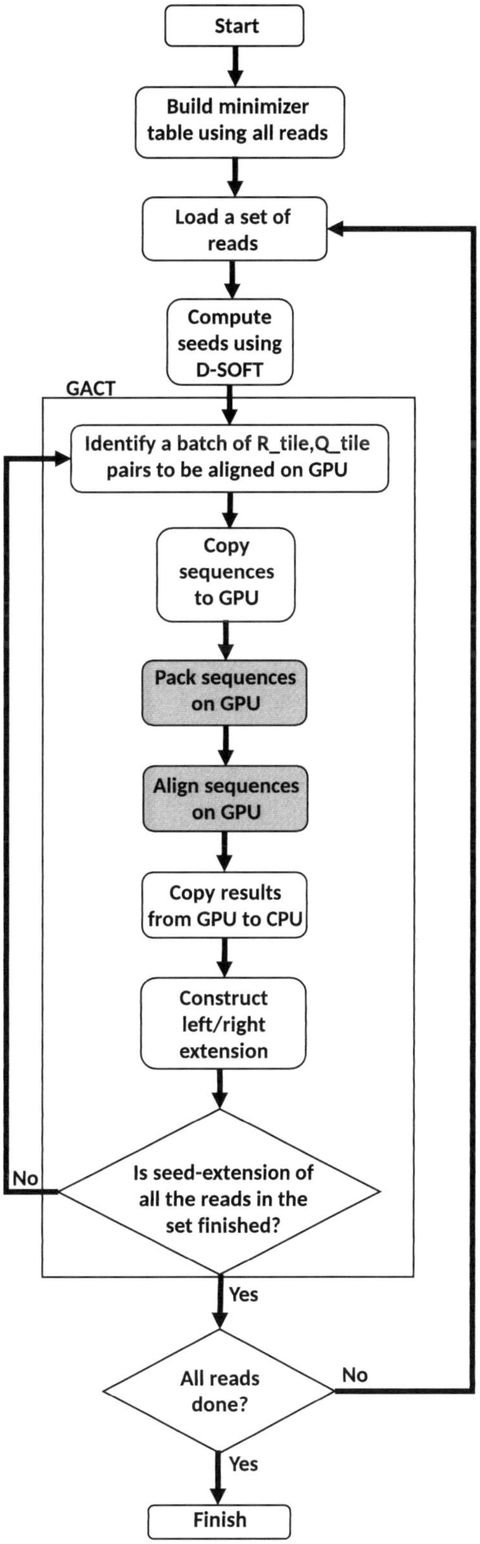


**Fig. 4** The flow graph of GPU accelerated Darwin. The green boxes are GPU kernels

shown by their number. The arrows in the submatrix number 4 show the order of computation of the dynamic programming cells in a submatrix. The figure shows that the *Q_tile* sequence is read multiple times. Hence, packing the sequence with 4 bits per base helps to keep it in the cache for faster access. It is clear from Eqs. 1-4 that to compute a column of the submatrix only the cells in the left column and the row above it are required. The required column and row are colored blue in Fig. 5. The column has only 8 elements, and hence can be stored in GPU registers. Therefore, the total amount of memory required is $\mathcal{O}\left(T + T^2\right)$, where $\mathcal{O}(T)$ is required for computing maximum alignment score $m$ and $\mathcal{O}\left(T^2\right)$ for storing the traceback matrix $D$. Algorithm 1 shows the GPU implementation of the *Align*() function. The pseudocode above Line 4 is for computing the position of maximum alignment score $(i_max, jmax)$ and the traceback matrix $D$. Observe that the all the writes in $D$ are coalesced to optimize the memory bandwidth and reduce the number of memory transactions. The pesudocode below Line 4 is for computing the traceback path *tb*. The GPU accelerated Darwin supports both linear as well as affine gap penalties. Algorithm 1 shows the alignment using the linear gap penalties. The algorithm with affine gap penalties has a similar layout and omitted here for brevity.

**Algorithm 1:** The GPU kernel for the *Align*() function of GACT (Listing 1) to compute a tile of size $T$x$T$

**Input:** Two sequences $R_{tile}\_pack$ and $Q_{tile}\_pack$ of length $T/8$ packed in 32 bit words where each word contains 4 bases (4 bits per base); A backtrack matrix $D$ of size $(T \cdot T) \cdot n_{threads}$ allocated in the GPU memory, where $n_{threads}$ is the number of GPU threads; tile number $t$; maximum number of bases to align $max_{off}$; number of GPU threads $n_{threads}$, GPU thread ID $tid$, score for match $match$, mismatch penalty $mismatch$ and gap penalty $gap$ are known by default

**Output:** Number of aligned bases in $R_{tile}$ and $Q_{tile}$ $(i_{off}, j_{off})$; position of the maximum alignment score $(i_{max}, j_{max})$ and the traceback arrows/pointers $tb$

```
1  Function AlignOnGPU(R_tile_pack , Q_tile_pack, D, t, max_off) begin
2      Initialize H array of length T containing zeros
3      max_sc = 0
4      for i = 0 to T/8 − 1 do
5          Initialize h and p arrays of length of 9 containing zeros
6          r = R_tile_pack[i] // r is a 32 bit word
7          r_idx = i·
8          q_idx = 0
9          for j = 0 to T/8 do
10             q = Q_tile_pack[i] // q is a 32 bit word
11             for k = 1 to 8 do
12                 qbase = (q >> (32 − (k · 4)))&15  /* C++ like shift operation followed by AND to extract a base from a 32 bit
                    word */
13                 h[0] = H[q_idx]
14                 for m = 1 to 8 do
15                     rbase = (r >> (32 − (m · 4)))&15    /* C++ like shift operation followed by AND to extract a base from a
                        32 bit word */
16                     if qbase == rbase then
17                         tmp = p[m] + match
18                     else
19                         tmp = p[m] − mismatch
20                     h[m] = max {tmp, p[m] − gap, h[m − 1] − gap, 0}
21                     if h[m] == 0 then
22                         D[(((r_idx + (m − 1)) · T) + q_idx) · n_threads) + tid] = 0
23                     else if h[m] == tmp then
24                         D[(((r_idx + (m − 1)) · T) + q_idx) · n_threads) + tid] =↖
25                     else if h[m] == p[m] − gap then
26                         D[(((r_idx + (m − 1)) · T) + q_idx) · n_threads) + tid] =←
27                     else if h[m] == h[m − 1] − gap then
28                         D[(((r_idx + (m − 1)) · T) + q_idx) · n_threads) + tid] =↑
29                     if h[m] > max_sc then
30                         max_sc = h[m]
31                         i_max, j_max = {r_idx + (m − 1), q_idx}
32                     p[m] = h[m − 1]
33                 H[q_idx] = h[m]
34                 q_idx = q_idx + 1
35     // the traceback starts below
36     if t==1 then
37         (start_i, start_j) = (i_max, j_max)
38     else
39         (start_i, start_j) = (T − 1, T − 1)
40     while D[((start_i · T) + start_j) · n_threads) + tid]! = 0 and i_off ≤ max_off and j_off ≤ max_off and start_i ≥ 0 and start_j ≥ 0
        do
41         tb.prepend(D[((start_i · T) + start_j) · n_threads) + tid])
42         if D[((start_i · T) + start_j) · n_threads) + tid] ==↖ then
43             start_i = start_i − 1
44             start_j = start_j − 1
45             i_off = i_off + 1
46             j_off = j_off + 1
47         else if D[((start_i · T) + start_j) · n_threads) + tid] ==↑ then
48             start_i = start_i − 1
49             i_off = i_off + 1
50         else if D[((start_i · T) + start_j) · n_threads) + tid] ==← then
51             start_j = start_j − 1
52             j_off = j_off + 1
53     return {(i_off, j_off), (i_max, j_max), tb}
```

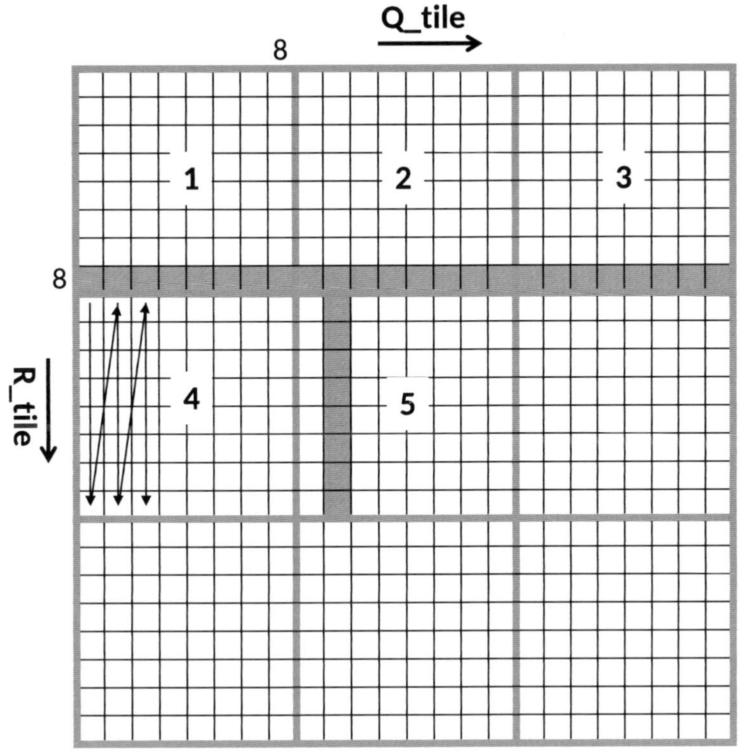


**Fig. 5** Layout of the tile computed on the GPU. Each green box is a 8x8 submatrix

Sequence alignment of long DNA sequence is a performance bottleneck in genome analysis algorithms. The results of Darwin alignment are same as for normal Smith-Waterman for reasonable values of $T$ and $O$. Hence, the Darwin algorithm can also be applied for Smith-Waterman alignment between two long DNA sequences and our GPU implementation of Darwin can be used to accelerate Smith-Waterman alignment (with traceback) for long DNA sequences.

## Results

We compared our GPU acceleration with the hand-optimized CPU version of Darwin [26] (commit: 16bdb81). Tests are performed on both IBM as well as Intel machines. The IBM machine (S824L) has 2 sockets with each socket containing a 10-core POWER8 @ 3.42 GHz processor. Each core has 8-way Simultaneous Multithreading. Hence, there are 160 logical cores in total. The machine has 256 GB of RAM and a Tesla K40m. The CUDA version is 7.5, and the operating system is Ubuntu 3.19.0-28-generic. The GCC version is 4.9.2.

The Intel machine has 2 sockets with each socket containing a 6-core Xeon E5-2620 @ 2.4 GHz processor. Each core has 2-way Hyperthreading. Hence, there are 24 logical cores in total. The machine has 32 GB of RAM and a Tesla K40c. The CUDA version is 9.2, and the operating system is CentOS 7.5. The GCC version is 4.8.5. The K40c and K40m have the same performance, the only difference lies in their cooling method.

We use Pacbio 54x Human sequencing data [27]. The data has a size of 172 gigabytes containing 21,856,161 reads. Since the runtime of the experiments has a quadratic relationship with the number of reads, we use first 50 megabytes (8566 reads) as the input dataset to finish the experiments in a reasonable time. Even with 50 megabytes input dataset, the CPU implementation takes more than 2 hours to run on 8 threads of the Intel machine (Fig. 8). The input dataset contains reads up to 33 kilo bases long with an average read length of 6 kilo bases. Darwin computes the overlaps between the reads in the input dataset. The settings for GPU and CPU implementation are as follows: $match, mismatch, gapo, gape = (1, -1, -1, -1)$, $N = 800$, $T = 320$, $O = 120$, $K = 14$, $h = 21$, $w = 1$.

Since our GPU implementation accelerates only the $Align()$ function on the GPU, everything else is executed on the CPU with multiple threads. Each CPU thread launches a batch of *R_tile* and *Q_tile* sequences to be aligned on the GPU. Since all these CPU threads share a single GPU, it is necessary to investigate how the choice of numbers of CPU threads, number of GPU blocks and the number of threads in a block affect the performance. Figure 6 shows the total execution size for various settings of these factors. The figure shows that the fastest execution time is obtained with 8 CPU threads running with the GPU launch parameters of 32 blocks and 64 threads per block. We performed a similar analysis for the Intel machine and found that 8/32/64 is the best setting in the case of the Intel machine as well. Therefore, we used the 8/32/64 ((Number of CPU threads) / (number of blocks) / (number of threads per block)) setting for running the GPU implementation in the remainder of the experimental results.

Figure 7 shows the total execution time of the CPU implementation of Darwin and compares it with the total execution time of the GPU accelerated Darwin, for the IBM machine. Note that the $y$-axis of the figure represents a logarithmic scale due to the high speedup achieved by the GPU implementation. The CPU implementation is running with 64 threads, which gives nearly the fastest execution time. Two GPU times are reported: "GPU" and "GPU-coalesced". "GPU" is the time without coalescing the accesses to the traceback matrix $D$. The figure shows that the GPU acceleration without coalescing is 2.4x faster than the CPU implementation. Coalescing further accelerates the GPU implementation by 10x to achieve an overall speedup of 24x.

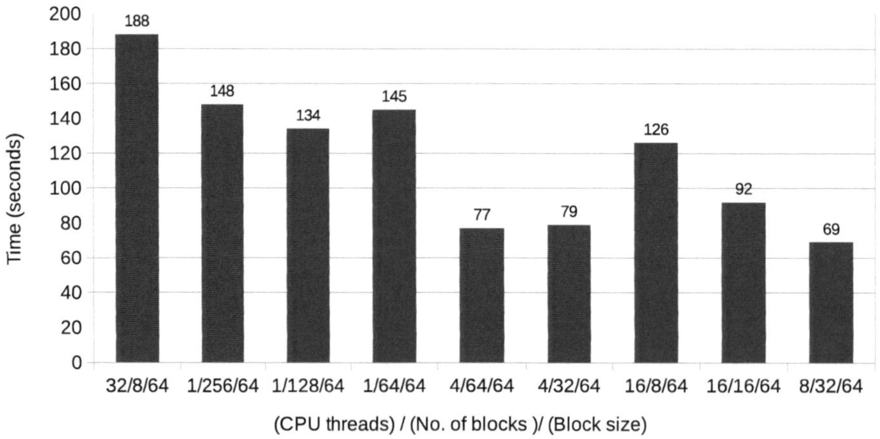


**Fig. 6** Total execution time of GPU accelerated Darwin for different values of number of CPU threads, numbers of blocks and block size on an IBM machine

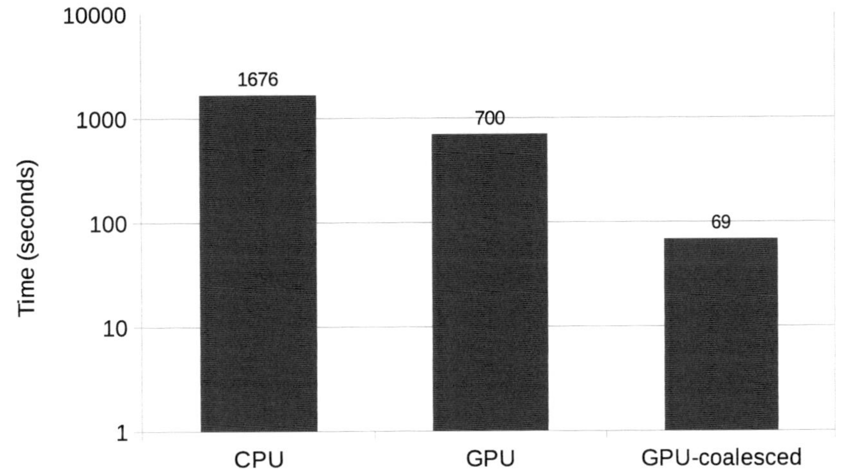


**Fig. 7** Total execution time of the Darwin's CPU implementation and GPU accelerated Darwin, on the IBM machine. The CPU implementation is running with 64 CPU threads

Figure 8 show the comparison of total execution times on the Intel machine, again with a logarithmic scale on the *y*-axis. The CPU implementation is running with 8 threads, which gives nearly the fastest execution time. The non-coalesced GPU implementation achieves a speedup of 11.8x over the CPU. With coalesced memory accesses the speedup becomes 109x. Figures 7 and 8 indicate that coalescing helps to improve the speedup by around 10x. This happens due to efficient utilization of large GPU global memory bandwidth.

The above results were obtained with the linear-gap penalty model which is also the default setting in Darwin. However, Darwin and hence our GPU acceleration also support the affine gap model. Table 1 shows the total execution time of the CPU and GPU implementation of Darwin for various values of *match*, *mismatch*, *gapo* and *gape* on the

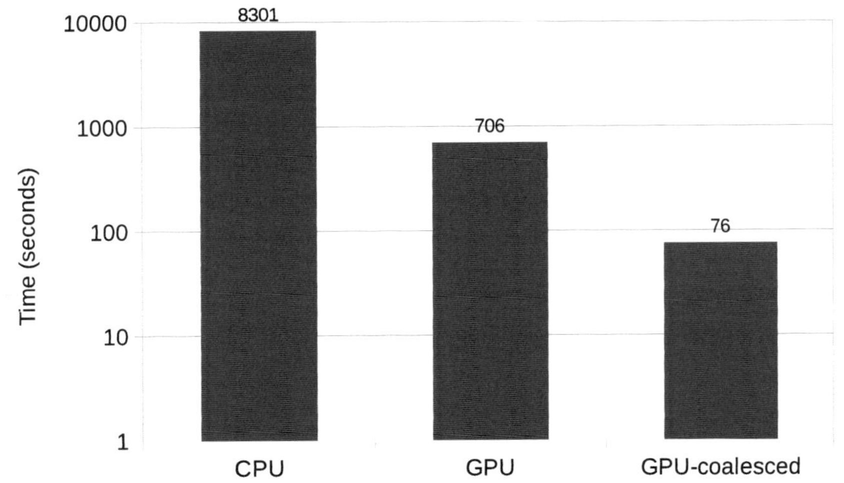


**Fig. 8** Total execution time of Darwin's CPU implementation and GPU accelerated Darwin, on the Intel machine. The CPU implementation is running with 8 threads

**Table 1** Runtimes and speedup for different scoring schemes on the IBM machine

| | (2,-1,-2,-2) | (1,-3,-1,-1) | (5,-4,-10,-1) |
|---|---|---|---|
| CPU | 31m15 | 21m28s | 31m27s |
| GPU-coalesced | 76.0 | 59.3 | 78.0 |
| speedup | 24.7 | 21.7 | 24.2 |

IBM machine. The CPU implementation is running with 64 threads. The table shows that the speedup nearly remains constant (20x-25x) regardless of the scoring scheme. Hence our GPU acceleration is equally effective for both linear and affine gap penalty scoring models.

## Conclusions

Read overlapping is an important step in OLC based de novo assemblers. Darwin is a fast and accurate read overlapper for assembly of long DNA reads. It is based on the seed-and-extend paradigm. It has two stages: 1) D-SOFT, to compute the seeds and 2) GACT, to extend the seed hits on both sides to compute the overlap between two reads. The ASIC implementation of Darwin is shown to be hundreds of times faster than software based read overlappers. GPUs are cost-effective and easily accessible processing units that are used to accelerate many high performance applications. In this paper, we have shown a GPU implementation of Darwin which accelerates the Smith-Waterman alignment with traceback computation used in the GACT stage. We pack the sequences on the GPU and compute the Smith-Waterman alignment matrix by dividing the matrix into 8x8 submatrices. This helps to reduce the GPU memory accesses. To further reduce the memory transactions, writing to the traceback matrix is coalesced. We tested our implementation against the hand-optimized CPU implementation of Darwin. The results show that using the real Pacbio dataset, our GPU implementation is 24x faster than 64 IBM Power8 threads and 109x faster than 8 Intel Xeon threads, regardless of the scoring scheme (linear or affine gap). The GPU implementation can also be used to accelerate generic Smith-Waterman alignment of long DNA sequences. The implementation is available at https://github.com/Tongdongq/darwin-gpu.

## Availability and requirements

**Project name:** darwin-gpu
**Project home page:** https://github.com/Tongdongq/darwin-gpu
**Operating system(s):** Linux
**Programming language:** C++, CUDA
**Other requirements:** CUDA toolkit version 8 or higher.
**License:** Apache 2.0
**Any restrictions to use by non-academics:** Not applicable

### Abbreviations

ASIC: Application Specific Integrated Circuit; CPU: Central Processing Unit; CUDA:Compute Unified Device Architecture; DNA: Deoxyribonucleic Acid D-SOFT: Diagonal-band Seed Overlapping based Filtration Technique; GACT: Genome Alignment using Constant memory Traceback; GASAL: GPU Accelerated Sequence Alignment Library; GPU: Graphical Processing Unit; NGS: Next Generation Sequencing; SM: Streaming Multiprocessor; SP: Streaming Processor

### Acknowledgments

We are thankful to Yatish Turakhia from Stanford University for his help during the research.

**About this supplement**
This article has been published as part of *BMC Bioinformatics Volume 21 Supplement 13, 2020: Selected articles from the 18th Asia Pacific Bioinformatics Conference (APBC 2020): bioinformatics*. The full contents of the supplement are available online at https://bmcbioinformatics.biomedcentral.com/articles/supplements/volume-21-supplement-13.

**Authors' contributions**
NA provided the initial GPU implementation, wrote the manuscript and performed some experiments. TDQ carried out the study, extended the GPU implementation to be integrated in Darwin, performed the experiments and wrote some parts of the manuscript. KB headed the research and revised the manuscript. ZA proposed the idea, supervised the research and helped in writing the manuscript. All author(s) read and approved the final manuscript.

**Funding**
This work is partly funded by the Faculty Development Program of the University of Engineering and Technology Lahore and by the FitOptiVis ECSEL Joint Undertaking project under grant number H2020 - ECSEL - 2017 - 2 - 783162. Publication costs are funded by the TUDelft library open access fund.

**Availability of data and materials**
Not applicable.

**Ethics approval and consent to participate**
Not applicable.

**Consent for publication**
Not applicable.

**Competing interests**
The authors declare that they have no competing interests.

Published: 17 September 2020

**References**

1. Kececioglu JD, Myers EW. Combinatorial algorithms for dna sequence assembly. Algorithmica. 1995;13(7):7–51.
2. Myers G, Tischler G, Cunial F, Pippel M. DAZZLER: Dresden Azzembler for Long Read DNA Projects. https://https://dazzlerblog.wordpress.com. Accessed 2 July 2019.
3. Simpson JT, Durbin R. Efficient de novo assembly of large genomes using compressed data structures. Genome Res. 2012;22(3):549–56.
4. Pevzner PA, Tang H, Waterman MS. An eulerian path approach to dna fragment assembly. Proc Natl Acad Sci U S A. 2001;98(17):9748–53.
5. Zerbino D, Birney E. Velvet: algorithms for de novo short read assembly using de bruijn graphs. Genome Res. 2008;18:074492.
6. Simpson JT, Wong K, Jackman SD, Schein JE. Abyss: a parallel assembler for short read sequence data. Genome Res. 2009;19:089532.
7. Luo R, Liu B, Xie Y, Li Z. Soapdenovo2: an empirically improved memory-efficient short-read de novo assembler. Gigascience. 2012;1(18):1–6.
8. Yatish Turakhia GB, Dally WJ. Darwin: genomics co-processor provides up to 15,000X acceleration on long read assembly. In: Proceedings of the Twenty-Third International Conference on Architectural Support for Programming Languages and Operating Systems. ASPLOS '18. Williamsburg: ACM; 2018. p. 199–213.
9. Ahmed N, Lévy J, Ren S, Mushtaq H, Bertels K, Al-Ars Z. GASAL2: a GPU accelerated sequence alignment library for high-throughput NGS data. BMC Bioinformatics. 2019;20(1):520.
10. Ren S, Ahmed N, Bertels K, Al-Ars Z. GPU accelerated sequence alignment with traceback for GATK HaplotypeCaller. BMC Genomics. 2019;20(2):184.
11. Houtgast EJ, Sima V-M, Bertels K, Al-Ars Z. Hardware acceleration of bwa-mem genomic short read mapping for longer read lengths. Comput Biol Chem. 2018;75:54–64.
12. Smith TF, Waterman MS. Identification of common molecular subsequences. J Mol Biol. 1981;147(1):195–7.
13. Ahmed N, Bertels K, Al-Ars Z. A comparison of seed-and-extend techniques in modern dna read alignment algorithms. In: 2016 IEEE International Conference on Bioinformatics and Biomedicine (BIBM). Piscataway: IEEE; 2016. p. 1421–8.
14. Altschul SF, Gish W, Miller W, Myers EW, Lipman DJ. Basic local alignment search tool. J Mol Biol. 1990;215(3):403–10.
15. Roberts M, Hayes W, Hunt BR, Mount SM. Reducing storage requirements for biological sequence comparison. Bioinformatics. 2004;20(18):3363–9.
16. Rucci E, Garcia C, Botella G, De Giusti A, Naiouf M, Prieto-Matias M. SWIFOLD: Smith-Waterman implementation on FPGA with OpenCL for long DNA sequences. BMC Syst Biol. 2018;12(5):96.
17. Farrar M. Striped smith–waterman speeds database searches six times over other SIMD implementations. Bioinformatics. 2007;23(2):156–61.
18. Hirschberg DS. A Linear Space Algorithm for Computing Maximal Common Subsequences. Commun ACM. 1975;18(6):341–3.
19. Chao KM, Pearson WR, Miller W. Aligning two sequences within a specified diagonal band. Comput Appl Biosci CABIOS. 1992;8(5):481–7.
20. Trapnell C, Schatz MC. Optimizing data intensive gpgpu computations for dna sequence alignment. Parallel Comput. 2009;35(8):429–40.

21. de O Sandes EF, de Melo ACMA. Smith-waterman alignment of huge sequences with gpu in linear space. In: 2011 IEEE International Parallel Distributed Processing Symposium. Piscataway: IEEE; 2011. p. 1199–211. https://doi.org/10.1109/IPDPS.2011.114. https://ieeexplore.ieee.org/document/6012857/.
22. Liu Y, Schmidt B. CUSHAW2-GPU: Empowering Faster Gapped Short-Read Alignment Using GPU Computing. Des Test IEEE. 2014;31(1):31–39.
23. Houtgast EJ, Sima VM, Bertels KLM, Al-Ars Z. An efficient gpu-accelerated implementation of genomic short read mapping with bwa-mem. In: Proc. International Symposium on Highly-Efficient Accelerators and Reconfigurable Technologies. Hong Kong, China: ACM; 2016.
24. Hasan L, Kentie MA, Al-Ars Z. Dopa: Gpu-based protein alignment using database and memory access optimizations. BMC Res Notes. 2011;4:1–11.
25. Ahmed N, Mushtaq H, Bertels KLM, Al-Ars Z. Gpu accelerated api for alignment of genomics sequencing data. In: Proc. IEEE International Conference on Bioinformatics and Biomedicine. Piscataway: IEEE; 2017. p. 510–515.
26. Turakhia Y. Darwin: A co-processor for long read alignment. https://github.com/yatisht/darwin. Accessed 5 Nov 2018.
27. Data release: 54x long-read coverage for PacBio-only de novo human genome assembly. 2014. https://www.pacb.com/blog/data-release-54x-long-read-coverage-for/. Accessed 2 July 2019.

## Publisher's Note

Lee et al. BMC Bioinformatics 2020, 21(Suppl 13):382
https://doi.org/10.1186/s12859-020-03686-0

BMC Bioinformatics

RESEARCH

Open Access

# Liver imaging features by convolutional neural network to predict the metachronous liver metastasis in stage I-III colorectal cancer patients based on preoperative abdominal CT scan

Sangwoo Lee[1†], Eun Kyung Choe[2,3†], So Yeon Kim[3,4], Hua Sun Kim[5], Kyu Joo Park[6†] and Dokyoon Kim[3,7*†]

*From* The 18th Asia Pacific Bioinformatics Conference
Seoul, Korea. 18-20 August 2020

* Correspondence: dokyoon.kim@pennmedicine.upenn.edu
†Sangwoo Lee, Eun Kyung Choe, Kyu Joo Park and Dokyoon Kim contributed equally to this work.
[3]Department of Biostatistics, Epidemiology & Informatics, Perelman School of Medicine, University of Pennsylvania, B304 Richards Building, 3700 Hamilton Walk, Philadelphia, PA 19104-6116, USA
[7]Institute for Biomedical Informatics, University of Pennsylvania, Philadelphia, PA 19104, USA
Full list of author information is available at the end of the article

## Abstract

**Background:** Introducing deep learning approach to medical images has rendered a large amount of un-decoded information into usage in clinical research. But mostly, it has been focusing on the performance of the prediction modeling for disease-related entity, but not on the clinical implication of the feature itself. Here we analyzed liver imaging features of abdominal CT images collected from 2019 patients with stage I – III colorectal cancer (CRC) using convolutional neural network (CNN) to elucidate its clinical implication in oncological perspectives.

**Results:** CNN generated imaging features from the liver parenchyma. Dimension reduction was done for the features by principal component analysis. We designed multiple prediction models for 5-year metachronous liver metastasis (5YLM) using combinations of clinical variables (age, sex, T stage, N stage) and top principal components (PCs), with logistic regression classification. The model using "$1^{st}$ PC (PC1) + clinical information" had the highest performance (mean AUC = 0.747) to predict 5YLM, compared to the model with clinical features alone (mean AUC = 0.709). The PC1 was independently associated with 5YLM in multivariate analysis (beta = – 3.831, $P <$ 0.001). For the 5-year mortality rate, PC1 did not contribute to an improvement to the model with clinical features alone. For the PC1, Kaplan-Meier plots showed a significant difference between PC1 low vs. high group. The 5YLM-free survival of low PC1 was 89.6% and the high PC1 was 95.9%. In addition, PC1 had a significant correlation with sex, body mass index, alcohol consumption, and fatty liver status.

(Continued on next page)

(Continued from previous page)

**Conclusion:** The imaging features combined with clinical information improved the performance compared to the standardized prediction model using only clinical information. The liver imaging features generated by CNN may have the potential to predict liver metastasis. These results suggest that even though there were no liver metastasis during the primary colectomy, the features of liver imaging can impose characteristics that could be predictive for metachronous liver metastasis.

**Keywords:** Radiomics, Colorectal cancer, Convolutional neural network, Artificial intelligence

## Background

In colorectal cancer (CRC) patients, liver is the most common site of metastasis due to its anatomical connection with portal circulation [1]. In advanced CRC, liver may be the sole site of metastasis in 30–40% of patients [2, 3]. In these patients, median survival is 5–20 months without intervention, and 5-year survival is extremely rare [4]. Thus, understanding the pathophysiology of the liver metastasis is one of the most critical subjects for CRC management.

Radiomics is an innovative technique that uses the large volume of imaging features to predict oncological features [5]. It consists of converting the medical images into high-dimensional features which is then used to predict clinical outcomes [6].Convolutional neural networks (CNNs) have recently manifested the ability to generate useful features from imaging data in various medical research [7]. There are multiple studies suggesting the CT imaging features extracted by CNNs have high predictive values in oncological outcomes [8, 9]. Machine learning approach is one of the major subfields of artificial intelligence which can be used for constructing prediction model in radiomics [6] and has shown promising performances for predicting various oncological subjects [10–13].

Nonetheless, there are challenges that to evaluate the values of each imaging feature in the prediction model for an oncological outcome is difficult [14]. In traditional approach for radiological research, to review the medical images, radiologists use the human visual interpretation based on the characteristics of the images systematized by accumulated experiences and researches in association with clinical manifestations [15] and the statistical models are used to find associations in these data that could enhance clinical reasoning [16]. This canonical way to use medical images for clinical research might be presenting challenges to radiomics, which uses the imaging features generated by CNNs and designs prediction models by machine learning to enhance clinical performances. If the approaches of radiomics study can encompass purposes focusing on not only the performance of the prediction modeling for disease-related entity, but also the clinical implication of the imaging features itself, its utility and contribution to the healthcare research would be in great value.

Here we propose a framework to extract the liver imaging features from CT scan using CNNs in CRC patients and analyze the imaging features using machine learning approaches to predict the metachronous liver metastasis. Additionally, we tried to elucidate its clinical implication in oncological perspectives using statistical analysis. As a proof of concept study, abdominal CT images were collected from 2019 patients who had colectomy for stage I–III CRC, preoperatively.

## Methods

### Patients and data acquisition

We performed a retrospective, cross-sectional study in patients who underwent colectomy for CRC from January 2008 to September 2013 at Seoul National University Hospital. Stage I-III CRC patients who had curative resection were included in the analyses. The exclusion criteria consisted of patients who had surgery less than 5 years; recurrence in or distant metastasis to other than liver after in less than 5 years after surgery; preoperative neoadjuvant chemotherapy; a history of liver resection; have liver lesion; and had a poor quality of preoperative CT scan. A total of 2019 patients were eligible, and their electronic medical records (EMR) and CT images were collected. The CRC staging description was based on the AJCC staging system, seventh edition, which is a classification system provided by the American Joint Committee on Cancer for describing the extent of cancer progression [17]. We used the clinical information turned into a dichotomized form such as T stages into T1, T2 vs. T3, T4, N stages into N0 vs. N1, N2 and patient ages into < 65 years vs. > = 65 years. 5-year metachronous liver metastasis (5YLM) rate was the primary endpoint. If the patient died or had recurrence other than in liver, they would not be counted as metachronous liver metastasis incidences. We used the abdominal CTs taken before the colectomy for clinical staging. For image acquisition, the non-contrast abdominal CT scan image of each patient was used. A physician (author EKC) acquired the images under the guidance of radiologist (author HSK). Cross-sectional images at the level where the caudate lob of the liver is most prominent were selected with abdominal view setting. The regions of interest (ROIs) were placed at the segment 7 of the Couinaud system [18], which is the right – posterior – superior segment of liver, with a size of 50 × 50 pixels.

### Feature generation

Figure 1 shows the overview of the analysis framework. First, feature extraction on the images was done by utilizing a pre-trained convolutional neural network, VGG16 [19], which do not require further training. For the feature extraction, we used the fully connected layers, which is a top-layer of the pre-trained model of VGGnet with 16 layers (VGG-16). The 4096 features were extracted as an output. Then, we preprocessed the extracted imaging features rather than using whole 4096 features based on the significance of association with 5YLM rate by performing univariate logistic regression analysis. The subsets of imaging features that passed the suggestive significance level ($P < 0.01$) were used for further analyses. Lastly, principal component analysis (PCA) was performed for reducing the feature dimensionality, and this generated new sets of features, such as 1st principal component (PC1) to 10th principal component (PC10), sequentially. At the input stage to the machine learning methods, we further standardized either the clinical features or the PC-transformed image features, by z normalization for each feature to have a mean of zero and unit variance.

### Prediction models

We trained two machine learning algorithms, including logistic regression classification (LR) and random forest classification (RFC) to predict 5YLM and compared the performances of each model. The models were designed by respective combination of features consisting of clinical features, which includes age, gender, T stage and N stage, and imaging features, which are the sequential summation of PC1 to PC10. The whole

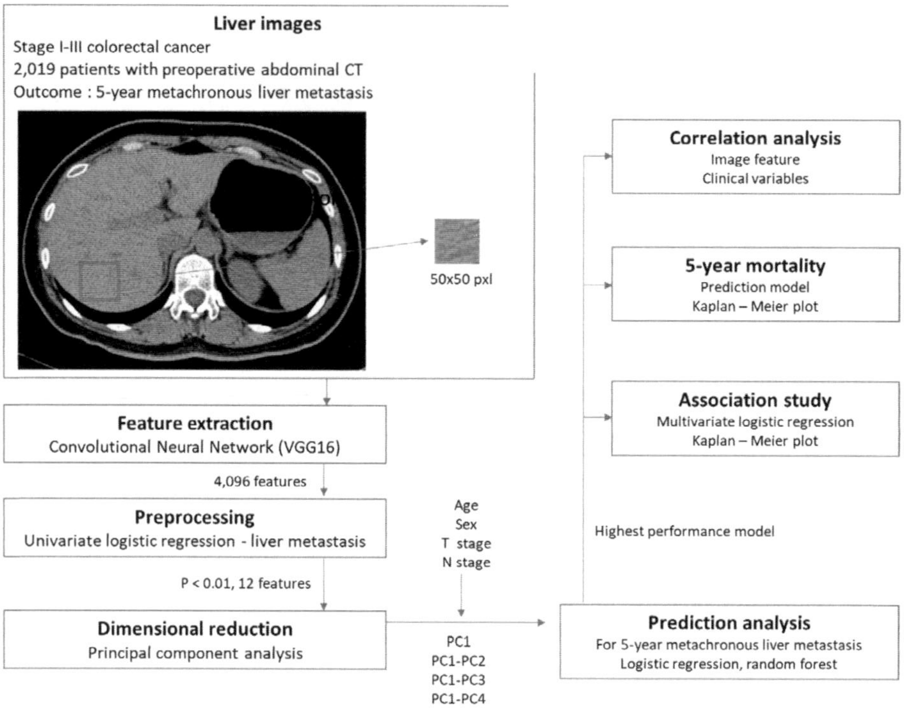


**Fig. 1** Study design. Overview of the analysis framework. Feature extraction on the abdominal CT, 50 × 50 pixel ROIs, was done by utilizing a pre-trained convolutional neural network. We preprocessed them based on the significance of association with 5-year liver metastasis (5YLM) rate by performing univariate logistic regression analysis. Principal component analysis (PCA) was done for feature reduction in dimensionality and this generated new sets of feature. We used two machine learning algorithms, such as logistic regression classification (LR) and random forest classification (RFC) to train prediction models for 5YLM and compared the performances of each model. Among the models to predict 5YLM, we used the highest AUC model to perform multivariate logistic regression to association between the image features and 5YLM statistically. Then Kaplan Meier analysis was done by the principal components (PCs) for metachronous liver metastasis free survival and overall survival. We done a correlation analysis between the significant PCs and the clinical variable in Table 1. We also applied the highest AUC model for 5YLM to predict 5-year mortality and observed whether the liver image feature could do a predictive role for 5-year mortality

data set was divided into a training set (70%) and a test set (30%), and a five-fold cross validation scheme was used on the training set for the parameter tuning of the classification models. Given our highly imbalanced data set, we adopted five-fold, considering that five can be reasonable to allow enough number of positive data samples in each fold. Due to the highly imbalanced data set, we dealt with the imbalance problem of the training set first by oversampling on the negative minority class to meet negative/positive ratio 0.2, followed by down-sampling on the positive majority class to meet negative/positive ratio 0.4 finally. For up-sampling, we used SMOTE (Synthetic Minority Over-Sampling Technique) [20], and for down-sampling, we used the random down-sampling technique. After performing five-fold cross-validation on the training set, the trained model is validated on the test set. We note that cross-validation procedure is used to prevent overfitting to the training set when fitting the prediction model (LR and RFC). To validate the trained model, we performed the procedure aforementioned (splitting the data set into a training set and a test set; SMOTE followed by down-sampling; a five-fold cross validation to find the optimal parameters; and performance evaluation on the test set) for 100 iterations. Using the trained model, we evaluated the performance of the test set. The performances were measured by the area under the curve (AUC) of the receiver operating characteristics (ROC) curve and presented as means and standard deviations.

We also applied the model with the highest AUC for 5YLM to predict 5-year mortality and observed whether the liver imaging feature contribute to a prediction of 5-year mortality. The AUC performance was compared with the model using clinical feature only for the prediction of 5-year mortality.

### Association study

Among the models to predict 5YLM, we used the model with the highest AUC to perform multivariate logistic regression to identify associations between the imaging features and 5YLM. Then, Kaplan-Meier analysis was conducted based on the principal components (PCs) that were significantly associated with metachronous liver metastasis-free survival and overall survival ($P < 0.05$). For survival analysis, the patients excluded by the previous exclusion criteria, such as recurrence in or distant metastasis to other than liver after in less than 5 years after surgery, were included. The optimal cut-off points for those PCs to divide patients into two groups were determined by MaxStat packages in R (Maximally selected Rank Statistics). MaxStat uses the maximally selected rank statistics to recommend the optimal cut-off point for the survival plot [21]. Univariate cox proportional hazard regression analysis was performed to compare the differences between PC-based groups using the time to event and the censoring data of it.

### Correlation study

To investigate the clinical relevance of the imaging features, we ran a correlation analysis between the significant PCs and the additional clinical variables in Table 1, such as age, sex, body mass index, tumor location, alcohol consumption, liver function test, fatty liver status, T stage, N stage, Angiolymphatic invasion venous invasion, and postoperative follow-up duration. We measured Pearson's correlation between numeric variables, and Spearman's correlation between categorical variables, respectively. $P$ value of the correlation coefficient between variables less than 0.05 was considered statistically significant. The results were visualized with "corrplot" R package.

All the statistical and computational analyses were done by R statistical software (version 3.5.3 R) and Python software (version 3.6.2). Associations between clinical information and 5YLM rate were assessed by Chi-square test, Student's t-test, and analysis of variance (ANOVA) for independent groups in Table 1.

### Ethics statement

The Institutional Review Board of Seoul National University Hospital approved the study protocol (IRB number 1902–088-1010), and the study was conducted in accordance with the Declaration of Helsinki. Informed consent was waived by the board.

## Results

### Patient demographics

Our study sample comprised 2019 patients (1269 males and 750 females) who had colectomy for stage I-III colorectal cancer. The mean patient age was 62.32 +/− 9.21 years.

**Table 1** Demographic features of the study population

| | Liver metastasis, no (***N*** = 1919, 95.04%) | Liver Metastasis, yes (***N*** = 100, 4.96%) | *P* value |
|---|---|---|---|
| Age (years) | 62.3 ± 9.2 | 63.1 ± 9.5 | 0.390 |
| Age > =65 years | | | 0.640 |
| No | 1092 (56.9%) | 54 (54.0%) | |
| Yes | 827 (43.1%) | 46 (46.0%) | |
| Sex | | | 0.727 |
| Male | 1204 (62.7%) | 65 (65.0%) | |
| Female | 715 (37.3%) | 35 (35.0%) | |
| BMI (kg/m2) | 23.9 ± 3.0 | 23.8 ± 3.1 | 0.805 |
| BMI (> = 25 kg/m2) | | | 0.648 |
| No | 1280 (66.7%) | 64 (64.0%) | |
| Yes | 638 (33.3%) | 36 (36.0%) | |
| Tumor location | | | 0.127 |
| Right | 501 (26.4%) | 19 (19.0%) | |
| Left | 1397 (73.6%) | 81 (81.0%) | |
| Heavy alcohol consumption | | | 1 |
| No | 1261 (65.7%) | 66 (66.0%) | |
| Yes | 658 (34.3%) | 34 (34.0%) | |
| GOT | 22.3 ± 9.0 | 21.8 ± 9.6 | 0.559 |
| GPT | 20.3 ± 12.8 | 19.8 ± 16.0 | 0.737 |
| Fatty liver | | | 0.283 |
| No | 1743 (95.0%) | 96 (98.0%) | |
| Yes | 91 (5.0%) | 2 (2.0%) | |
| T stage | | | < 0.001 |
| T1 stage | 359 (18.7%) | 4 (4.0%) | |
| T2 stage | 352 (18.3%) | 6 (6.0%) | |
| T3 stage | 1106 (57.6%) | 74 (74.0%) | |
| T4 stage | 102 (5.3%) | 16 (16.0%) | |
| N stage | | | < 0.001 |
| N0 | 1293 (67.4%) | 28 (28.0%) | |
| N1 | 463 (24.1%) | 39 (39.0%) | |
| N2 | 163 (8.5%) | 33 (33.0%) | |
| Lymph node metastasis | | | < 0.001 |
| Absent | 1293 (67.4%) | 28 (28.0%) | |
| Present | 626 (32.6%) | 72 (72.0%) | |
| Overall stage | | | < 0.001 |
| Stage 1 | 581 (30.3%) | 6 (6.0%) | |
| Stage 2 | 658 (34.3%) | 18 (18.0%) | |
| Stage 3 | 680 (35.4%) | 76 (76.0%) | |
| Angiolymphatic invasion | | | < 0.001 |
| Absent | 1434 (77.0%) | 57 (57.6%) | |
| Present | 429 (23.0%) | 42 (42.4%) | |
| Venous invasion | | | < 0.001 |
| Absent | 1729 (92.8%) | 75 (75.8%) | |

**Table 1** Demographic features of the study population *(Continued)*

| | Liver metastasis, no (***N*** = 1919, 95.04%) | Liver Metastasis, yes (***N*** = 100, 4.96%) | *P* value |
|---|---|---|---|
| Present | 134 (7.2%) | 24 (24.2%) | |
| Postoperative follow up duration | 1893.5 ± 767.7 | 1554.5 ± 784.9 | < 0.001 |
| 5-year Mortality | | | |
| Alive | 1919 (100.0%) | 61 (61.0%) | < 0.001 |
| Dead | 0 (0.0%) | 39 (39.0%) | |

There were 100 cases (4.96%) of metachronous liver metastasis during the follow. Patient characteristics are shown in Table 1.

### 5-year metachronous liver metastasis prediction based on principal components and clinical information

We preprocessed the extracted imaging features from CNN based on the significance of association with 5YLM rate by performing univariate logistic regression analysis. Twelve features passed the suggestive significance level ($P < 0.01$). From the 12 features, using the PCA, we generated new sets of features notated as PC1 to PC10.

For 5YLM prediction models, we used PCs and clinical features as inputs to our models, in which we not only incremented the number of PCs one by one (for example, PC1, PC1-PC2, PC1-PC3 and so on), but also adopted various combinations (for example, clinical feature only model, PCs only model, clinical feature plus PC model) in order to validate which combination of features more contribute to the prediction performance. Two supervised machine learning methods, LR and RFC, were used for the performance evaluations.

The performances of each model are shown in Table 2. Model with 1st PC (PC1) showed the highest performance among other PCs combinations both in LR (mean AUC = 0.606) and RFC (AUC = 0.557). In the combination of clinical and imaging features, LR model trained with PC1 and clinical features showed the best performances (mean AUC = 0.747), which imply that the model using the imaging features in combination with the clinical features improved the prediction performance rather than the model using clinical feature only (mean AUC = 0.709).

### Association study with the oncological and clinical variables using the designed model

For the model with PC1 and clinical features (age, sex, T stage, and N stage), which is the best performed model, an association study was done to investigate the association between the features and 5YLM. By multivariate logistic regression analysis, PC1 showed independent association with 5YLM, significantly (beta = − 3.831, $P < 0.001$) (Table 3).

For the PC1, Kaplan-Meier plots were generated for metachronous liver metastasis free survival (Fig. 2a). The patients were divided by the optimal cut offs for each PC1 based on MaxStat. (− 0.135 for the PC1 score). The results are shown in Fig. 2 with the result of univariate cox proportional hazard regression. The 5YLM-free survival of low group (PC1 score below − 0.135) was 88.7% and the high group (PC 1score above − 0.135) was 95.6%. We also evaluated with the Kaplan-Meier plot for overall survival

**Table 2** Performances of the prediction models in the test set for 5-year mortality and 5-year metachronous liver metastasis

| Predictors | Logistic regression classification AUC (mean +/− standard deviation) | Random forest classification AUC (mean, standard deviation) |
|---|---|---|
| Prediction model for 5-year metachronous liver metastasis | | |
| Clinical* | 0.709 +/− 0.038 | 0.692 +/− 0.038 |
| PC1 | 0.606 +/− 0.044 | 0.557 +/− 0.043 |
| PC1-PC2 | 0.600 +/− 0.042 | 0.536 +/− 0.042 |
| PC1-PC3 | 0.588 +/− 0.040 | 0.503 +/− 0.046 |
| PC1-PC4 | 0.580 +/− 0.040 | 0.520 +/− 0.042 |
| **Clinical + PC1** | **0.747 +/− 0.036** | 0.697 +/− 0.038 |
| Clinical + PC1-PC2 | 0.744 +/− 0.036 | 0.676 +/− 0.043 |
| Clinical + PC1-PC3 | 0.740 +/− 0.038 | 0.668 +/− 0.042 |
| Clinical + PC1-PC4 | 0.736 +/− 0.038 | 0.691 +/− 0.042 |
| Prediction model for 5-year mortality | | |
| Clinical* | 0.704 +/− 0.028 | 0.679 +/− 0.030 |
| PC1 | 0.482 +/− 0.031 | 0.511 +/− 0.030 |
| Clinical + PC1 | 0.695 +/− 0.031 | 0.647 +/− 0.033 |

*Clinical: Age, Sex, T stage, N stage

and the results were similar that low group had poor overall survival compared to high group, significantly ($P < 0.001$) (Fig. 2b).

In the correlation analysis between the significant PC1 and the clinical variable in Table 1, sex, body mass index, alcohol consumption and fatty liver status had significant correlation with PC1 (Fig. 3).

### Performance of the prediction model in 5-year mortality prediction

We applied the 5YLM prediction model with the highest AUC model, which combined clinical features with PC1, with respect to the prediction of 5-year mortality. In the 5 year mortality prediction model, the liver imaging features did not have an additional predictive power (mean AUC =0.690), when compared with clinical only models (mean AUC = 0.700), in both models, trained with LR and RFC.

## Discussion

Currently, clinical features, such as age, sex, T stage, and N stage are most commonly used to predict the prognosis of colorectal cancer survival [22]. Comparing the

**Table 3** Multivariate logistic regression analysis for 5-year metachronous liver metastasis

| | Beta (standard error) | P value |
|---|---|---|
| **Using PCs from 12 features** | | |
| Age (> = 65 years) | 0.119 (0.213) | 0.213 |
| Gender (Female) | −0.232 (0.223) | 0.297 |
| T3, T4 stage | 1.276 (0.345) | < 0.001 |
| N1, N2 stage | 1.467 (0.234) | < 0.001 |
| **PC1** | **−3.831 (1.012)** | **0.0001** |

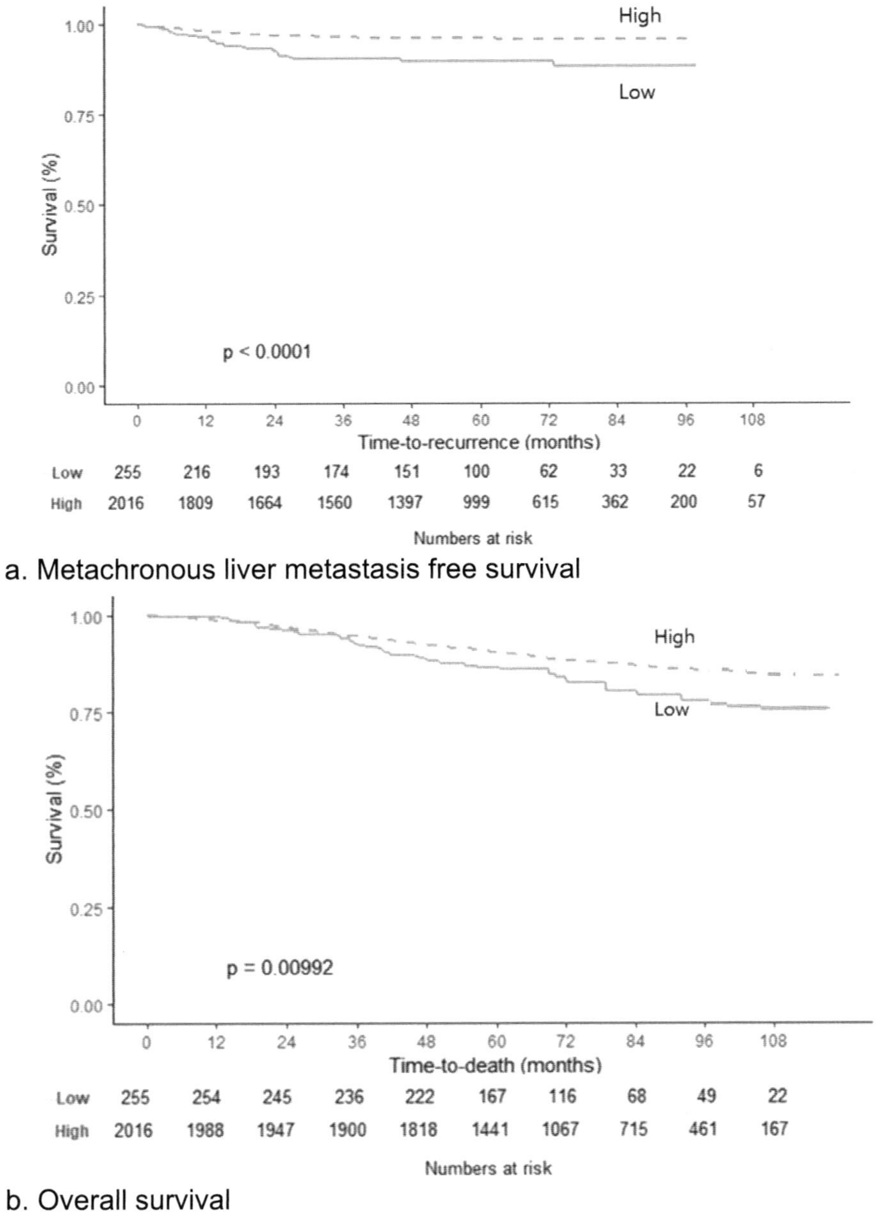


**Fig. 2** Kaplan Meier plots for metachronous liver metastasis free survival and overall survival using 1st PC of image features. **a**. Metachronous liver metastasis free survival The populations were divided by the optimal cut offs for PC1 score based on MaxStat (− 0.135). The difference between two group was compared by univariate cox proportional hazard regression. The 5-year metachronous liver metastasis free survival of low group (PC1 score below − 0.135) was 89.6% and the high group (PC 1score above − 0.135) was 95.9%. **b**. Overall survival Using the same PC1 group, K-M plot was visualized for overall survival

prediction performance of these clinical features, the integrative model that integrates clinical information and imaging features by convolutional neural network significantly improved performance of prediction for 5YLM rate.

This is quite interesting result because adding the 50 × 50 pixel liver imaging features would contain extremely limited information, but still, we demonstrated the additional imaging features when combined with clinical features improved prediction performance

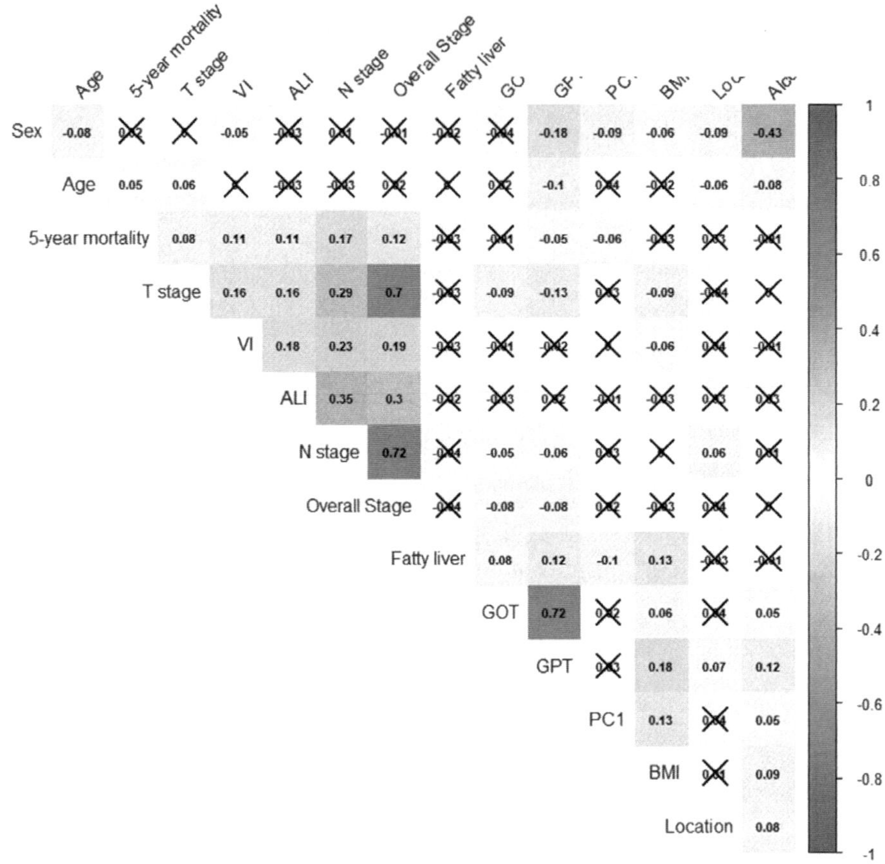


**Fig. 3** Correlation plots for 1st PC and clinical variables. Correlations with *p*-value > 0.05 are considered as insignificant. In this case the correlation coefficient values are leaved blank or crosses are added. 1st PCA had significant correlation with sex, body mass index, alcohol consumption and fatty liver status

of liver metastasis. The underlying mechanism would be that, though at the time of surgery the liver does not seem to impose liver metastasis on gross finding of CT images, there could be underlying molecular changes in liver that could be predictive of liver metastasis. For 5-year mortality, the liver feature did not have additive predictive power in prediction, but there was a significant association with overall survival in K-M plot. In our study, we applied both statistical association analysis and computational predictive analysis for the outcome. Association studies would focus on understanding a relationship between the variables and outcomes, while prediction studies train the model with the training data to obtain the variables to investigate their predictive power for the corresponding outcome. Association studies might provide explanation for the relationship but might not have predictive power. On the contrary, prediction studies might have high performance but hard to interpret it. This might imply that the liver image features alone are not enough to precisely predict the outcome of 5-year mortality, however they are associated with the 5-year mortality based on the results of Kaplan-Meier survival analysis with significant *p*-values.

Metachronous liver metastasis significantly influence the prognosis of CRC patients who had curative colectomy [16, 23], and it is reported that in 20–30% of patients, it will be detected following primary colectomy [24–26]. Well-known risk factors for liver metastasis are N stage, vascular invasion and preoperative carcinoembryonic antigen (CEA) level [23, 27, 28]. Additionally, there are also suggested factors predisposing to the development of metachronous liver metastasis in CRC, such as tissue micro-environmental changes and chronic inflammation [29]. In the present study, we focused on investigating the clinical relevance of this imaging features performing association and correlation studies with patient's demographical and oncological information. The representative imaging feature, the 1st principal component (PC1), had a significant association with 5YLM-free survival, and the results were shown in Kaplan-Meier plot. In the correlation study, PC1 had significant correlations with sex, body mass index, alcohol consumption and fatty liver status. Sex [30], body mass index [29], alcohol consumption [31] and fatty liver status [32] are reported to be correlated with liver metastasis in CRC. Obesity determined by body mass index change or sex difference and hepatosteatosis derivative of fatty liver or alcoholism are suggestive predisposing factors in the function of tissue microenvironment change and chronic inflammation [29]. It can be postulated that the imaging features of liver ROIs impose various heterogeneity of predisposing factors for metachronous liver metastasis comprehensively.

In the study design, we additionally preprocessed the 4096 imaging features extracted by training the convolutional neural network, VGG16, based on the significance of association with 5YLM rate. VGG16 is a convolutional neural network model which achieved 92,7% accuracy in ImageNet, a dataset of over 14 million images belonging to 1000 classes, and it used RGB images for training [19]. This model can capture complex features like human faces, natural scenes and showed human level performance [33]. The preprocessing step was introduced because designing the prediction model with the whole 4096 imaging features from CNN did not provide good results for 5YLM (data not shown). This might come from the fact that pretrained CNN basically trained RGB images which has chromatic color, shape and size variance but in CT scan image, it is achromatic color, fixed square shape and fixed $50 \times 50$ pixel sized. Thus, preprocessing with statistical association threshold will remove the noises and include only the effective features to be introduced.

With the preprocess features, principal component analysis was performed for reducing the feature dimensionality, and new sets of features, such as 1st principal component (PC1) to 10th principal component (PC10) as an imaging features for the prediction model and showed that PC1 had improved the performance of prediction model by clinical information to predict 5YLM. We used the PC transformed as imaging features rather than the ones after preprocessing, which is primarily generated by CNN, because these did not have promising results in the prediction model (data not shown). This might reflect the fact that the respective imaging feature by itself do not have meaningful contribution for prediction but the selected features should be aggregated to have a predictive role in the model.

In our study, we used logistic regression and random forest classification, which are very common and widely studied machine learning models [34]. For the prediction of 5YLM, the model trained by logistic regression showed the improvement by adding imaging features to the clinical features compared with the ones only consisting of clinical features. However, the improvement of performances was modest by the models trained by random forest classification. Logistic regression is a classical machine learning classifier and it has advantage of having fast speed to train the inputs and making it more interpretable [6, 35]. In a recent study, comparing logistic regression and random forest classification for binary outcomes, when increasing the variance in the explanatory and noise variables, logistic regression consistently better performed as compared to random forest classification [34]. However, true positive rate was higher in random forest classification compared to logistic regression. When machine learning is applied to the clinical fields, the characteristics of the data set and the purpose of using machine learning should be considered carefully before choosing which algorithms to be applied.

This study has several advantages. First, we comprehensively analyzed the imaging feature by respective combination of image features and clinical information. By this way, we were able to find that the best-performing combination of imaging features, preprocessed features by association analysis and PCs from PCA analysis, which showed improved prediction performance combined with clinical information. Second, we performed both the prediction model based on computational analysis and association and correlation study based on statistical analysis. This will help to interpret the complex nature of the liver CT scan image features more intuitively, provide the clinical relevance of it and support an evidence for the results of prediction model. Third, all the patients enrolled in the analyses had at least 5 years of follow-up after primary colectomy. This will make the results of analyses more reliable since the information involves the long-term observations.

Despite demonstrating the validity of our proposed approach, there might be a couple of remaining potential limitations. First, as we used the retrospectively collected CT scan, we could not collect the parameters and the product information of the CT scanner, which could influence the imaging features as a batch effect. But since, the CT images were taken on the purpose of preoperative clinical staging in colorectal cancer, the quality of the CT images was well controlled. Second, we simply concatenated the clinical factors and imaging features level for integration. Transformational integration could be applied in a larger set of samples [36]. Third, since there are no open source databases that includes both abdominal CT images and clinical information, we could not replicate the results in another data set.

## Conclusion

By analyzing the liver image features by convolutional neural network in stage I-III colorectal cancer patients based on preoperative abdominal CT scan, we were able identify the contribution of imaging features to predict the metachronous liver metastasis. The 1st PC from imaging feature combined with clinical information improved the performance of standardized prediction model using only clinical information such as age and TNM stage. The preoperative liver imaging features generated by CNN might have the potential to predict liver metastasis after 5 years. This result suggests that even though there were no metastasis or liver lesion during the primary colectomy for stage I-III CRC, the features of liver in preoperative CT scan can impose characteristics that could be predictive for liver metastasis during the postoperative follow-up. The findings might conceptualize the importance of the imaging features in liver which could be applied for clinical practice.

## Abbreviations

CRC: Colorectal cancer; CNN: Convolutional neural network; 5YLM: 5-year metachronous liver metastasis; PCs: Principal components; AUC: Area under the curved; VGG-16: VGGnet with 16 layers; PCA: Principal component analysis; PC1: 1st principal component; PC10: 10th principal component; LR: Logistic regression; RFC: Random forest classification; SMOTE: Synthetic Minority Over-Sampling Technique

## About this supplement

This article has been published as part of *BMC Bioinformatics Volume 21 Supplement 13, 2020: Selected articles from the 18th Asia Pacific Bioinformatics Conference (APBC 2020): bioinformatics.* The full contents of the supplement are available online at https://bmcbioinformatics.biomedcentral.com/articles/supplements/volume-21-supplement-13 .

## Authors' contributions

SL, EKC, KJP, DK conceptualized and designed the study; EKC, KJP provided the study materials and patients; SL, EKC, SYK, KJP, DK analyzed and interpreted the patient data regarding the colorectal cancer; EK, HSK performed evaluation of the data quality and processing the raw data set; and SL, EKC, KJP, DK were a major contributor in writing the manuscript. All authors read and approved the final manuscript and are accountable for all aspects of the work.

## Funding

The support for this research in the design of the study, and analysis, interpretation of data and in writing the manuscript was provided by NLM R01 LM012535. Publication costs are funded by NLM R01 funding (LM012535).

## Availability of data and materials

The datasets generated and/or analyzed during the current study are not publicly available due to restrictions (the institution policy to protect the privacy of research participants) but are available from the corresponding author on reasonable request.

## Ethics approval and consent to participate

The Institutional Review Board of Seoul National University Hospital approved the study protocol (IRB number 1902-088-1010), and the study was conducted in accordance with the Declaration of Helsinki. Informed consent was waived by the board.

## Consent for publication

Not applicable.

## Competing interests

The authors declare that they have no competing interests.

## Author details

[1]Division of Future Convergent, The Cyber University of Korea, Seoul 03051, South Korea. [2]Department of Surgery, Seoul National University Hospital Healthcare System Gangnam Center, Seoul 06236, South Korea. [3]Department of Biostatistics, Epidemiology & Informatics, Perelman School of Medicine, University of Pennsylvania, B304 Richards Building, 3700 Hamilton Walk, Philadelphia, PA 19104-6116, USA. [4]Department of Software and Computer Engineering, Ajou University, Suwon 16499, South Korea. [5]Department of Radiology, Seoul National University College of Medicine, Seoul 03080, South Korea. [6]Department of Surgery, Seoul National University College of Medicine, Seoul 03080, South Korea. [7]Institute for Biomedical Informatics, University of Pennsylvania, Philadelphia, PA 19104, USA.

Published: 17 September 2020

## References

1. Sheth KR, Clary BM. Management of hepatic metastases from colorectal cancer. Clin Colon Rectal Surg. 2005; 18(3):215–23.
2. Garden OJ, Rees M, Poston GJ, Mirza D, Saunders M, Ledermann J, et al. Guidelines for resection of colorectal cancer liver metastases. Gut. 2006;55(Suppl 3):iii1–8.
3. Hadden WJ, de Reuver PR, Brown K, Mittal A, Samra JS, Hugh TJ. Resection of colorectal liver metastases and extra-hepatic disease: a systematic review and proportional meta-analysis of survival outcomes. HPB (Oxford). 2016;18(3):209–20.
4. Valderrama-Trevino AI, Barrera-Mera B, Ceballos-Villalva JC, Montalvo-Jave EE. Hepatic metastasis from colorectal cancer. Euroasian J Hepatogastroenterol. 2017;7(2):166–75.
5. Choi ER, Lee HY, Jeong JY, Choi YL, Kim J, Bae J, et al. Quantitative image variables reflect the intratumoral pathologic heterogeneity of lung adenocarcinoma. Oncotarget. 2016;7(41):67302–13.
6. Forghani R, Savadjiev P, Chatterjee A, Muthukrishnan N, Reinhold C, Forghani B. Radiomics and artificial intelligence for biomarker and prediction model development in oncology. Comput Struct Biotechnol J. 2019;17:995–1008.
7. Litjens G, Kooi T, Bejnordi BE, Setio AAA, Ciompi F, Ghafoorian M, et al. A survey on deep learning in medical image analysis. Med Image Anal. 2017;42:60–88.
8. Meng Y, Zhang Y, Dong D, Li C, Liang X, Zhang C, et al. Novel radiomic signature as a prognostic biomarker for locally advanced rectal cancer. J Magn Reson Imaging. 2018;48(3):605.
9. Li Q, Kim J, Balagurunathan Y, Liu Y, Latifi K, Stringfield O, et al. Imaging features from pretreatment CT scans are associated with clinical outcomes in nonsmall-cell lung cancer patients treated with stereotactic body radiotherapy. Med Phys. 2017;44(8):4341–9.

10. Lubner MG, Stabo N, Lubner SJ, del Rio AM, Song C, Halberg RB, et al. CT textural analysis of hepatic metastatic colorectal cancer: pre-treatment tumor heterogeneity correlates with pathology and clinical outcomes. Abdom Imaging. 2015;40(7):2331–7.
11. Wu W, Parmar C, Grossmann P, Quackenbush J, Lambin P, Bussink J, et al. Exploratory study to identify Radiomics classifiers for lung cancer histology. Front Oncol. 2016;6:71.
12. Chamming's F, Ueno Y, Ferre R, Kao E, Jannot AS, Chong J, et al. Features from computerized texture analysis of breast cancers at pretreatment MR imaging are associated with response to neoadjuvant chemotherapy. Radiology. 2018; 286(2):412–20.
13. Kickingereder P, Burth S, Wick A, Gotz M, Eidel O, Schlemmer HP, et al. Radiomic profiling of glioblastoma: identifying an imaging predictor of patient survival with improved performance over established clinical and radiologic risk models. Radiology. 2016;280(3):880–9.
14. Deo RC. Machine learning in medicine. Circulation. 2015;132(20):1920–30.
15. Krupinski EA. Current perspectives in medical image perception. Atten Percept Psychophysiol. 2010;72(5):1205–17.
16. Sendak M, Gao M, Nichols M, Lin A, Balu S. Machine learning in health care: a critical appraisal of challenges and opportunities. EGEMS (Wash DC). 2019;7(1):1.
17. Edge SB, Compton CC. The American joint committee on cancer: the 7th edition of the AJCC cancer staging manual and the future of TNM. Ann Surg Oncol. 2010;17(6):1471–4.
18. Couinaud C. Liver lobes and segments: notes on the anatomical architecture and surgery of the liver. Presse Med. 1954; 62(33):709–12.
19. Simonyan K, Zisserman A. Very deep convolutional networks for large-scale image recognition. ICLR. 2015;2015:1–14.
20. Nakamura M, Kajiwara Y, Otsuka A, Kimura H. LVQ-SMOTE - learning vector quantization based synthetic minority over-sampling technique for biomedical data. BioData Min. 2013;6(1):16.
21. Hothorn T, Lausen B. On the exact distribution of maximally selected rank statistics. J Comput Stat Data Anals. 2003; 43(2):121–37.
22. Oliveira T, Silva A, Satoh K, Julian V, Leao P, Novais P. Survivability prediction of colorectal cancer patients: a system with evolving features for continuous improvement. Sensors (Basel). 2018;18:9.
23. Assumpcao L, Choti MA, Gleisner AL, Schulick RD, Swartz M, Herman J, et al. Patterns of recurrence following liver resection for colorectal metastases: effect of primary rectal tumor site. Arch Surg. 2008;143(8):743–9 discussion 749-750.
24. Fegiz G, Ramacciato G, D'Angelo F, Barillari P, Indinnimeo M, Gozzo P, et al. Patient selection and factors affecting results following resection for hepatic metastases from colorectal carcinoma. Int Surg. 1991;76(1):58–63.
25. Abdalla EK. Resection of colorectal liver metastases. J Gastrointest Surg. 2011;15(3):416–9.
26. Scheele J, Altendorf-Hofmann A. Resection of colorectal liver metastases. Langenbeck's Arch Surg. 1999;384(4):313–27.
27. Cromheecke M, de Jong KP, Hoekstra HJ. Current treatment for colorectal cancer metastatic to the liver. Eur J Surg Oncol. 1999;25(5):451–63.
28. Minsky B, Mies C. The clinical significance of vascular invasion in colorectal cancer. Dis Colon Rectum. 1989;32(9):794–803.
29. Pathak S, Pandanaboyana S, Daniels I, Smart N, Prasad KR. Obesity and colorectal liver metastases: mechanisms and management. Surg Oncol. 2016;25(3):246–51.
30. Sorski L, Levi B, Shaashua L, Neeman E, Benish M, Matzner P, et al. Impact of surgical extent and sex on the hepatic metastasis of colon cancer. Surg Today. 2014;44(10).1925–34.
31. Mohr AM, Gould JJ, Kubik JL, Talmon GA, Casey CA, Thomas P, et al. Enhanced colorectal cancer metastases in the alcohol-injured liver. Clin Exp Metastasis. 2017;34(2):171–84.
32. Wu K, Zhai MZ, Weltzien EK, Cespedes Feliciano EM, Meyerhardt JA, Giovannucci E, et al. Non-alcoholic fatty liver disease and colorectal cancer survival. Cancer Causes Control. 2019;30(2):165–8.
33. Xu Y, Jia Z, Wang LB, Ai Y, Zhang F, Lai M, et al. Large scale tissue histopathology image classification, segmentation, and visualization via deep convolutional activation features. BMC Bioinformatics. 2017;18(1):281.
34. Kirasich K, Smith T, Sadler B. Random Forest vs logistic regression: binary classification for heterogeneous datasets. SMU Data Sci Rev. 2018;1(3):1–24.
35. Christodoulou E, Ma J, Collins GS, Steyerberg EW, Verbakel JY, Van Calster B. A systematic review shows no performance benefit of machine learning over logistic regression for clinical prediction models. J Clin Epidemiol. 2019;110:12–22.
36. Ritchie MD, Holzinger ER, Li R, Pendergrass SA, Kim D. Methods of integrating data to uncover genotype-phenotype interactions. Nat Rev Genet. 2015;16(2):85–97.


## Publisher's Note

Springer Nature remains neutral with regard to jurisdictional claims in published maps and institutional affiliations.

# AUTHOR INDEX

**BioMed Central Ltd**
236 Gray's Inn Road
London WC1X 8HB
UK

ISBN 978-1-7138-1953-0

CURRAN ASSOCIATES INC.
proceedings.com